FUNDAMENTALS OF MACHINING AND MACHINE TOOLS

FUNDAMENTALS OF MACHINING AND MACHINE TOOLS

R.K. Singal
Ph.D., FIE
Professor and Consultant
rk.singal@gmail.com

Mridul Singal
B.Tech (IIT, Kanpur), M.Tech (IISc, Bangalore)
I/T Architect, IBM India Pvt. Ltd.

Rishi Singal
B.Tech (PEC Chandigarh), Ph.D. (PMI, USA)
Project Manager, Accentire India Pvt. Ltd.

TECHSAR™

NEW DELHI

Published and Printed by
Techsar Pvt. Ltd.
4435-36/7, Ansari Road, Daryaganj
New Delhi-110 002 (India)
E-mail: tech_info@techsarworld.com
Website: www.techsarworld.com

ISBN: 978-81-89866-66-2

Reprint 2025

PREFACE

Fundamentals of Machining and Machine Tools deals with analytical modeling techniques of machining processes, modern cutting tool materials and their effects on the economics of machining.

The book thoroughly illustrates the causes of various phenomena and their effects on machining practice. It includes description of machining processes outlining the merits and demerits of various modeling approaches.

Spread in 22 chapters, the book is broadly divided in five sections:

1. Machining Processes
2. Cutting Tools
3. Machine Tools
4. Automation
5. Annexures

Data on cutting parameters for machining operations and main characteristics of machine tools have been separately provided in Annexures.

In Addition to exhaustive theory, a number of numerical examples have been solved and arranged in various chapters. Question Bank has been given at the end of every chapter.

The book is a must for anyone involved in metal cutting, machining, machine tool technology, machining applications, and manufacturing processes.

The authors are grateful to the publisher for bringing out this book in a very short time. Although the authors are in touch with the students and their difficulties and requirements have been taken care during the preparation of this book, any constructive criticism will be welcome with gratitude.

AUTHORS

CONTENTS

Part I

Machining Processes

1

MACHINING PROCESSES

Introduction; Principle of machining, Estimation of parameters for important machining processes; Summary of equations for machining operation; Question Bank.

1.1 INTRODUCTION

Machining process is the gradual removal of metal from a workpiece. This includes metal cutting using single-point and multi-point tools and also grinding using abrasive wheels. These processes are carried out on machines called machine tools to achieve desired shape, size and surface finish on workpiece.

The first machine tool was invented by John Wilkinson which was called horizontal-boring machine. The cutting tool was mounted on a boring bar supported on bearings located outside the cylinder to be machined. The boring bar could be rotated as well as fed through the cylinder. A rough casting of cylinder was machined for the steam engine used by James Watt. This boring machine was the first effective machine tool.

Metal cutting started with the introduction of Wilkinson's machine tool. Today, machine tools form the basis of present industrial civilization and all products are manufactured by the direct or indirect use of some machine tool for machining processes.

Before industrial revolution, machine tools were made slowly by skilled craftsmen. The developments in machine tool technology became faster after industrial revolution. However, most rapid growth of machine tool industry and technology occurred in twentieth century. Special purpose machines for flow line production, machining heads, transfer machines, numerically-controlled machine tools, robots, automatic guided vehicles and flexible manufacturing systems are notable achievements of the previous century. A new type of machine called machining centre has been developed which can perform several different operations by choosing and picking desired tool from an array of tools and loading it on the machine.

A complete automatic factory will be in place in a very near future. This will be completely managed by computer with the help of robots and automatic guided vehicles. There would be no worker, no lighting, no ventilation, no canteens, no rest rooms and other facilities required by the workers.

1.2 PRINCIPLE OF MACHINING

The principle of machining is the generation of required surface by suitable relative motions between the cutting tool and the workpiece. The cutting edge (or edges) on the cutting tool removes a layer of work material. The removed material is called a shaving or a chip. The simplest required surfaces can be a flat surface and internal or external cylindrical surfaces.

1. **Generation of flat surface:** The tool is given to and fro motion in a straight line and a workpiece is incrementally fed beneath the tool in a direction at right angle to the motion of the tool.
2. **Generation of cylindrical surface:** The workpiece is rotated and tool is fed parallel to the axis of the workpiece rotation.

1.2.1 Classification of Machining Processes

The following are the basic machining processes:

1. **Turning and Boring:** Turning is a machining operation for generating external surfaces of revolution while in boring, internal surfaces of revolution are generated. The workpiece or tool is rotated about its axis and the cutting tool or workpiece is also given a feed motion in a direction normal to the cutting motion.
2. **Shaping, Planning and Slotting:** Shaping is a machining operation for generating flat surface by means of single-point cutting tools. The workpiece is kept stationary and the cutting tool is given a reciprocating motion. The feed motion is intermittent and imparted to the workpiece at the end of each stroke of the cutting tool. The planning operation is similar to the shaping operation. Here, the workpiece is reciprocated to impart the primary cutting motion instead of the cutting tool. The cutting tool is imparted the feed motion intermittently after each stroke of the table holding the workpiece.
3. **Drilling and Reaming:** Drilling is a machining process in which a hole is produced or enlarged by a multi-point cutting tool called the drill. The primary cutting motion is a rotary motion usually imparted to the drill and feed motion is translation and imparted to the drill. The cutting action is done by the cutting edge of the end face of the drill. Reaming is a hole finishing process. The tool motions are similar to drilling process. However, the cutting edges which may be straight or helical are on the periphery of the reamer.
4. **Milling:** Milling is a machining process in which flat as well as curved surfaces are generated by rotating multi-edged cutting tools called milling cutters. The milling cutter is rotated and the feed is given to the

workpiece. The main milling processes are slab milling, face milling, slot milling, form milling, angular milling, etc.

5. **Broaching:** Broaching is a machining process in which a multi-teeth cutter called a broach is pushed or pulled over the workpiece surface. Broaching is used for production of external and internal gears, internal key slots, splines and finishing of internal bores, etc.
6. **Thread Cutting:** Tapping and die cutting are machining operations to produce internal and external screw threads by the helical (cutting) motion of taps and dies which are multi-point tools.
7. **Grinding:** Grinding is a machining process to remove excess material from workpiece by rotation of multi-edged abrasive tool called grinding wheel. Flat as well as curved surfaces may be ground. Different grinding processes are cylindrical grinding, face grinding, internal grinding, taper grinding, etc.
8. **Fine Finishing Process:** (i) **Honing** is the fine finishing operation where abrasive sticks are used as tools. The tools are rotated and simultaneously reciprocated on the workpiece surface. There is slow abrasion of workpiece for getting desired finish and accuracy.

   (ii) **Lapping:** It is a fine finishing operation in which a material softer than workpiece lightly rubs the abrasive particles against the workpiece. Flat as well as curved surfaces may be lapped.
9. **Gear Cutting:** Gears are produced by the following processes:

   (i) Gear milling

   (ii) Gear hobbing, shaping and planning.

   (iii) Gear rolling

   Gear finishing operations are:

   (i) Burnishing

   (ii) Shaving

   (iii) Grinding

   (iv) Lapping

1.2.2 Relative Motion Generation

In general, two types of relative motion must be provided for machining processes.

1. **Primary Motion or Cutting Motion:** It is the main motion provided by the machine tool or manually to cause relative motion between the tool and workpiece so that face of the tool approaches the workpiece material. Primary motion accounts for the major part of power requirements for machining operation.
2. **Feed Motion:** It is the motion to be provided to the tool or workpiece when added to the primary motion helps the repeated or continuous chip

removal for creation of a machined surface with the desired geometry. The motion may be continuous or in steps. The feed motion absorbs a small portion of the total power required.

1.2.3 Axes System of Motion by ISO

A system of machine tool axes has been recommended by International Organization for Standards (ISO) for programming numerically controlled machine tools. However, it is also suitable as a general system for all machine tools. The system is based on the right hand coordinate system as shown.

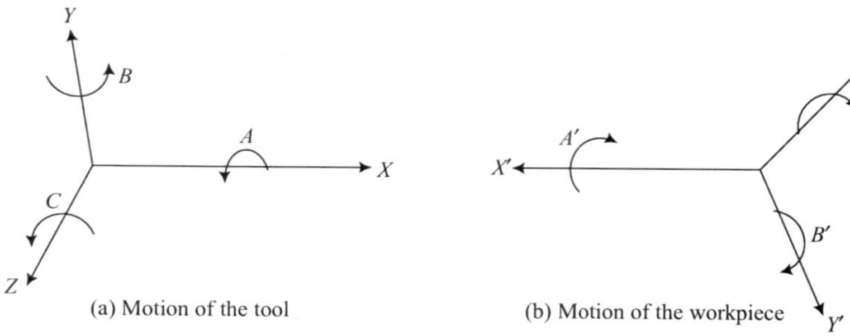

Fig. 1.1 Coordinate system for machine tool motions.

1. **Motion of the Tool:** The three coordinate axes *x*, *y*, and *z* indicate possible linear motions of the tools, and the motions *A*, *B* and *C* indicate possible rotary motions of the tool about these axes, respectively. If a right-hand screw nut is rotated about an axis, the movement of the nut in positive direction of axis (away from origin) indicates positive rotary motion.

   The primary axes of the machine tools are the three mutually perpendicular axes *x*, *y*, and *z*. The spindle provides the primary motion and its axis is taken as *Z*-axis. Positive *Z* motion increases the distance between the workpiece and tool holder. If the machining has no spindle, the Z-axis is arranged perpendicular to the work-holding surface.

   The larger table movement, which is generally along the length of the machine bed, is taken as the *X*-axis. The *Y*-axis corresponds to direction of saddle travel normal to the *X* and *Z* axes.

   (i) **Machines with no Main Spindle:** Where possible, the *X*-axis is horizontal and parallel to work-holding surface. On machines with no main spindle, the *X*-axis is parallel to and positive in the principle direction of cutting (primary motion).

   (ii) **Machines with rotating workpieces:** The *X* motion is radial and parallel to the cross slide. Positive *X* motion of the tool recedes it from the axis of the rotation of the workpiece.

   (iii) **Machines with rotating tools:** (a) ***Z* axis horizontal:** Positive *X* motion is to the right when viewed from the main spindle toward the workpiece. (b) ***Z* axis vertical:** Positive *X* motion is to the right when viewed from the main spindle toward the machine column.

2. **Motion of Workpiece:** Figure 1.1(b) defines the set of axes and motions employed for motion of the workpiece in a machine tool.
   (i) All the letters are primed as X′, Y′, Z′ and A′, B′ and C′.
   (ii) The directions of the linear and rotational axes are reversed.

If the tool can be made to approach a workpiece by motion in the possitive *X* direction, the same effect can be obtained by moving the workpiece in the positive *X*′ direction.

1.2.4 Cutting Tools

There are three types of cutting tools:

(1) Single-point tools
(2) Multi-point tools
(3) Abrasive wheels.

1. **Single-point Tools:** These are cutting tools with one cutting part or chip-producing element and one shank. These tools are used in lathes, turret lathes, planers, shapers, boring mills and similar machine tools. The important features are shown in Fig. 1.2 and defined as follows:

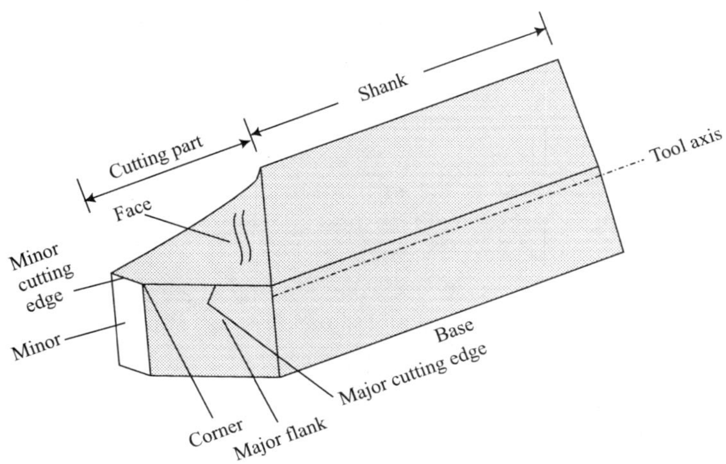

Fig. 1.2 Single-point tool.

   (i) The face is the surface or surfaces over which the chip flows.
   (ii) The flank is the tool surface or surfaces over which the surface produced on the workpiece passes.
   (iii) The cutting edge performs the cutting. The major cutting edge is responsible for the transient surface on the workpiece. The minor cutting edge is the remainder part of the cutting edge.

(iv) The corner is the intersection of major and minor cutting edges. The corner may be curved or straight.

2. **Multi-point tools:** A multi-point tool has a series of two or more cutting parts or chip-producing elements secured to a common body. The important features such as face, flank, and cutting edges are similar to that defined for single-point tools. The multi-point tools are used as milling cutters, drills, broaching tools, taps and dies. These tools are rotated for machining. The shank may be conical, cylindrical for holding or it may be bored for inserting a spindle or arbor.
3. **Abrasive wheels:** Abrasive wheels or grinding wheels are generally cylindrical, disc-shaped or cup-shaped and are mounted on high speed rotating spindles. In grinding machines, the primary motion is the rotation of abrasive wheels and workpiece is given two or more feed motions to produce the desired surface shape.

A grinding wheel essentially consists of a large number of abrasive particles (grains) held together by a suitable bond. It may be regarded as a multi-point cutting tool with cutting action similar to a milling cutter but the cutting edges are irregularly shaped.

1.2.5 Machining Parameters

There are many terms and definitions applicable to all types of machining operations. There is international terminology and equations to estimate machining parameters or variables. Important parameters are defined as follows:

1. **Cutting speed (*v*):** It is the relative velocity between the workpiece and cutting edge of tool responsible for cutting action. It is expressed as m/min.
2. **Feed (*f*):** The feed is the displacement of the tool relative to the workpiece in the direction of feed motion. It is expressed as mm per stroke or per revolution of workpiece or tool.
3. **Uncut Chip Thickness (*t*):** It is the thickness of the layer of the material being removed by the cutting tool in the direction of the feed motion.
4. **Depth of cut (*d*):** It is the normal distance between the machined and unmachined surfaces measured along a normal to the machined surface.
5. **Area of uncut chip (*Ac*):** It is the cross-sectional area of the layer of the workpiece being machined away. It is the product of feed rate and depth of cut.

$$Ac = f.d.$$

6. **Metal Removal Rate (*Zw*):** It is the volume of material being removed per unit time from the workpiece. It is the product of the mean cutting speed and cross-sectional area of the uncut chip.

$$\begin{aligned} Zw &= Ac\ \upsilon_{av}. \\ &= 1000.f.d.\upsilon_{av}.\ [\text{mm}^3/\text{min}]. \end{aligned}$$

where, f = feed rate (mm)

υ_{av} = average cutting speed (m/min)

7. **Machining Time (t_m):** It is the time required to machine a length of the surface or hole, l_w.

$$t_m = \frac{l_w}{f N} \text{ (min).}$$

where, N = spindle speed (rpm)

8. **Machining Power (P_m):** The power required to perform any machining operation.

$$P_m = P_s . Z_w$$

where, P_s = energy required to remove a unit volume of material.

= specific cutting energy.

9. **Electrical Power (P_e):** The electrical power consumed by the machine tool,

$$P_e = \frac{P_m}{\eta_m}$$

where, η_m = overall efficiency of the machine tool motor and drive systems.

1.3 ESTIMATION OF PARAMETERS FOR IMPORTANT MACHINING PROCESSES

1.3.1 Turning and Boring

The important turning operations carried out on a lathe are:

1. Cylindrical turning
2. Facing
3. Boring
4. External Threading
5. Cutoff.

These operations are shown in Fig. 1.3. The primary motion ($-C'$), feed motion ($-X$, $-Z$) and other terms and dimensions are also indicated. The workpiece has three important surfaces.

1. **Work surface:** The surface on the workpiece to be removed by machining.
2. **Machined surface:** The desired surface produced by the action of the cutting tool.
3. **Transient surface:** The part of surface formed on the workpiece by the cutting edge and removed during the subsequent cutting stroke or pass of the tool.

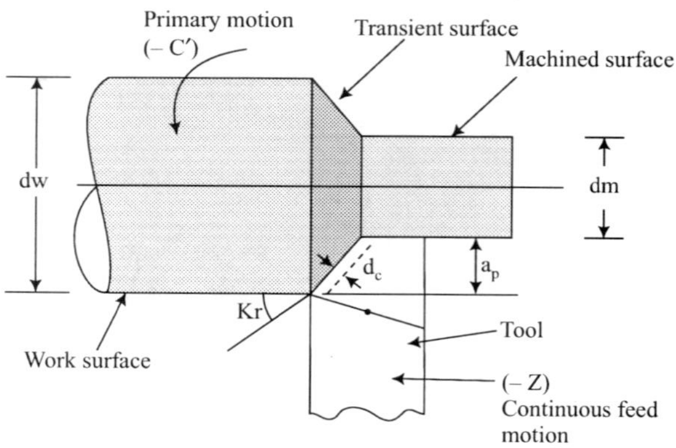

Key : d_m = Diameter of machined surface
d_w = Diameter of work surface
a_p = Back engagement (depth of cut)
K_r = Major cutting edge angle.
a_c = Undeformed chip thickness.

Fig. 1.3(a) Cylindrical Turning

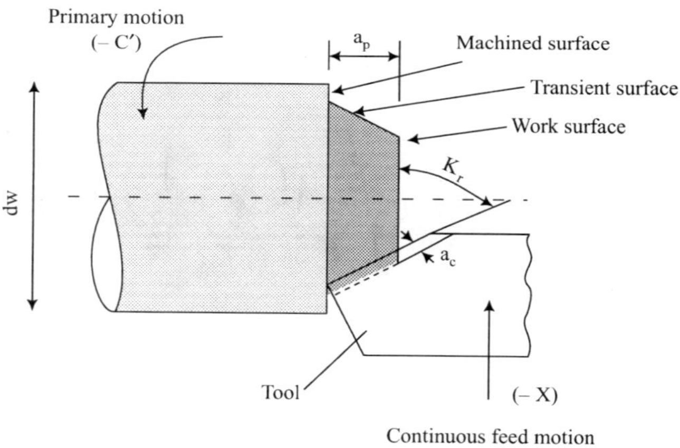

Fig. 1.3(b) Facing.

1.3.1.1 Result and cutting motion

When a tool is applied to a workpiece, its motion relative to workpiece has two components:

1. The primary motion
2. The feed motion

The resultant cutting motion is the vector sum of the above two motions.

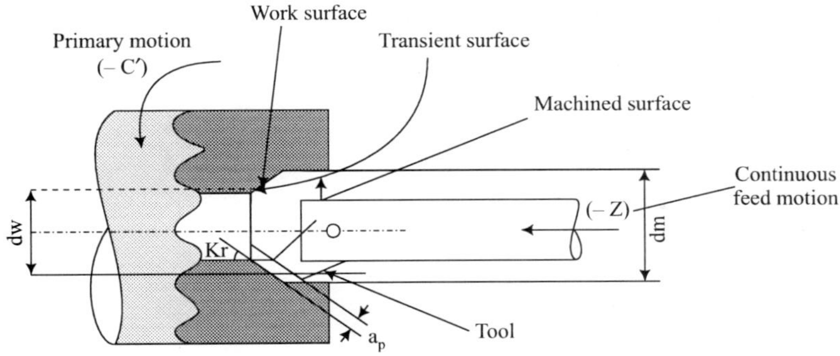

Fig. 1.3(c) Boring.

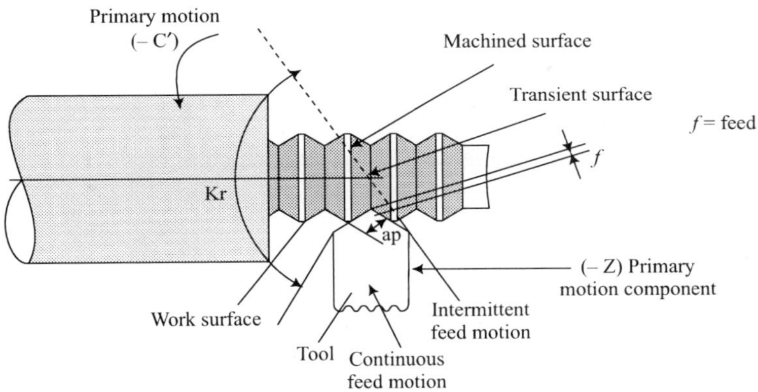

Fig. 1.3(d) External Threading

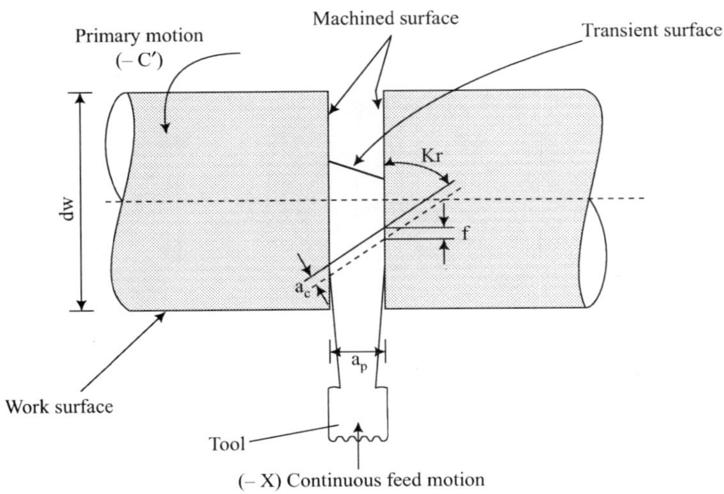

Fig. 1.3(e) Parting and cutoff.

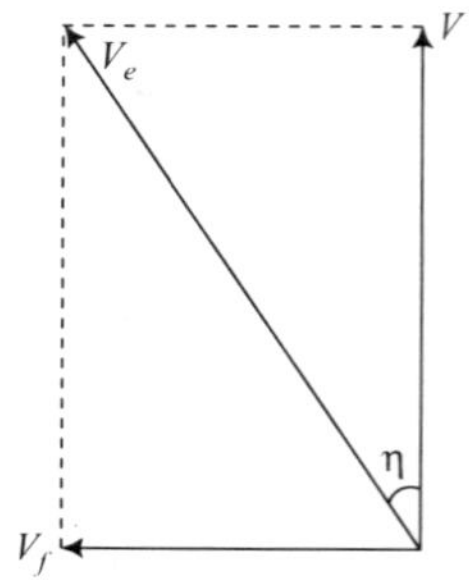

Fig. 1.3(f) Resultant cutting motions.

The resultant cutting speed,

$$V_e = \frac{V}{\cos \eta}$$

where, V = Cutting speed

V_f = Feed speed

η = Angle of resultant cutting speed with the cutting speed

In most practical operations, η is very small

$$\therefore \quad V_e \approx V$$

1.3.1.2 Geometry of single-point tool operation

The geometry of single-point tool operation in cylindrical turning is shown in Fig. 1.4. The various dimensions shown are:

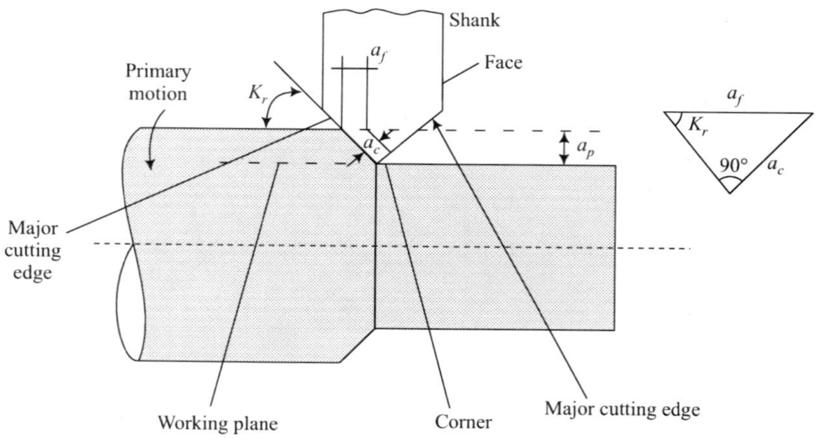

Fig. 1.4 Geometry of single-point tool operations.

K_r = Major cutting edge angle
a_c = Undeformed chip thickness
a_p = Depth of cut (back engagement)
a_f = Feed engagement in the direction of feed motion
f = feed

The undeformed chip thickness (a_c) is the thickness of the layer of material being removed. The power consumption mainly depends upon a_c. This dimension is nearly same when measured normal to the cutting edge and normal to the resultant cutting direction because angle η is very small.

$$a_c = a_f \sin K_r$$

But $a_f = f$ for single-point cutting operations

The cross-sectional area of the uncut chip,

$$A_c = f\, a_p$$

The back engagement or depth of cut (a_p) is the depth of material removed from the workpiece in a single-point cutting operation.

1.3.1.3 Machining parameters in cylindrical turning

1. Cutting Speed (V)

The cutting speed at the tool corner,

$$V = \pi\, d_m\, n_w \text{ (m/min)}$$

where, d_m = diameter of machined surface (m)

n_w = rotational speed of the workpiece (rpm)

The average or mean cutting speed,

$$V_{av} = \frac{\pi n_w (d_w + d_m)}{2}$$

where, d_w = diameter of work surface (m).

2. Metal removal rate (Z_w)

The metal removal rate is the volume of material being removed per unit time from the workpiece. It is the product of mean cutting speed and cross-sectional area of the uncut chip (A_c).

$$Z_w = A_c\, V_{av}.$$

$$= \frac{\pi a_f\, a_p\, a_w (d_w + d_m)}{2}$$

or, $$Z_w = \pi f\, a_p\, n_w\, (d_m + a_p)$$

For operations where depth of cut (a_p) is small as compared to the diameter of the machined surface (d_m), the material removal rate,

$$Z_w \approx \pi f a_p n_w d_m.$$

3. Power Consumption

The power (P_m) required to perform any machining operation,

$$P_m = p_s Z_w$$

where, p_s = specific cutting energy

Specific cutting energy is the energy required to remove a unit volume of material. This depends mainly on the properties of work material and undeformed chip thickness. The electrical power consumed by the machine tool,

$$P_e = \frac{P_m}{\eta_m}$$

where, η_m = The overall efficiency of the machining tool motor and drive system.

1.3.1.4 Machining parameters in facing and groove cutting

The tool is fed at right angles (–*X*) to the axis of the workpiece rotation.

The cutting speed at the tool corner varies from maximum at the beginning of the cut to zero when the tool reaches the centre of the workpiece.

The maximum cutting speed,

$$V_{max} = \pi n_w d_m.$$

The undeformed chip thickness,

$$a_c = f \sin K_r$$

The cross-sectional area of the uncut chip.

$$A_c = f a_p$$

The machining time

$$t_m = \frac{d_m}{2 f n_w}$$

The maximum metal removal rate,

$$Z_{w\,max} = \pi f a_p n_w d_m.$$

1.3.1.5 Machining parameters in boring operation

Boring operation is carried out on a lathe to enlarge an existing hole in the workpiece.

The mean cutting speed,

$$V_{av} = \frac{\pi n_w (d_w + d_m)}{2}$$

where, d_w = diameter of work surface (initial hole).

d_m = diameter of machined surface (final bore).

The metal removal rate,

$$Z_w = \pi f a_p n_w (d_w - a_p)$$

The machining time,

$$t_m = \frac{l_w}{f n_w}$$

where, l_w = length of the hole to be bored.

1.3.1.6 Machining parameters in parting operation

Parting is the operation used to separate the finished workpiece from the bar gripped in the chuck. Two machined surfaces are produced simultaneously. The cutting speed varies from a maximum at the beginning of the cut to zero at the centre of workpiece.

The machining time,

$$t_m = \frac{d_m}{2 f n_m}$$

The maximum removal rate,

$$Z_{w\,\max} = \pi f a_p n_w d_m$$

1.3.1.7 Machining parameters in external threading

A combination of motions $(-C')$ and $(-Z)$ is used to generate a helix on the workpiece for screw cutting. The required pitch of threads is obtained by setting suitable gears to drive the lead screw.

The machining of threads requires several passes of tool along the workpiece. A thin layer of metal is removal from one side of thread in each pass. The feed is applied in increments parallel to machined surface after every pass of tool. The calculation of production time for screw cutting should include:

(1) the time taken for return of tool to the start of cut,

(2) the time taken to increment the feed,

(3) the time taken to engage the lathe carriage with the lead screw.

The ratio of pitch of lead screw and pitch of thread to be cut is called gear ratio.

$$\text{The gear ratio} = \frac{N_s}{N_i} = \frac{l}{p}$$

where, N_s = speed of lathe spindle (rpm)

N_i = speed of lead screw (rpm)

p = pitch of thread to be cut

l = pitch of the lead screw

The cutting speed,

$$V = \pi\, d_m\, n_w \text{ (m/min)}$$

The feed

$$f = p\, n_w$$

The machining time,

$$t_m = \frac{l_w}{f}$$

where, l_w = length of screw threads to be cut.

Example 1.1 Estimate the machining time for cylindrical turning of a 50 mm diameter copper bar to a final diameter of 42 mm over a length of 50 mm. The spindle speed is 450 rpm. The feed is set at 0.2 mm per revolution, the depth of cut is limited to 3 mm, the side edge angle of tool is 30°. Take approach allowance of 3 mm.

Solution: Number of passes required = 2

Depth of cut in roughing pass = 3 mm

Depth of cut in finishing pass = 1 mm

Spindle speed, n_w = 450 rpm

Side cutting edge angle, K_r = 30°

Uncut ship thickness, a_c $= f \sin K_r$

$= 0.2 \sin 30°$

$= 0.1$ mm

Cutting speed for roughing pass, $V = \pi\, d_w\, n_w$

$= \pi \times 50 \times 450 = 70685.835$ mm/min

$= 70.9$ m/min

Cutting speed for finishing pass, $V = \pi\, d_w\, n_w$

$= \pi \times 44 \times 450 = 62203535$ (mm/min)

$= 62.2$ m/min

Length of tool travel, l_m $= 50 + 3 = 53$ mm.

The machining time, t_m $= \dfrac{2l_m}{f \times n_w} = \dfrac{2 \times 53}{0.2 \times 450}$

$= 1.1778$ minutes. **Ans.**

1.3.2 Shaping, Planing and Slotting

(a) Shaping and Slotting

The single-point tool moves backward and forward (*X* motion). The forward stroke is the cutting stroke with lower speed and backward stroke is idle stroke and has quicker speed. The length of stroke can be adjusted to suit the workpiece. The feed is applied to the workpiece in increments at the end of return stroke.

The different motions of shaping operation are shown in Fig. 1.5 along with geometrics of shaping horizontally, vertically and inclined flat surfaces.

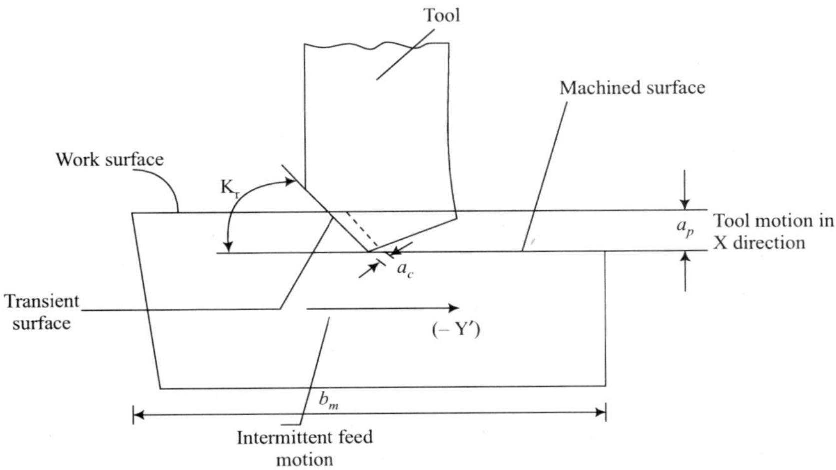

Fig. 1.5(a) Horizontal surface.

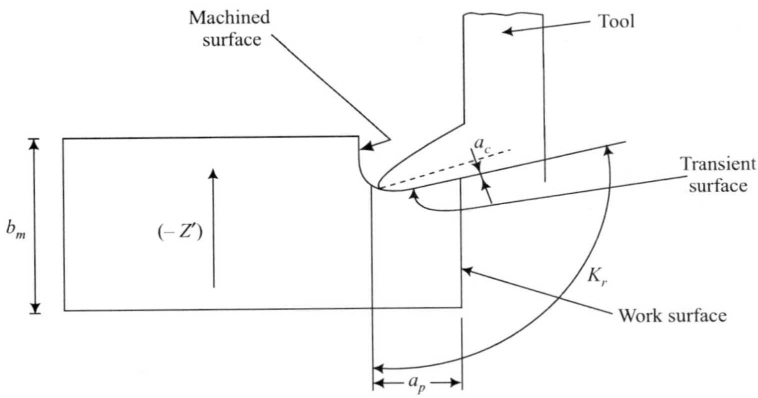

Fig. 1.5 (b) Vertical surface.

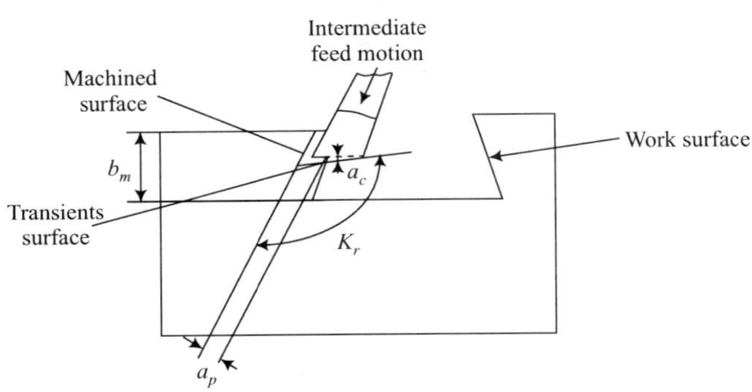

Fig. 1.5(c) Inclined surface.

The machining time,

$$t_m = \frac{b_m}{fn_r}$$

where, b_m = width of the surface to be machined

f = feed

n_r = frequency of cutting strokes.

The metal normal rate,

$$Z_w = a_c\ V = f\,a_p\ V$$

where, V = cutting speed

a_p = back engagement

The underformed chip thickness

$$a_c = f \sin K_r.$$

Example 1.2 A cast iron block of face 100 mm × 150 mm is fixed on a shaping machine with dimension 150 mm along the stroke. Calculate the machining time and metal removal rate if the mean cutting speed is 20 m/min, depth of cut 3 mm and table feed 0.3 mm/stroke. Take ratio of forward speed to return speed as 0.7.

Solution: Take stroke length, $l_m = 150 + 2 \times 5 = 160$ mm where 5 mm is tool clearance length on each side.

$$\text{Time for forward travel of tool} = \frac{l_m}{V} = \frac{160}{1000} \times \frac{60}{20} = 0.48 \text{ sec.}$$

$$\text{Time for return stroke} = 0.7 \times 0.48 = 0.336 \text{ sec.}$$

$$\text{Time for complete stroke} = 0.48 + 0.336 = 0.816 \text{ sec.}$$

$$\text{Machining time, } t_m = \frac{b_m \times n_r}{f} = \frac{100}{0.3} \times 0.816 = 272 \text{ sec.}$$

$$\text{Metal removal rate, } Z_w = \frac{fa_p b_m}{n_r} = \frac{0.3 \times 3 \times 150 \times 60}{0.816}$$

$$= 9926.47 \text{ mm}^3\text{/min.}$$

(b) Planing

A shaping machine is not suitable for generating flat surfaces on very large parts. There are limitations on the stroke and over hung of the ram. In a planing machine the linear primary motion (X') is given to the workpiece and the tool is fed at right angles to primary motion (Y or Z). The machining time, metal removal rate, and the undeformed chip thickness is estimated using the equations of shaping process.

Example 1.3: A surface plate 1 m × 1.2 m is machined on a planer with an average cutting speed of 24 m/min. The ratio of forward to return speed is 0.75. Determine the machining time if the feed per stroke is 0.5 mm and time for reversal of work table is 1.2 sec. Also determine metal removal rate if depth of cut is 4 mm.

Solution: Take stroke lengths, $l_m = 1200 + 2 \times 10 = 1220$ mm

where 10 mm is tool clearance length on each side of the length of workpiece.

$$\text{Time for one complete stroke} = \frac{l_m}{V}$$

$$= \frac{1220 \times (1 + 0.75)}{24 \times 1000} \times 60$$

$$= 5.3375 \text{ sec.}$$

$$\text{Depth of cut } a_p = 4 \text{ mm}$$

$$\text{Feed } f = 0.5 \text{ mm/stroke}$$

$$\text{Cutting speed } V = 24 \text{ m/min}$$

$$\text{Machining time, } t_m = \frac{1000 \times 1220\,(1 + 0.75) \times 60}{0.5 \times 24 \times 1000} + \frac{1000}{0.5} \times 1.2$$

$$= 217.9 \text{ min.}$$

$$\text{Metal removal rate, } Z_w = \frac{f a_p b_m}{n_r}$$

$$= \frac{0.5 \times 4 \times 1200}{5.3375} \times 60$$

$$= 26979 \text{ mm}^3\text{/min.}$$

1.3.3 Drilling

The tool is rotated (motion *C*) and fed along its axis of rotation (motion *Z*). The workpiece remains stationary and a twist drill generates an internal cylindrical surface. The geometry of drilling operation is shown in Fig. 1.6.

The tool has two cutting edges, each removes part of work material. Feed per tooth is the depth of material removed by each tooth and measured parallel to the direction of feed motion.

The feed engagement,

$$a_f = \frac{f}{2},$$

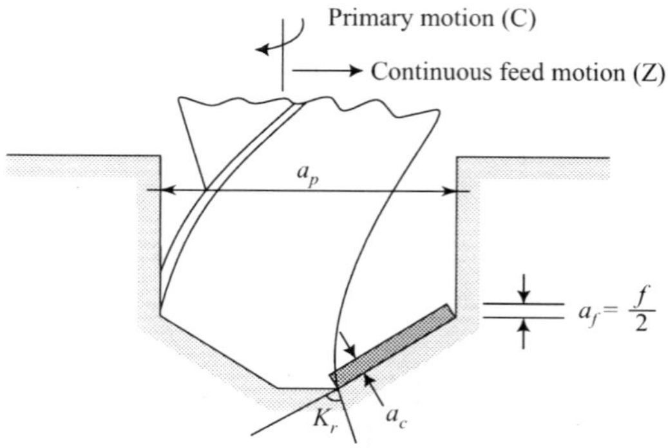

Fig. 1.6 Drilling operation.

The thickness of undeformed chip,

$$a_c = \frac{f}{2} \sin K_r$$

where, K_r = major cutting edge angle.

The machining time,

$$t_m = \frac{l_w}{fn_t}$$

where, l_w = length of hole produced.

n_t = rotational frequency of the tool.

In case of blind holes, length of machining is depth of holes, l_w.

For through holes

$$l_w = t_w + \frac{a_p}{2 \tan K_r}$$

$$= t_w + \frac{d_m}{2 \tan K_r}$$

where, t_w = thickness of workpiece

a_p = work engagement

= drill diameter

= d_m.

The metal removal rate Z_w is the product of feed speed, V_f and the cross-sectional area of the hole produced.

$$Z_w = \frac{\pi}{4} d_m^2 \; V_f = \frac{\pi}{4} d_m^2 f n_t$$

For enlargement of an existing hole, the metal removal rate,

$$Z_w = \frac{\pi f (d_m^2 - d_w^2) n_t}{4}$$

where, d_w = Diameter of existing hole.

The cutting speed is maximum at outer radius of the major cutting edges and zero at the tip of the drill.

Some other drilling operations are shown in Fig. 1.7.

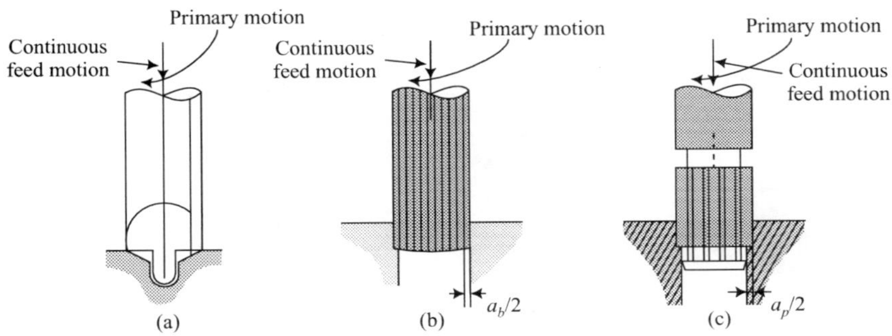

Fig. 1.7 (a) Centre drilling. **(b)** Reaming. **(c)** Spot facing.

1.3.4 Horizontal Milling

The tools are rotated (motion *C*) on a horizontal axis and the work is fed continuously (motion X' or Y'). The geometry of slab milling is shown in Fig. 1.8.

Feed (f) is the distance moved by workpiece during one revolution of the tool.

$$f = \frac{V_f}{n_t}$$

where, V_f = feed speed of workpiece

n_t = rotational frequency of the cutter

The feed engagement (a_f) is equal to the thickness of the chip removed by one tooth and measured parallel to the feed direction or feed per tooth

$$a_f = \frac{f}{N}$$

where, N = number of cutter teeth.

The maximum undeformed chip thickness ($a_{c\,max}$) is measured normal to the direction of primary motion.

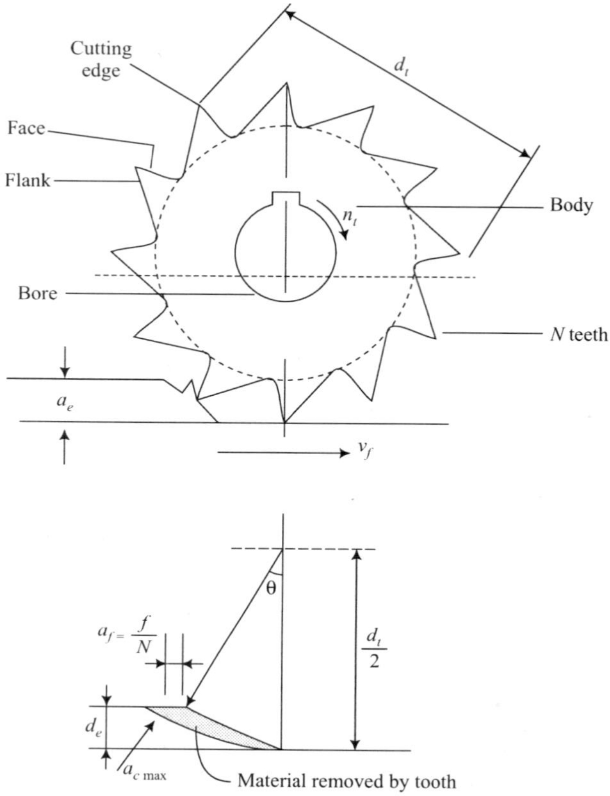

Fig. 1.8 Geometry of Slab-milling operation.

$$a_{c\,max} = \frac{V_f \sin\theta}{N n_t} \qquad ...(1)$$

where, $\cos\theta = 1 - \frac{2a_e}{d_t}$

and, d_t = diameter of the cutter.

The working engagement (a_e) is the instantaneous engagement of the tool with the workpiece and is measured in the plane containing the directions of primary and feed motions and perpendicular to feed motion.

In slab milling, a_e is equal to the depth of cut.

Now, $$\sin\theta = \sqrt{1-\cos^2\theta} = 2\sqrt{\frac{a_e}{d_t} - \left(\frac{a_e}{d_t}\right)^2}$$

From equation (1)

$$a_{c_{max}} = \frac{2V_f}{Nn_t}\sqrt{\frac{a_e}{d_t} - \left(1 - \frac{d_e}{d_t}\right)}$$

But $\frac{a_e}{d_t}$ is very small

$$\therefore \qquad a_{c_{max}} = \frac{2V_f}{Nn_t}\sqrt{\frac{a_e}{d_t}}$$

The distance travelled by the cutter in slab milling is illustrated in Fig. 1.9 and is equal to $l_w + \sqrt{a_e(d_t - a_e)}$.

where, l_w = length of workpiece.

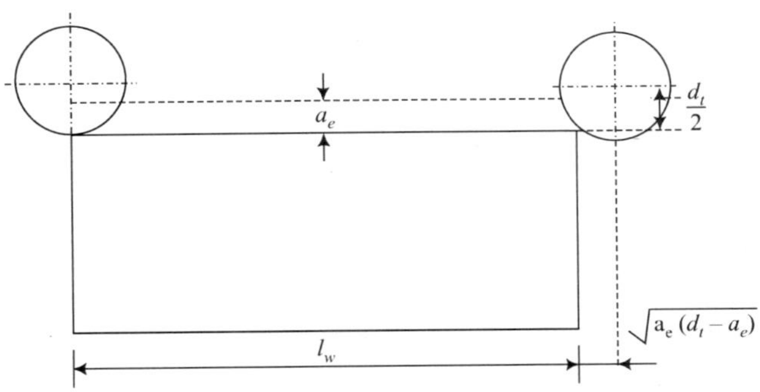

Fig. 1.9 Distance travelled by cutter in slab milling

Therefore, the machining time,

$$t_m = \frac{l_w + a_e\sqrt{d_t - a_e}}{v_f}$$

The metal removed rate (Z_w) is the product of feed speed and cross-sectional area of the metal removal, measured in the direction of feed motion.

$$Z_w = a_c\, a_p\, v_f$$

Since the back-engagement a_p is equal to the width of the workpiece.

Some additional horizontal-milling operations are shown in Fig. 1.10.

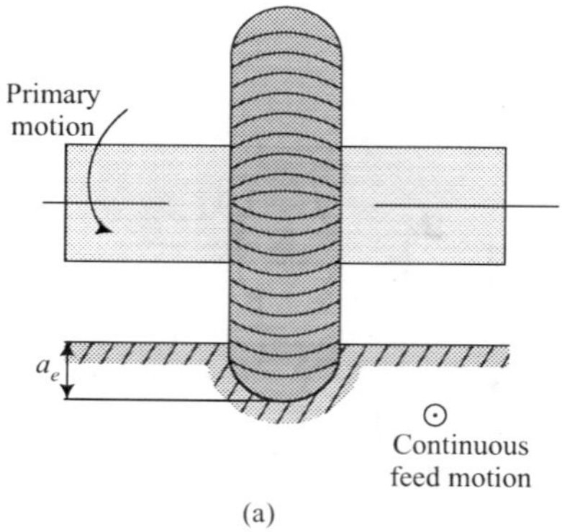

Fig. 1.10(a) Form cutting.

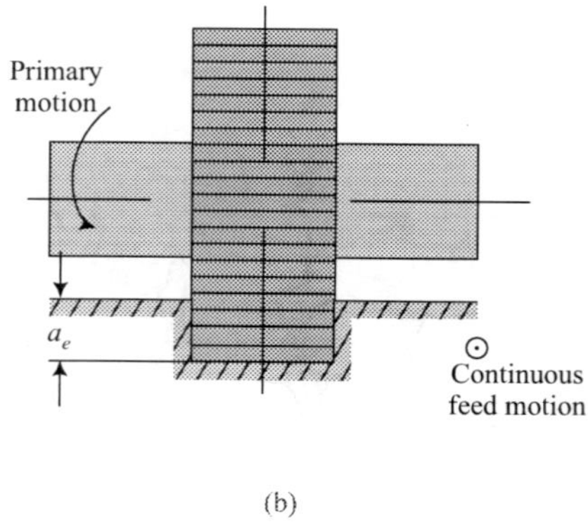

Fig. 1.10(b) Slotting

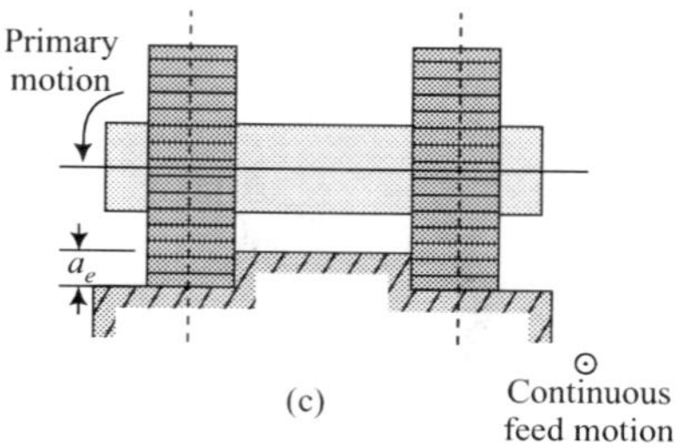

Fig. 1.10(c) Straddle milling.

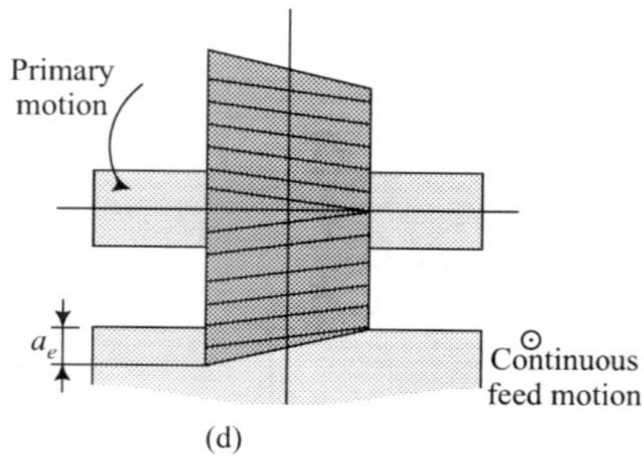

Fig. 1.10(d) Angular milling.

1.3.5 Vertical Milling

Vertical milling includes machining of horizontal, vertical and inclined surfaces. End milling tools can be used to mechanic slots, peripheral steps, profiles, etc. Drilling, tapping, reaming and boring operations can also be carried out on vertical milling machines.

Milling cutter mounted on a vertical spindle is rotated and feed motion to workpiece can be given either.

1. Along vertical axis (Z' motion) by raising or lowering of knee.
2. Along horizontal axis (Y' motion) by moving the saddle.
3. Along horizontal axis (X' motion) by moving the table across the saddle.

The geometry of face milling is shown in Fig. 1.11.

Feed (f) is the distance moved by the cutter across workpiece during its one revolution.

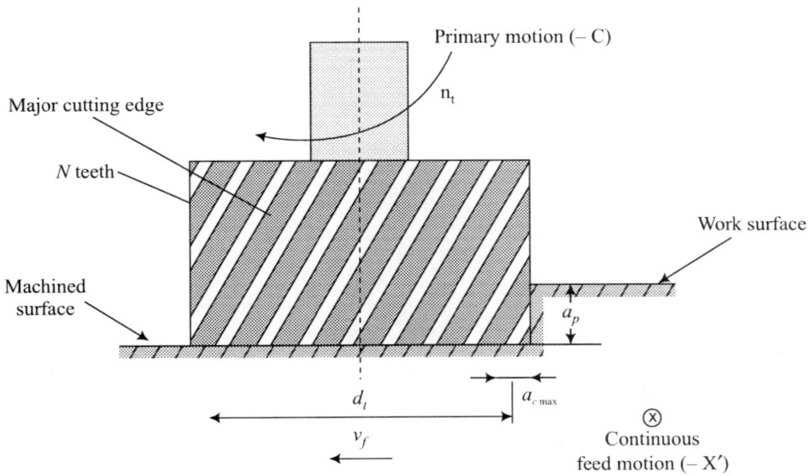

Fig. 1.11 Geometry of face milling.

$$f = \frac{V_f}{n_t}$$

where, V_f = feed speed

n_t = rotational speed of cutter.

The maximum value of undeformed chip thickness is equal to the feed engagement.

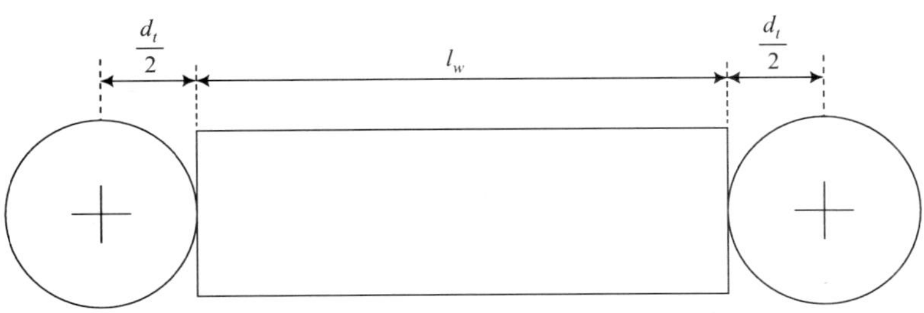

Fig. 1.12 Distance moved by cutter.

$$a_{c_{max}} = \frac{V_f}{Nn_t}$$

where, N = number of teeth on the cutter.

The total distance moved by the cutter is equal to $(l_w + d_t)$ and is illustrated in Fig. 1.12.

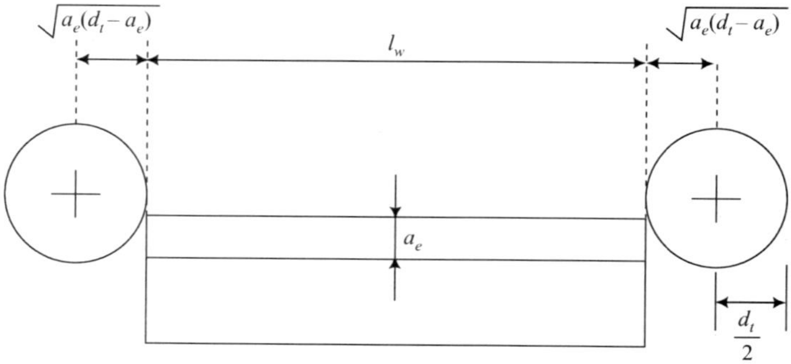

Fig. 1.13 Distance moved by workpiece.

The machining time,

$$t_m = \frac{(l_w + d_t)}{V_f}$$

When the path of tool axis does not pass over the workpiece, the distance moved by workpiece is illustrated in Fig. 1.13.

The machining time,

$$t_m = \frac{l_w + 2\sqrt{a_e(d_t - d_e)}}{V_f}$$

where, a_e = working engagement

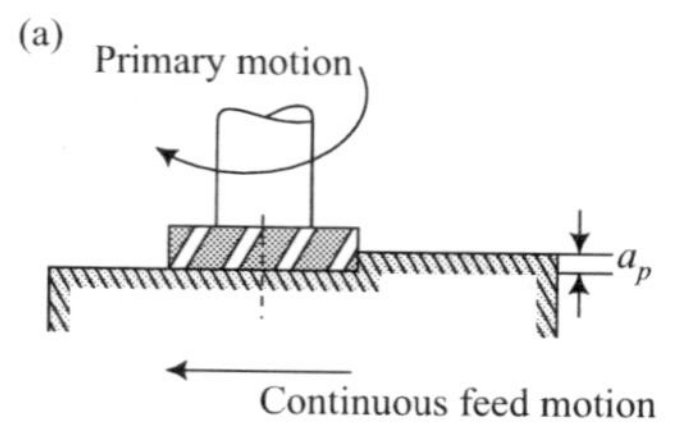

Fig. 1.14(a) Horizontal surface

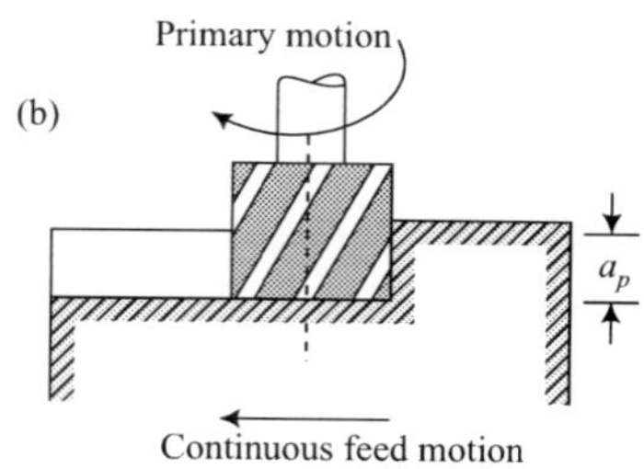

Fig. 1.14(b) Slotting

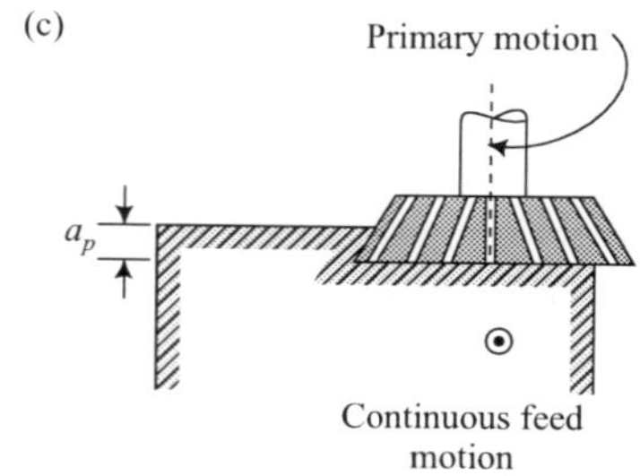

Fig. 1.14(c) Dovetailing

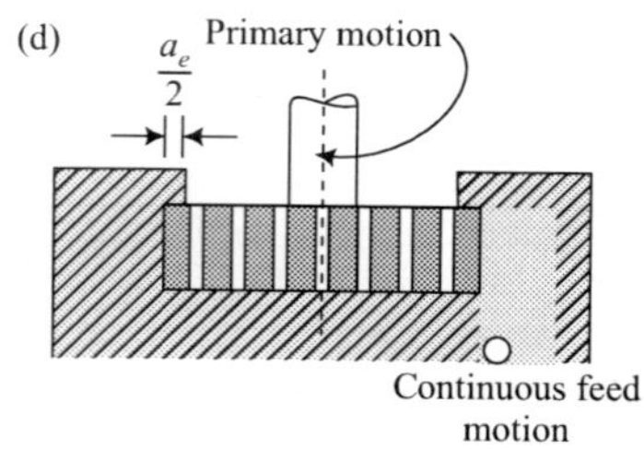

Fig. 1.14(d) T-slot.

The maximum value of undeformed chip thickness,

$$a_{c_{max}} = \frac{2V_f}{Nn_t}\sqrt{\frac{a_e}{d_t}\left(1-\frac{a_e}{d_t}\right)}$$

The metal removal rate,

$$Z_w = a_c\, a_p\, V_f.$$

Additional vertical-milling operations are shown in Fig. 1.14.

Example 1.4 A key-way slot of 10 mm × 4 mm × 100 mm long is to be machined on 75 shafts using an end mill. Assume a cutting speed of 22 m/min and time for loading and unloading 0.3 min., calculate the time for milling of slot.

Solution: Assume the following data:

Diameter of end mill = 10 mm

Number of cutter teeth = 4

Feed per tooth = 0.1 mm

Cutting speed = 22 m/min

The rotational speed of cutter, $n_t = \frac{22 \times 1000}{\pi \times 10} = 700$ rpm.

Length of travel of cutter $= l_w + d_t = 100 + 10 = 110$ mm

Feed of cutter, $f = 4 \times 0.1 \times 700 = 280$ mm/min.

Time required for cutting one slot $= \frac{110}{280} = 0.4$ min.

Total time of machining = cutting time + unloading and loading time

$= 0.4 + 0.3 = 0.7$ min.

Time required for milling slots on 75 shafts = 0.7 × 75 = **52.5 min.**

1.3.6 Broaching

A vertical broaching machine is used for machining shaped slots in the workpiece. The primary motion is provided between the tool and workpiece (motion *X*) and feed is provided by staggering the teeth on the broach, each tooth removes a thin layer of material. The machined surface is produced during one pass of the tool.

The machining time,

$$t_m = \frac{l_t}{V}$$

where, l_t = length of broach

V = cutting speed.

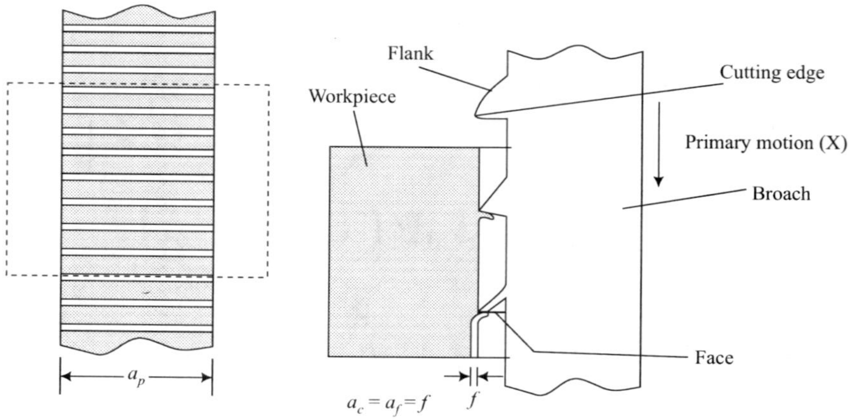

Fig. 1.15 Geometry of Broaching.

The feed (f) is the motion of an imaginary single cutting edge which would produce the same result as the array of cutting edges provided on the broach. The undeformed chip thickness (a_c) is equal to the feed at a given point on cutting edge.

The average metal removal rate (Z_w) is the ratio of total volume of metal removed and the machining time.

1.3.7 Grinding

The important grinding operations are:

1. Horizontal surface grinding
2. Vertical surface grinding
3. Cylindrical grinding
4. Internal grinding
5. Centreless grinding

1. Horizontal surface grinding

The primary motion is given to the abrasive wheel mounted on a horizontal rotating spindle (motion C). The principal feed motion is called transverse motion (motion X') and is given to workpiece mounted on a reciprocating worktable. The feed motion may be given to either:

(i) The wheel head by moving downwards (motion $-Y$) and is called *infeed*, or to

(ii) The worktable by moving it parallel to machine spindle (motion Z') and is called *cross-feed*.

The geometry of traverse grinding and plunge grinding is shown in Fig. 1.16.

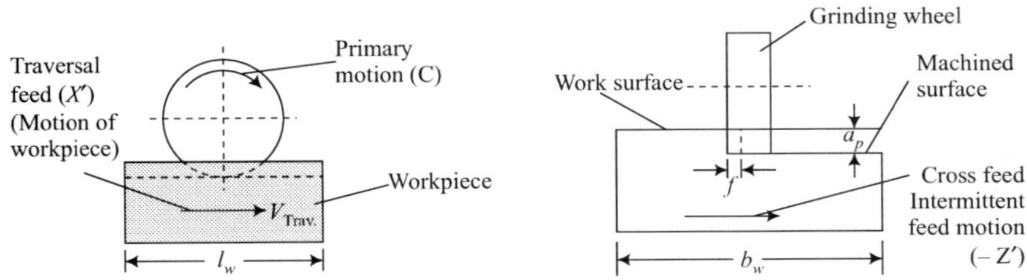

(a)

Fig. 1.16(a) Traverse Grinding.

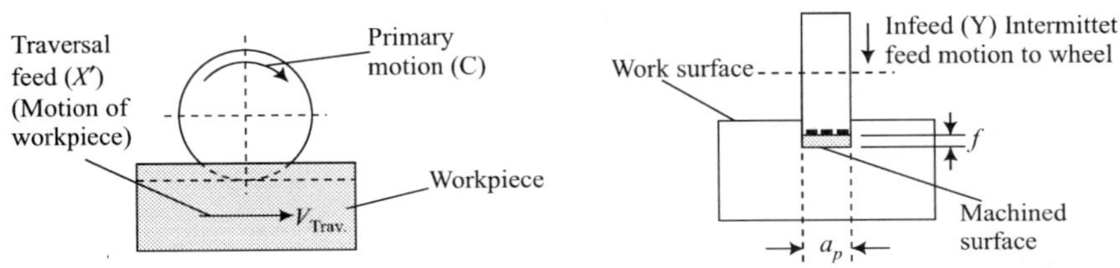

(b)

Fig. 1.16(b) Plunge Grinding.

(a) The metal removal rate in *traverse grinding*,

$$Zw = fa_p V_{\text{trav}}.$$

where, $f =$ cross-feed per cutting stroke

a_p = back engagement

V_{trav} = traverse speed

The machining time,

$$t_m = \frac{b_w}{2fn_r}$$

where, n_r = frequency of reciprocation of worktable.

b_w = width of workpiece.

The frequency of reciprocation depends on traverse speed setting and stroke setting.

(b) The metal removal rate in *plunge grinding*,

$$Z_w = f\,a_p\,V_{\text{trav}}$$

The machinery time in plunge grinding,

$$t_m = \frac{a_t}{2fn_r} + t_s.$$

where, a_t = total depth of work material to be removed

t_s = time for sparking out, i.e., time for continuation of grinding with no further feed for finishing operation when no more sparks appear.

2. Vertical surface grinding

The cup-shaped grinding wheel is mounted on a vertical rotating wheel and forms the primary motion (C). The worktable is reciprocated (X motion) and the tool is fed intermittently downwards ($-Z$ motion). These motions are called *traverse* and *infeed* respectively. The geometry of vertical surface grinding is shown in Fig. 1.17.

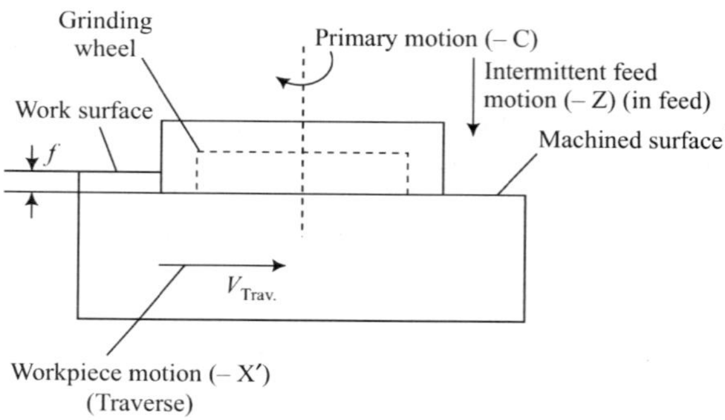

Fig. 1.17 Geometry of vertical surface grinding.

The metal removal rate,

$$Z_w = f a_p V_{trav}.$$

where, a_p = back engagement

= width of workpiece

V_{trav} = traverse speed.

The machining time, $t_m = \dfrac{a_t}{2 f n_r} + t_s.$

The machining time for vertical surface grinding with rotary worktable on which many workpieces are mounted for machining,

$$t_m = \frac{a_t}{f n_w} + t_s.$$

where, n_w = rotational frequency of worktable.

3. Cylindrical grinding

The grinding wheel is mounted on a rotating horizontal spindle which is parallel to workpiece (primary motion C'). The cylindrical workpiece is driven slowly (C' motion) and is mounted on centres. The workpiece is mounted on a worktable that is reciprocated horizontally (Z' motion). The grinding wheel can be given a feed (infeed) normal to the axis of workpiece ($-X$ motion).

The geometries of traverse and plunge cylindrical grinding are shown in Fig. 1.18.

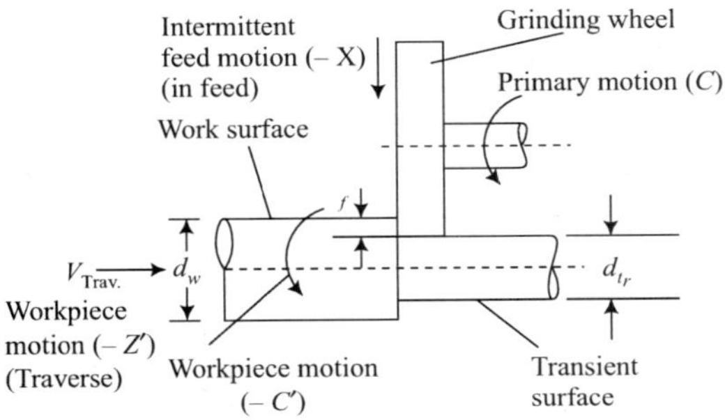

Fig. 1.18(a) Traverse grinding.

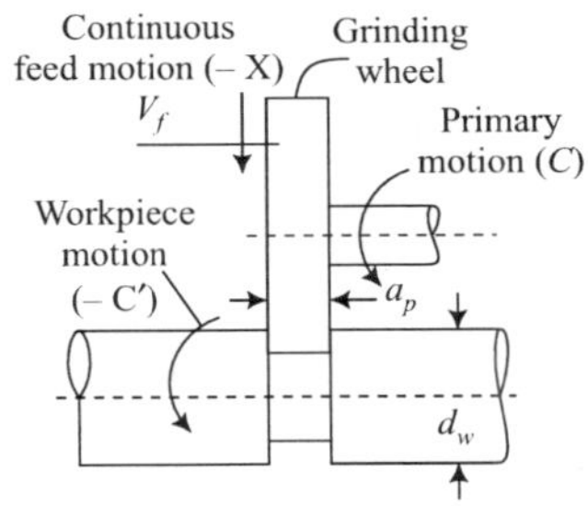

Fig. 1.18(b) Plunge grinding.

(a) The maximum metal removal rate for *traverse grinding,*

$$Z_{W_{max}} \cong \pi d_w f.\ V_{trav}.$$

where, d_w = diameter of work surface.

f = feed of machine table per stroke

f is very small as compared to d_w.

The machining time,

$$t_m = \frac{a_t}{2fn_r} + t_s.$$

(b) A groove is formed in a cylindrical workpiece by *plunge grinding*.

The maximum metal removal rate,

$$Z_{W_{max}} = \pi d_m f V_f$$

where, V_f = feed speed of grinding wheel

d_m = diameter of machined surface

a_p = back engagement

= width of grinding wheel

The machining time,

$$t_m = \frac{a_t}{V_f} + t_s.$$

where, a_t = total depth of metal to be removed

t_s = sparking-out time.

4. Internal grinding

A small cylindrical grinding wheel is rotated at a very high speed (primary motion $-C$). The horizontal spindle of wheel can be traversed parallel to workpiece (Z motion). The workpiece mounted on a chuck or magnetic faceplate is rotated (motion C'). The horizontal feed (infeed) is given to wheel spindle (motion X).

The geometries of traverse and plunge grinding are shown in Fig. 1.19.

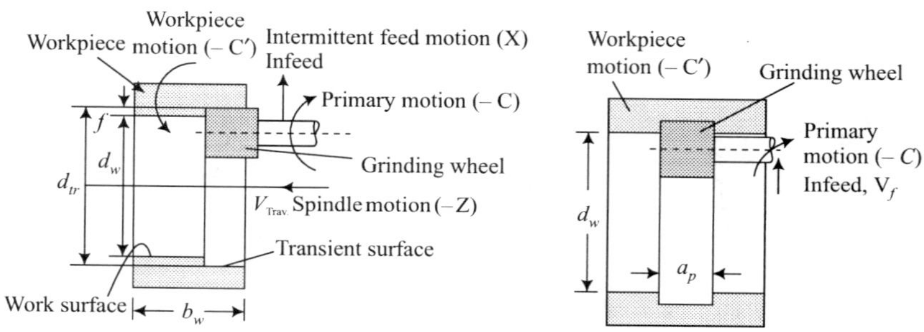

Fig. 1.19(a) Traverse grinding. **Fig. 1.19(b)** Plunge grinding.

(a) Maximum metal removal rate in *traverse grinding*,

$$Z_{W_{\max}} = \pi\, d_m f\, V_{\text{trav}}.$$

where, f = feed

V_{trav} = traverse speed

d_m = diameter of machined surface.

The machining time,

$$t_m = \frac{b_w}{2 f n_r}$$

(b) Maximum metal removal rate in *plunge grinding*,

$$Z_{W_{\max}} = \pi\, a_p\, d_m\, V_f$$

The machining time,

$$t_m = \frac{a_t}{V_f} + t_s.$$

5. Centreless grinding

Centreless grinding can be used for external grinding as well as internal grinding.

(a) External Centreless Grinding

The workpiece is supported by a rest blade and two wheels contact the workpiece. A conventional grinding wheel removes the material. A rubber bonded regulating wheel rotates at slow speed and causes the workpiece to rotate slowly so that whole of the external cylindrical surface is machined.

The geometry of external centreless grinding is shown in Fig. 1.20.

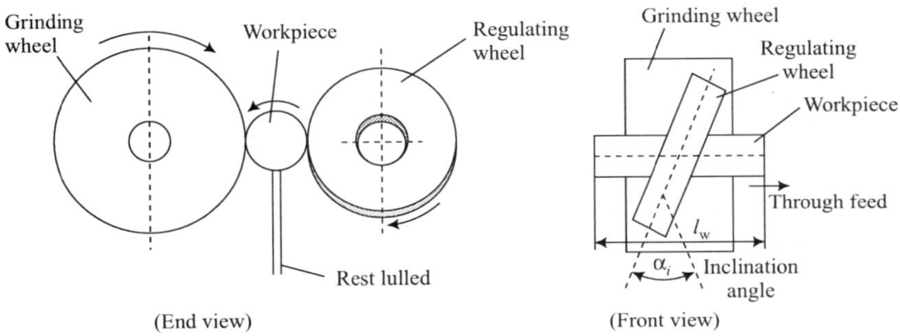

Fig. 1.20 External Centreless Grinding.

The axis of the regulating wheel is inclined at a small angle a_i to the axis of grinding wheel. This induces are axial feed velocity (V_f) to the workpiece.

where, $V_f = \pi\, d_r\, n_r \sin \alpha_i$

d_r = diameter of regulating wheel

n_r = rotational frequency of regulating wheel.

a_i = Inclination angle.

The machining time,

$$t_m = \frac{l_w}{V_f}$$

where, l_w = length of workpiece

(b) Internal Centreless Grinding

The geometry of internal centreless grinding is shown in Fig. 1.21. The workpiece is supported on two support rolls. The metal is removed by grinding wheel internally. The regulating wheel contacts the outside diameter of workpiece.

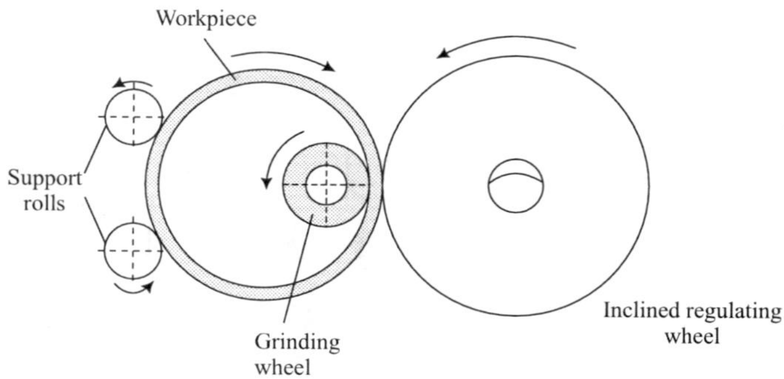

Fig. 1.21 Internal centreless grinding.

1.4 SUMMARY OF EQUATIONS FOR MACHINING OPERATIONS

The equations for machining operations are required for estimation of machining times and power consumption. The machining time is an important factor for considering the economics and production rate of an operation. The total production time should also include loading and unloading of workpieces on machining tools, changing tools, positioning of workpiece or tools before engaging the feed.

The estimation of power requirement is necessary for rough machining operations to remove metal at the highest rate possible although cost of power consumption in machining operations is not an important economic factor.

The equations for various machining operations have been summarized in Table 1.1 and equations for grinding operations have been summarized in Table 1.2. The following symbols have been used.

a_e = working engagement
a_p = back engagement
b_w = width of surface to be machined or workpiece
d_m = diameter of machined surface
d_t = diameter of tool
f = feed
l_t = length of tool
l_w = length of surface or hole to be machined
n_r = frequency of reciprocation
n_t = rotational frequency of tool
n_w = rotational frequency of workpiece
N = number of teeth on cutter
V = cutting speed
V_f = feed speed
K_r = major cutting edge angle
a_t = total depth of material to be removed
t_s = spark-out time
V_{trav} = traverse speed.

Table 1.1 Equations for Machining Operations.

Sl. No.	*Operation*	*Machining time* (t_m)	*Metal removal rate* (Z_w or Z_{wmax})	*Mean undeformed chip thickness* a_{Cav}
1.	Turning	$l_w/f n_w$	$\pi f a_p n_w (d_m + a_p)$	$f \sin K_r$ (items 1–4)
2.	Boring	$l_w/f n_w$	$\pi f a_p n_w (d_m - a_p)$	
3.	Facing/parting	$d_m/2fn_w$	$\pi f a_p n_w d_m$	
4.	Shaping & Planing	$b_w f n_t$	$f a_p V$	
5.	Drilling	$l_w/f n_t$	$\frac{\pi f_m^2 nt}{4}$	$\frac{f}{2} \sin k_r$.
6.	Slab milling	$l_w + \frac{\sqrt{a_e(d_t - d_e)}}{V_f}$	$a_e a_p V_f$ (items 6–8)	$\frac{V_f}{Nn_t}\sqrt{\frac{a_e}{d_t}}$
7.	Face milling	$\frac{l_w + d_t}{V_f}$		V_f/Nn_t.
8.	Side & face milling	$l_w + \frac{2\sqrt{a_e(d_t - d_e)}}{V_f}$		$\frac{V_f}{Nn_t}\sqrt{\frac{a_e}{d_t}\left(1 - \frac{a_e}{d_t}\right)}$
9.	Broaching	l_t/V	$f a_p V$ per engaged cutting edge	f

Table 1.2 Equations for Grinding Operations.

Sl. No.	*Machine*	*Operations*	*Machining time (tm)*	*Metal removal rate* (Z_w or $Z_{w\ max}$)
1.	Horizontal surface grinder	Traverse	$\frac{b_w}{2fn_r}$	$f a_p V_{trav}$. (items 1–2)
		Plunge	$\frac{a_t}{2fn_r} + t_s$	
2.	Vertical surface grinder - traverse	Reciprocating work table		
		Rotary work table	$\frac{a_t}{fn_w} + t_s$	
3.	Cylindrical grinder	Traverse	$\frac{a_t}{2fn_r} + t_s$	$\pi d_w f V_{trav}$
		Plunge	$\frac{a_t}{V} + t_s$	$\pi d_w a_p V_f$
4.	Internal grinder	Traverse	$\frac{a_t}{2fn_r} + t_s$	$\pi d_w f V_{trav}$
		Plunge	$\frac{a_t}{V} + t_s$	$\pi d_w a_p V_f$

QUESTION BANK NO. 1

1. What is the importance of machining in modern industry?
2. Classify different machining operations.
3. Explain axes system of motion of I.S.O.
4. What are different types of cutting tools?
5. Define different machining parameters.
6. What are the processes that can be carried out on a lathe?
7. Explain geometry of single-point tool operations.
8. How are different machining parameters estimated in cylindrical turning?
9. How is planing different from shaping?
10. Explain the geometry of drilling operation?
11. Describe different milling operations using geometries and machining equations.
12. What are different grinding operations?
13. What is broaching?
14. What is the difference between traverse grinding and plunge grinding?
15. Compare external grinding and internal grinding with suitable sketches and equations.
16. What is the importance of machining time, metal removal rate and power consumption in machining and grinding operations?

2

MECHANICS OF METAL CUTTING

Introduction; chip formation; orthogonal cutting; cutting forces in orthogonal cutting, theory of Ernst and Merchant; Theory of Lee and Shaffer, stress distribution on rake face; ploughing force; chip velocity; Machining with variable uncut chip thickness; oblique cutting; mechanics of turning process; Mechanics of milling process; Mechanics of drilling process; Question Bank.

2.1 INTRODUCTION

Mechanics of metal cutting consists of study of machining process and accurate estimation of dynamic cutting forces by the use of suitable analytical models. Different scientists have prepared different models and advanced their own theory and analysis of metal cutting action. There are certain basic concepts used by most of them. A wedge-shaped tool with a straight cutting edge is made to move relative to the workpiece and a layer of metal called chip is removed. The model of cutting process is shown in Fig. 2.1. The chip is formed by a continuous shearing action of workpiece and there is friction between the flowing chip and face of the tool.

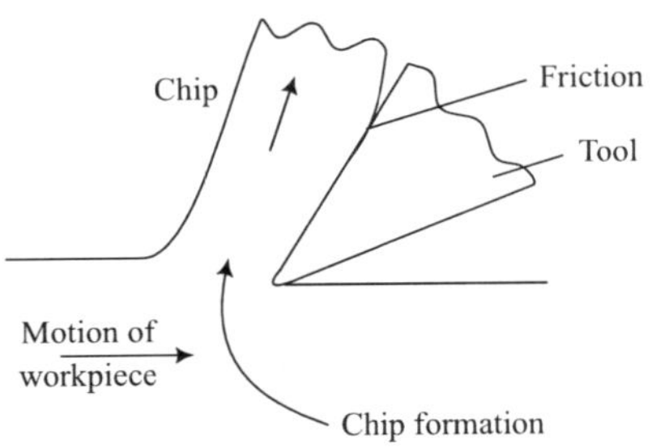

Fig. 2.1 Model of cutting process.

Different machining processes may be classified into two categories:

1. **Orthogonal cutting:** A special case of metal cutting, where the cutting edge of the fool is arranged perpendicular to the direction of relative work-tool motion is called orthogonal cutting. It represents a two-

dimensional problem and prone to research investigation. Many of the independent variables can be eliminated. This arrangement is used for theoretical and experimental work.

2. **Oblique cutting:** It is a general case of metal cutting where the cutting edge of the tool is not arranged perpendicular to the direction of relative woke-tool motion. It is a three-dimensional problem and more difficult to analyze. Only recently, a few attempts have been made to understand the mechanics of oblique cutting. However any analysis of the oblique cutting process also applies to orthogonal cutting which is a particular case of general oblique cutting.

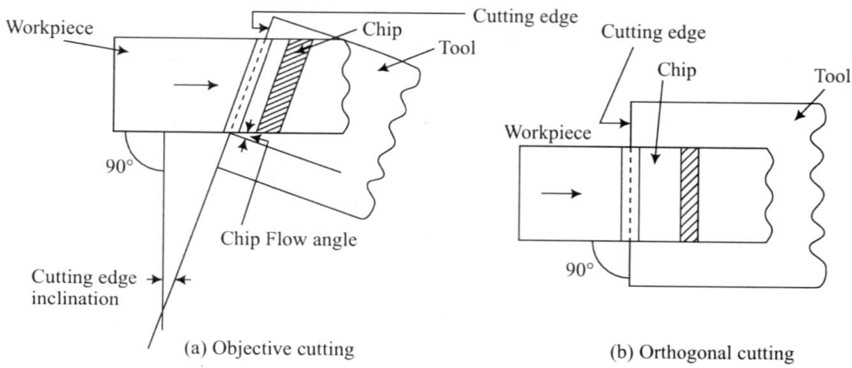

Fig. 2.2 Oblique and orthogonal cutting.

When the straight edge of tool moves relative to workpiece, the metal ahead of cutting edge deforms plastically before it goes to form the chip. The depth of the individual layer of material removed by the action of the tool is called *undeformed chip thickness* or uncut chip thickness. This dimension varies as cutting proceeds. However, due to complexity of an exact analysis, most scientists assume that this thickness remains constant for the simplicity of research work.

The wedge-shaped cutting tool basically consists of two surfaces intersecting to form the cutting edge:

1. **Face or rake face:** The chip flows along the rake face.
2. **Flank:** The tool surface ground back to clear the machined surface of workprice. The clearance between the flank of the tool and new machined surface of workpiece is called clearance *crevice.*

The mechanics of machining process depends upon the *rake angle* or slope of tool face. In orthogonal cutting, rake angle is the angle between the tool face and a line perpendicular to the new machined surface of workpiece. In ISO terminology, the rake angle is called the *working normal rake*.

The angle between tool face and normal to the cut surface is positive ($+\gamma_{ne}$) when the wedge angle is decreased. Negative rake angle ($-\gamma_{ne}$) results in stronger cutting edge and are used for rough machinery.

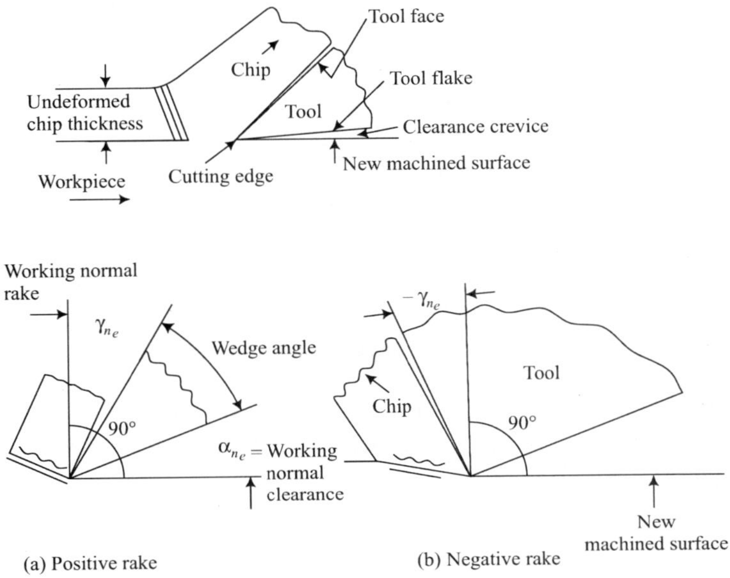

Fig. 2.3 Rake Angles.

The angle between the flank and the new machined surface (α_{ne}) is called *working normal clearance* or clearance angle. This angle does not play any part in the process of chip removal. However, this angle can significantly affect the wear rate of cutting tool.

The sum of rake, clearance and wedge angles is equal to 90 degrees.

2.2 CHIP FORMATION

All machining processes involve formation of chips by deforming the workpiece on its surface with the help of a cutting tool. The amount of plastic deformation of the material being machined depends on material of workpiece and cutting conditions used.

Following information can be determined from the above deformation.

1. Type of chip
2. Quality of machined surface
3. Cutting forces
4. Dimensional accuracy of the job
5. Temperature developed.

A variety of chip shapes and sizes are produced during different machining operations depending upon tool geometry, cutting conditions and work material Mainly following types of chips are formed:

1. Continuous chips
2. Continuous chips with built-up edge
3. Discontinuous chips
4. Segmented chips

2.2.1 Continuous Chip

A continuous ribbon-type chip is produced during cutting of ductile materials such as low carbon steel, copper, brass and aluminium alloys. The tool presses against the material which is deformed plastically. The material is subjected to both compression and shear. The chip slides over the tool rake face for some distance and then leaves the tool. Friction between the chip and tool may produce additional deformation of the chip material called secondary deformation. Similarly, the plastic zone ahead of tool edge is called the primary zone of deformation and the deformation zone on the rake face is called secondary zone of deformation, see figure 2.4.

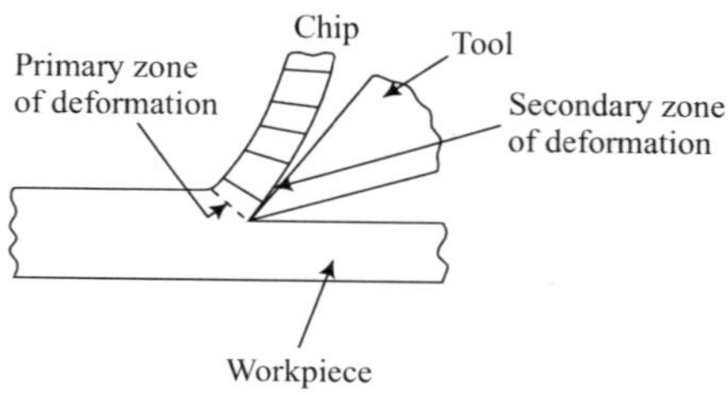

Fig. 2.4 Continuous chip.

Heat is generated in the two deformation zones. Heat is also generated due to sliding of chip on the rake face. The heat generated results in increase of temperature of tool-chip interface and that of tool.

The size of primary zone of deformation depends upon the following factors:

1. Rake angle of tool
2. Cutting speed
3. Properties of work material
4. Friction on rake face

1. Rake angle

The transition of work material into chip is gradual and the material suffers less overall deformation if rake angle of tool is large. The cutting forces are small. If rake angle is small or negative, the material suffers more severe deformation. The cutting forces are also large.

2. Cutting speed

The thickness of primary zone of deformation reduces and zone becomes narrower with the increase of cutting speed.

3. Work material

The size of the primary zone depends upon the following properties of work material.

1. Strength
2. Strain hardening
3. Strain rate
4. Heat conductivity

4. Friction on rake face

The size of both primary and secondary zones of deformation increases due to increase in friction between chip and tool rake face.

2.2.2 Continuous Chip with Built-up Edge

The chip slides under heavy pressure on the rake face. Also there is high temperature between the chip and tool. Due to high temperature and pressure existing, the chip may stick to the rake face of the tool. The heat of chip is conducted to the tool at close contact area. The chip becomes stronger and more of deforming work material is attracted. The size of built-up edge (*BUE*) goes on increasing till a "critical size" is reached when it breaks. The broken portion of *BUE* gets embedded into the machined surface or gets attached to the under side of the flowing chip. The building up and breaking of *BUE* is cyclic and there is a regular embedding of broken *BUE* into machined surface. The surface friction is affected.

With the increase of cutting speed, the temperature of cutting zone increases. The built-up edge is softened at high temperature and critical size of *BUE* reduces. At very high cutting speeds, the *BUE* disappears completely.

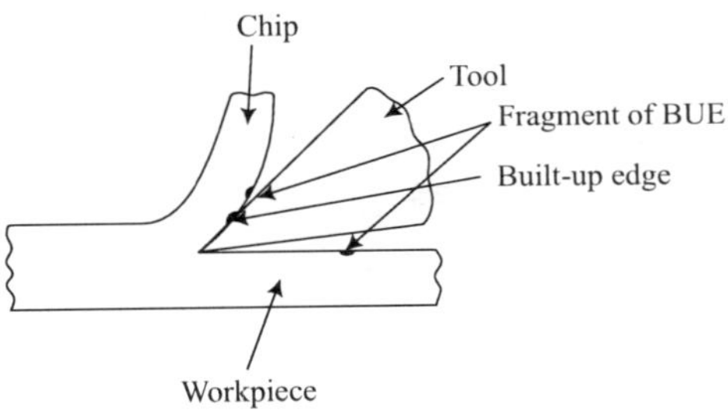

Fig. 2.5 Continuous chip with *BUE*.

2.2.3 Discontinuous Chip

Discontinuous chips are produced during machining of brittle materials such as cast iron and brasses. There is crack formation in the deforming zone ahead of cutting edge. The crack travels with further advancement of tool and results in formation of small lumps of chip. These lumps start moving up the rake face.

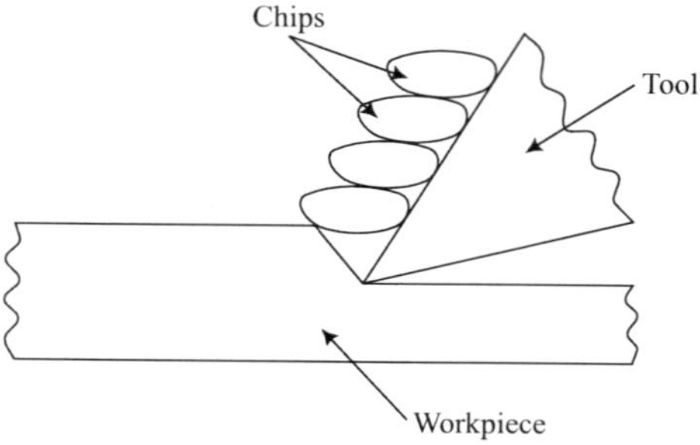

Fig. 2.6 Discontinuous chips.

The friction force obstructs the motion of the lump and crack propagates towards the surface. The fragment of chip gets detached. The heat generated in the cutting zone is mostly carried by the chip and tool life increases due to its lower temperature.

2.2.4 Segmented Chip

During machining of most engineering materials, segmented chips are produced which are a combination of continuous chips and discontinuous chips. The shape of these chips depends upon rake angle, cutting speed and material properties.

2.3 ORTHOGONAL CUTTING

The cutting edge of the tool is perpendicular to the direction of cutting speed. See Fig. 2.7.

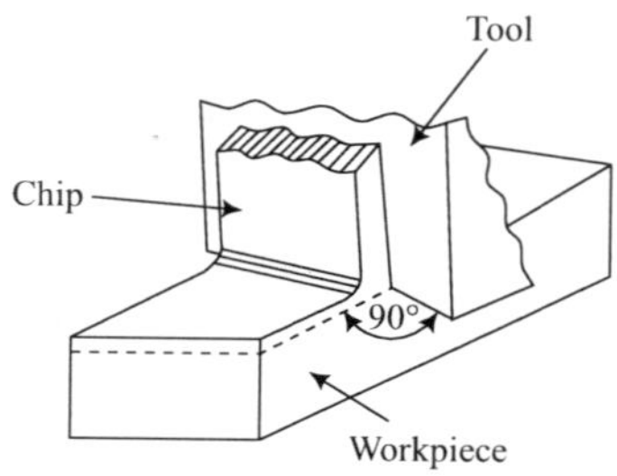

Fig. 2.7 Orthogonal cutting.

2.3.1 Shear Plane

The work material deforms plastically ahead of cutting tool edge. It slides on the rake face of tool and forms a chip. The region between the start of the chip and undeformed (elastically deformed) workpiece is called zone of plastic deformation.

The size of zone of plastic deformation depends upon cutting parameters. The size of this zone decreases with the increase of cutting speed. In the analysis of thin zones, it is assumed that the work material shears across a plane and forms the chip. This plane is called shear plane.

2.3.2 Shear Plane Angle

The angle between the cutting velocity vector and shear plane is called angle of shear plane ϕ. The chip is formed by plastic deformation of work material and material flow is continuous.

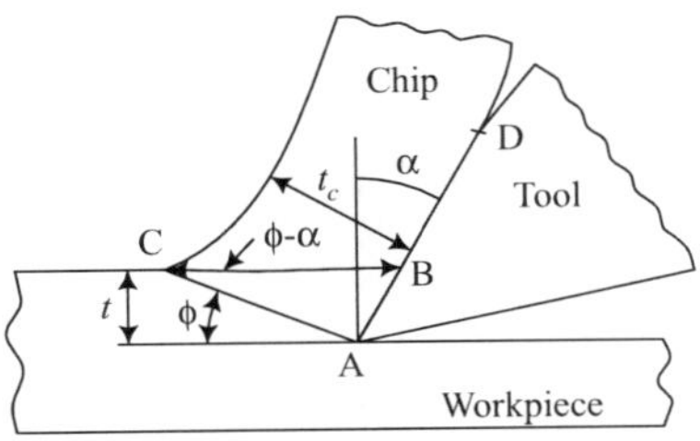

Fig. 2.8 Shear plane angle.

From equation of continuity,

$$t.b.v = t_c.b_c.v_c \quad ...(1)$$

where, t = deprth of cut.

b = width of cut

v = cutting velocity

t_c = chip thickness

b_c = width of chip

v_c = chip velocity

In most cutting processes, $b \approx b_c$

$$\therefore \quad tv = t_c v_c \quad ...(2)$$

The chip thickness ratio,

$$r_t = \frac{t}{t_c} = \frac{v_c}{v} = \frac{L_c}{L} \quad ...(3)$$

where, L = uncut chip length

L_c = length of chip formed.

From Fig. 2.8, the length of shear plane,

$$AC = \frac{t}{\sin\phi} = \frac{t_c}{\cos(\phi - \alpha)} \quad ...(4)$$

where, α = rake angle of tool.

$$\therefore \quad r_t = \frac{t}{t_c} = \frac{\sin \phi}{\cos(\phi - \alpha)} \qquad ...(5)$$

$$\therefore \quad \tan \phi = \frac{r_t \cos \alpha}{1 - r_t \sin \alpha} \qquad ...(6)$$

The shear angle ϕ can be experimentally found out from r_t and rake angle α.

A pipe of ductile material is turned with the help of tool with side rake angle (condition for orthogonal cutting). Measure L and L_c and calculate r_t. Measure rake angle α.

The tool should have only back rake angle and side zero rake angle for orthogonal cutting.

Example 2.1 The end of a pipe was orthogonally cut with a tool of 20° rake angle. The chip length was measured as 85 mm whereas uncut chip length was 202 mm. Determine shear plane angle and chip thickness if depth of cut was 0.5 mm.

Solution: Assume, the chip width b_c = width of uncut chip, b

Chip length $L_c = 85$ mm

Length of uncut chip $L = 202$ mm

$$\therefore \quad \text{Chip thickness ratio } r_t = \frac{L_c}{L} = \frac{85}{202} = 0.42$$

Now,

$$\tan \phi = \frac{r_t \cos \alpha}{1 - r_t \sin \alpha}$$

$$= \frac{0.42 \cos 20^\circ}{1 - 0.42 \sin 20^\circ}$$

$\therefore$ Shear plane angle, ϕ = **27.4° Ans.**

Chip thickness, $t_c = \frac{t}{r_t} = \frac{0.5}{0.42}$ = **1.19 mm Ans.**

Example 2.2 A specimen of 100 mm length along the stroke of a shaper is machined with 15° rake angle tool. Determine the shear plane angle and chip thickness if uncut chip thickness is 1.5 mm and chip length obtained is 40 mm.

Solution: Assuming that there is no change in the width of chip during machining,

$$t \times L = t_c \times L_c.$$

$$\therefore \quad \text{Chip thickness ratio,} \quad r_t = \frac{t}{t_c} = \frac{L_c}{L} = \frac{40}{100} = 0.4.$$

$$\tan \phi = \frac{r_t \cos \alpha}{1 - r_t \sin \alpha}$$

$$= \frac{0.4 \cos 15°}{1 - 0.4 \sin 15°}$$

Shear plane angle,

$$\phi = 25.8°$$

Chip thickness, $t_c = \frac{t}{r_t} = \frac{1.5}{0.4} = \mathbf{3.75\ mm}$ **Ans.**

2.4 CUTTING FORCES IN ORTHOGONAL CUTTING

In orthogonal cutting the resultant force $\vec{R}$ applied by the tool to the chip lies in a plane normal to the tool cutting edge. This can be determined experimentally by measuring its orthogonal components in the direction of cutting (known as cutting force F_h) and other normal to the direction of cutting (known as thrust force F_v) with the help of dynamometers. The magnitude of resultant force may be found out as follows

$$R = \sqrt{F_h^2 + F_v^2}$$

The component of R in the direction of width b is zero.

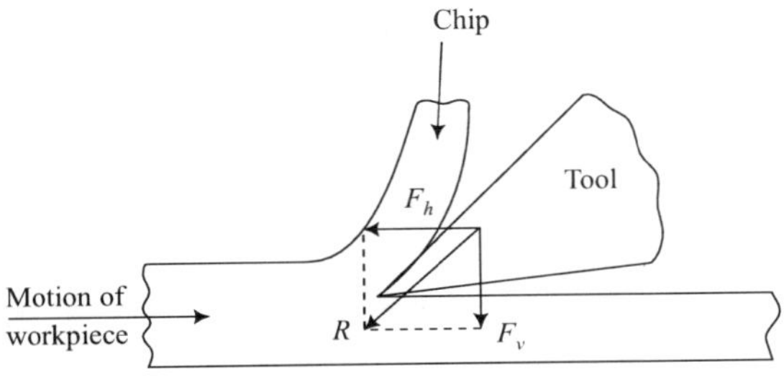

Fig. 2.9 Orthogonal component of cutting force.

The geometrical relationships of components of resultant force in other directions can be found out with the help of Fig. 2.10 which shows the following angles and directions.

Shear plane angle $= \phi$

Angle of friction $= \beta$

Tool rake angle $= \alpha$.

The following pairs of components are shown in selected directions.

1. F_h = Horizontal force component along cutting velocity vector.
   F_V = Vertical force component normal to cutting velocity vector.
2. F_s = Force component parallel to shear plane (AC)
   F_p = Force component normal to shear plane (AC)
3. F_t = Force component parallel to tool rake face.
   F_n = Force component normal to tool rake face.

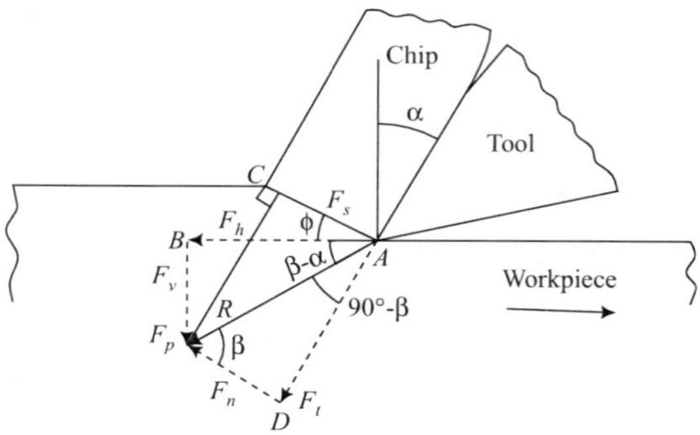

Fig. 2.10 Cutting forces.

$$\therefore \quad R = \sqrt{F_h^2 + F_v^2}$$

$$R = \sqrt{F_s^2 + F_p^2}$$

$$R = \sqrt{F_t^2 + F_n^2}$$

If average coefficient of friction between the chip and tool is μ.

$$\therefore \quad \mu = \tan \beta = \frac{F_t}{F_n} \qquad \text{...(1)}$$

$$\text{Also} \quad \tan(\beta - \alpha) = \frac{F_v}{F_h} \qquad \text{...(2)}$$

Equations (1) and (2) can be used to find friction angle β.

Expressing different force components in term of R, φ, β and α,

$$\left.\begin{aligned} F_s &= R\cos(\phi + \beta - \alpha) \\ F_P &= R\sin(\phi + \beta - \alpha) \end{aligned}\right\}$$

$$\left.\begin{aligned} F_h &= R\cos(\beta - \alpha) \\ F_v &= R\sin(\beta - \alpha) \end{aligned}\right\}$$

$$\left.\begin{aligned} F_n &= R\cos\beta \\ F_t &= R\sin\beta) \end{aligned}\right\}$$

2.5 THEORY OF ERNST AND MERCHANT

Merchant's model for orthogonal cutting is based on the assumption that shear plane angle ϕ should minimize the work done during cutting.

Following assumptions are made

1. The tool edge is sharp.
2. The shear plane is thin.
3. The deformation is in two dimensions only.
4. The normal and shear stresses are distributed uniformly on the shear plane.
5. The work material is rigid and perfectly plastic.

From Fig. 2.10, following relations were obtained:

$$F_h = R\cos(\beta - \alpha)$$

$$F_s = R\cos(\phi + \beta - \alpha)$$

Let yield stress of material in shear = k

$$\begin{aligned} F_s &= \text{Area} \times k \\ &= (\text{AC}) \times b \times k \\ &= \frac{t \times b}{\sin\phi} k. \end{aligned}$$

$$\therefore \quad R = \frac{F_s}{\cos(\phi + \beta - \alpha)} = \frac{t \times b \times k}{\sin\phi\cos(\phi + \beta - \alpha)}$$

$$\therefore \quad F_h = \frac{t \times b \times k\cos(\beta - \alpha)}{\sin\phi\cos(\phi + \beta - \alpha)} \qquad \text{...(1)}$$

The energy consumption during machining.

$$P_m = F_h.v$$

Assume that ϕ and β are not functions of cutting velocity v.

$$\therefore \quad \frac{dP_m}{d\phi} = \frac{dF_h}{d\phi} = 0$$

Differentiating equation (1) w.r.t ϕ and putting equal to zero for minimi-zation of F_h.

$$\therefore \quad \frac{t \times b \times \cos(\beta - \alpha)[\cos\phi\,(\phi + \beta - \alpha) - \sin\phi \sin(\phi + \beta - \alpha)]}{\sin^2\phi \cos^2(\phi + \beta - \alpha)}$$

$$\sin\phi \neq 0 \text{ and } \cos(\phi + \beta - \alpha) \neq 0$$

$$\therefore \quad \cos(\phi + \beta - \alpha) = 0$$

$$\cos(\beta - \alpha) \neq 0$$

$$\therefore \quad \cos(2\phi + \beta - \alpha) = 0$$

Hence, optimum value,

$$\phi = \frac{\pi}{4} - \frac{(\beta - \alpha)}{2}.$$

Example 2.3 Determine the shear plane angle, cutting force component and resultant force on the tool for orthogonal cutting of a material with yield stress of 250 N/mm^2. Following are the machining parameters.

Tool rake angle = 15°

Uncut chip thickness = 0.25 mm

Chip width = 2 mm

Chip thickness ratio = 0.46

Angle of friction = 40°

Solution:

$$\tan\phi = \frac{r_t \cos\alpha}{(1 - r_t \sin\alpha)}$$

$$= \frac{0.46 \cos 15°}{(1 - 0.46 \sin 15°)}$$

$$\therefore \quad \phi = 29.57°$$

Shear force along the shear plane,

$$F_s = \frac{t.b.K}{\sin\phi} = \frac{0.25 \times 2 \times 250}{\sin 29.57°}$$

$$= 279.04 \text{ N}$$

The resultant force on cutting tool,

$$R = \frac{F_s}{\cos(\phi + \beta - \alpha)} = \frac{279.04}{\cos(29.57 + 40° - 15°)}$$

$$= \frac{279.04}{\cos 54.57°} = 426.3 \text{ N}.$$

The cutting force component,

$$F_h = R\cos(\beta - \alpha)$$

$$= 426.3\cos(40° - 15°) = 426.3\cos 25°$$

$$= \mathbf{393.85\ N} \quad \textbf{Ans.}$$

Example 2.4 Determine the shear plane angle, resultant force on the tool and cutting force component for orthogonal cutting operation of a material with shear yield strength of 200 N/mm^2. The machining data is as follows:

Uncut chip length = 100 mm
Length of cut length = 50 mm
Rake angle of tool = 10°
Width of cut = 1.5 mm
Uncut chip thickness = 0.2 mm
Coefficient of friction = 0.8

Solution: Chip thickness ratio, $r_t = \dfrac{L_c}{L} = \dfrac{50}{100} = 0.5$

$$\tan\phi = \frac{r_t\cos\alpha}{1 - r_t\sin\alpha}$$

$$= \frac{0.5\cos 10°}{1 - 0.5\sin 10°}$$

∴ Shear plane angle,

$$\phi = 28.34°.$$

Shear force along shear plane,

$$F_s = \frac{tbK}{\sin\phi} = \frac{0.2 \times 1.5 \times 200}{\sin 28.34°}$$

Resultant force on cutting tool,

$$R = \frac{tbK}{\sin\phi.\cos(\phi + \beta - \alpha)}$$

$$\tan\beta = \mu = 0.8$$

∴ $$\beta = 38.66°$$

∴ $$R = \frac{126.3}{\cos(28.34° + 38.66° - 10°)} = \frac{126.3}{\cos 57°}$$

$$= \mathbf{232\ N} \quad \textbf{Ans.}$$

Cutting force component

$$F_h = R \cos(\beta - \alpha) = 232 \cos(38.66° - 10°)$$

$$= \mathbf{203.58\ N} \quad \textbf{Ans.}$$

2.6 THEORY OF LEE AND SHAFFER

The theory of Lee and Shaffer is based on slip line field theory and applies simplified plasticity analysis to the problem of orthogonal metal cutting. The following assumptions are made on the behaviour of work material under stress.

1. The material is rigid plastic. The stress-strain curve is shown in Fig. 2.11.
2. The behaviour of material is independent of rate of deformation.
3. The effect of temperature increase during deformation are negligible.
4. The inertia effects resulting from acceleration of material during deformation are negligible.

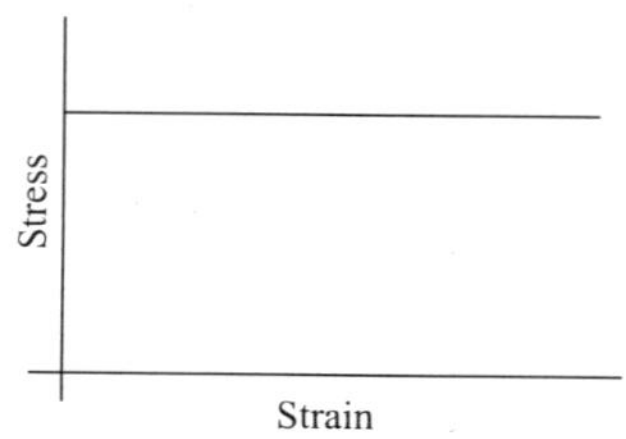

Fig. 2.11 Stress-stain curve for a rigid-plastic material.

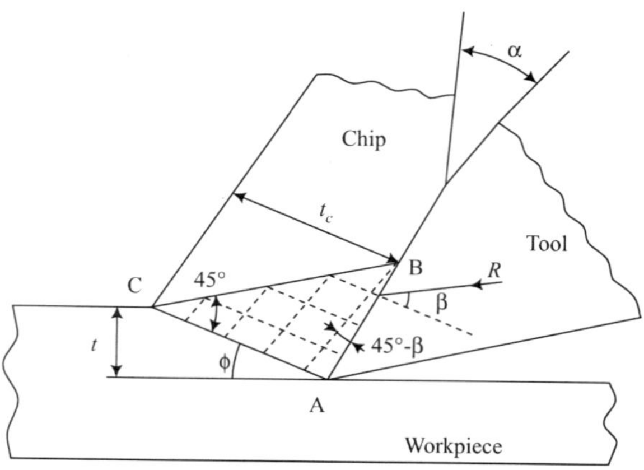

Fig. 2.12 Slip-line field for orthogonal cutting.

The slip-line field for orthogonal cutting is shown in Fig. 2.12.

R = resultant tool force

β = mean friction angle on tool face

ϕ = shear plane angle

α = rake angle

t = undeformed chip thickness

t_c = chip thickness

Triangle *ABC* contains the zone of deformation. The plane *CB* is stress free and slip lines meet *CB* at 45°. *AC* is the shear plane and there is velocity discontinuity along *AC*. A set of slip lines are parallel to *AC* and other set perpendicular to *AC* and inclined at an angle (45° – β) with the tool rake face.

From Fig. 2.12, $\angle CAB = 45° + \beta$

$$\therefore \quad \phi + \angle \text{CAB} = 90° + \alpha$$

$$\phi = 45° - (\beta - \alpha).$$

or, $$\phi + \beta - \alpha = 45° \quad ...(1)$$

Equation (1) is the required shear-angle solution.

Case 1: Zero rake angle

If $\beta = 45°$ and $\alpha = 0°$, $\phi = 0°$

High friction and low rake angle leads to formation of built-up edge.

Case 2: Negative rake angle

If $\beta = 35°$ and $\alpha = -10°$, $\phi = 0°$.

This solution is again impracticable.

2.7 STRESS DISTRIBUTION ON RAKE FACE

In the analysis of orthogonal cutting as per theory of Lee and Shaffer, shear and normal stresses are uniformly distributed in the plastic region and also there is uniform distribution of stress on the rake face. However, it is found that this assumption is not true. Both the stresses are found experimentally to vary along the contact length.

Photo-elastic apparatus was used and stress distribution pattern on the rake face were determined by the analysis of fringe patterns observed. The forces were measured by dynamometers. The distribution of normal and shear stresses on the rake face found out experimentally are shown in Fig. 2.13.

2.7.1 Idealized Stress Distribution

Idealized stress distribution of normal stress (*CB*) along the rake face is shown by an exponential curve in Fig. 2.14. The shear stress is shown constant upto a portion of contact length from tool tip. This zone is called "**sticking zone**". The shear stress exponentially decreases to zero beyond sticking zone. This constant stress zone is called "slipping zone" and Coulomb's laws of friction are applicable.

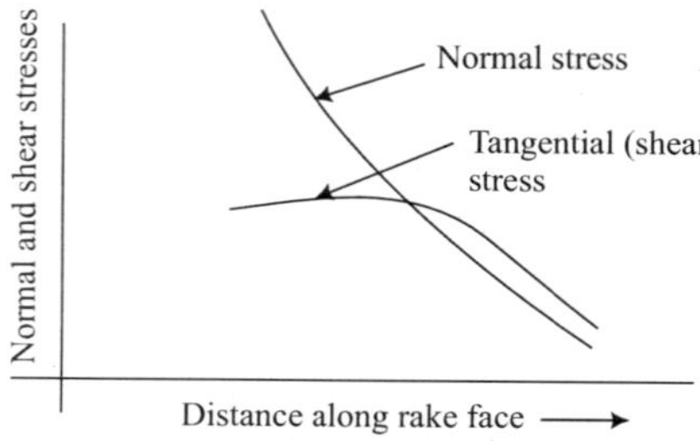

Fig. 2.13 Normal and shear stress distribution on rake face.

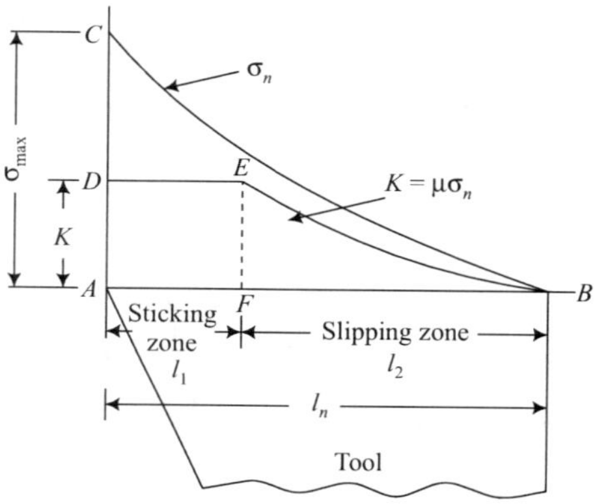

Fig. 2.14 Idealized stress distribution on rake face.

2.7.2 Stress Relationships

The distribution function of normal stress

$$\sigma_n = Ax^m$$

where, A = constant

m = constant

x = distance from end of chip-tool contact.

If σ_n is the normal stress on the rake face at a distance x from the end of contact length.

$$\sigma_n = \sigma_{max}\left(\frac{x}{l_n}\right)^m$$

where, σ_{max} = max. normal stress at the cutting edge

l_n = natural contact length between tool and chip.

Total normal force on the rake face,

$$F_n = b\int_0^{\ln} \sigma_{\max}\left(\frac{x}{l_n}\right)^m dx$$

$$= \frac{b\sigma_{\max} l_n}{m+1} \quad ...(1)$$

Shear stress distribution

(i) In sticking zone, shear stress is constant and is equal to shear yield stress of deforming material.

(ii) In slipping zone, shear stress is proportional to normal stress. As per Coulomb's law, shear stress,

$$\tau = \mu\sigma_n$$

where, μ = average coefficient of friction between chip and tool.

$$\therefore \quad K = \mu\sigma_{\max}\left(\frac{l_2}{l_n}\right)^m$$

$$\therefore \quad l_2 = \left(\frac{K}{\mu\sigma_{\max}}\right)^{1/m} .l_n$$

Total tangential force on the rake face,

$$F_t = F_{t_1} + F_{t_2}\,.$$

where, F_{t_1} = tangential force in sticking zone

F_{t_2} = tangential force in slipping zone

Now F_{t_1} = Kb ($l_n - l_2$)

and

$$F_{t_2} = \int_0^{l_2} b\mu\sigma_n dx$$

$$= \int_0^{l_2} b\mu\sigma_{\max}\left(\frac{x}{l_n}\right)^m dx$$

$$= \frac{\mu b.\sigma_{\max}}{(m+1)}\,\frac{(l_2)m+1}{(l_n)^m} \quad ...(2)$$

Average coefficient of friction,

$$\mu = \tan\beta = \frac{F_t}{F_n} = \frac{(F_{t_1} + F_{t_2})}{F_n}$$

Substituting the values of F_{t_1}, F_{t_2} and F_n,

$$\mu = \frac{K.(l_n - l_2)(m+1) + \mu\sigma_{\max}(l_2)^{m+1}}{\sigma_{\max}(l_n)^{m+1}}$$

2.8 PLOUGHING FORCE

The resultant tool force in metal cutting is distributed over the areas of the tool that contact the chip and workpiece. No cutting tool is perfectly sharp and the cutting edge can be represented by a cylindrical surface joining the tool flank and the tool face. For a freshly ground high-speed steel tool, the radius of the edge varies from 0.005 to 0.03 mm. Deformation of tool material occurs in this region due to high stresses on the cutting edge. There is contact between the tool and the new workpiece surface over a small area of tool flank. A frictional force may act in the tool flank region. This force does not contribute to the removal of the chip. This force is called ploughing force.

The total resultant tool force,

$$R = F_c + F_p$$

where, F_c = force required to remove the chip

F_p = ploughing force acting on the tool edge and work-tool interface region.

The specific cutting energy increases especially at low values of undeformed chip thickness.

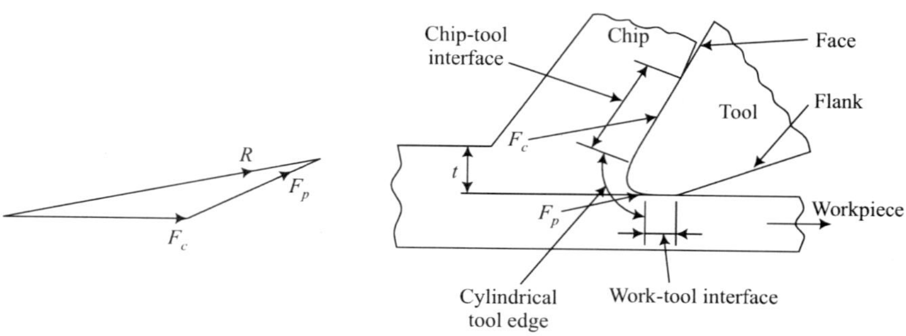

Fig. 2.15 Contact regions in a cutting tool.

2.9 CHIP VELOCITY

The chip velocity along the tool rake face, V_c is the vector sum of velocity of uncut chip V and the velocity discontinuity along the shear plane.

$$\vec{V}_c = \vec{V}_t + \vec{V}_s$$

The velocity of the uncut chip is the velocity of workpiece with respect to cutting tool. The surface layer shears across the shear plane AC and becomes a part of the chip. Therefore, surface layer suffers a velocity discontinuity parallel to shear plane as shown in Fig. 2.16.

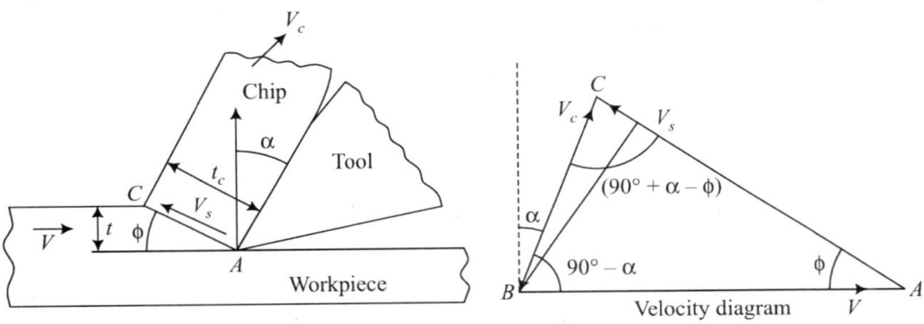

Fig. 2.16 Velocity relationship in orthogonal cutting.

From velocity triangle BAC, using Lami's theorem,

$$\frac{V}{\sin(90° + \alpha - \phi)} = \frac{V_s}{\sin(90° - \alpha)} = \frac{V_c}{\sin\phi}$$

The chip velocity,

$$V_c = \frac{V \sin\phi}{\sin(90° + \alpha - \phi)} = \frac{V \sin\phi}{\cos(\phi - \alpha)}$$

The shear velocity,

$$V_s = \frac{V \sin(90° - \alpha)}{\sin(90° + \alpha - \phi)} = \frac{V \cos\alpha}{\cos(\phi - \alpha)}.$$

2.10 MACHINING WITH VARIABLE UNCUT CHIP THICKNESS

In most machining processes such as milling, grinding, gear cutting, etc., the chip thickness varies during cutting. The uncut chip thickness may change from maximum to zero or zero to maximum. The vibration and chatter of machine tools also lead to chip thickness variation.

The following two cases are shown in Fig. 2.17.

1. The workpiece with positive surface slope. The uncut chip thickness varies from maximum to zero.

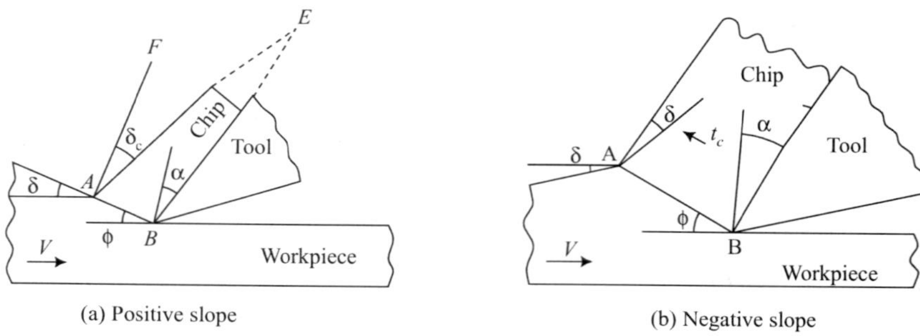

Fig. 2.17 Variable uncut chip thickness.

2. The workpiece with negative surface slope. The thickness of uncut chip varies from zero to maximum.

Following relationship may be used to find shear plane angle ϕ in case of variable or unsteady chip thickness cutting. The instantaneous shear plane angle.

$$\phi = \phi_0 + c\delta$$

where, ϕ_0 = shear plane angle when slope of work surface is zero, i.e., $\delta = 0$

δ = slope of workpiece surface ahead of tool edge.

c = constant

= 0.50 Merchant

= 1.00 Shaw and Sanghoni

= 0.75 Wallace and Andrew

= 0.20 Oxleg

= 1.00 Kebayashi and Shabaik.

In most cases, c can be taken to unity. From so many values of c by various authors, it can be suggested that it is not a constant and may be a function of α, δ and ϕ_0.

2.11 OBLIQUE CUTTING

The rake and other tool angles required for ideal orthogonal cutting cannot be achieved due to practical considerations. Therefore, most machining processes approach oblique cutting with orthogonal cutting as a particular case.

2.11.1 Chip Flow Direction

The angle between the normal to the cutting edge and chip velocity vector is called chip flow angle. In orthogonal cutting, chip flows in a direction normal to the cutting edge and chip flow angle is zero.

The chip flow angle can be found out by the following methods:

1. **Direction of scratches on the tool:** A dye is smeared on the rake face of the tool and direction of scratches produced by chip flow can be observed.
2. **Photography of cutting process:** The photographs can be analyzed to find the chip flow direction.
3. **Deformation of chip:** The chip flow angle can be simply and accurately found out by the analysis of deformation of chip.
4. **Analysis of cutting forces:** The shear force component on the rake face is the direction of chip flow.

The chip flow angle,

$$\psi = \tan^{-1} (\tan I. \tan \alpha_n)$$

where, I = angle of obliquity

= angle of cutting edge with normal to cutting velocity vector

α_n = normal rake angle

$$\psi \approx 1 \text{ for } a_n = 0$$

$$> 1 \text{ for } a_n < 0$$

$$< 1 \text{ for } a_n > 0.$$

The relationships between angle of obliquity and chip flow angle are plotted in Fig. 2.18 for different values of normal rake angle. There are based on experimental results of Brown and Armarego.

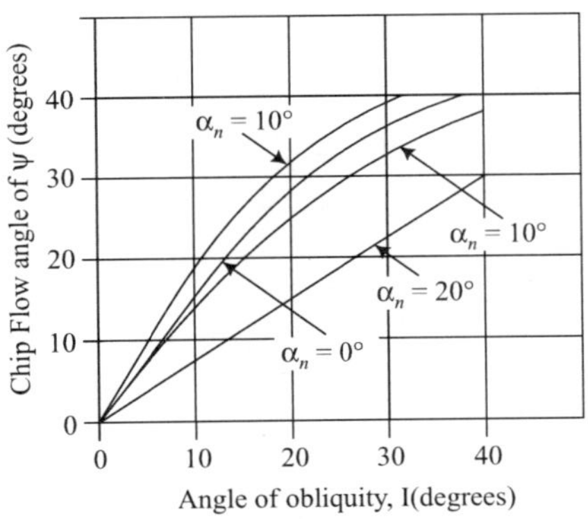

Fig. 2.18 Chip flow angle.

2.11.2 Rake Angle

Various angles in oblique cutting depend upon tool rake angle and angle of obliquity. The rake angle is measuresd in different planes.

1. **Normal rake angle or primary rake angle, α_n:** It is the angle between the rake face and a line perpendicular to cutting velocity vector and measured in a plane normal to the tool cutting edge.
2. **Velocity rake angle, α_v:** It is the angle between the rake face and a line perpendicular to the cutting velocity and measured in a plane parallel to cutting velocity and normal to machined surface.
3. **Effective rake angle, α_e:** It is the angle between the rake face and a line normal to the cutting velocity and measured in a plane containing cutting velocity vector and chip flow velocity vector.s

   The velocity rake angle,

$$\alpha_v = \tan^{-1}\left(\frac{\tan \alpha}{\tan I}\right)$$

   Effective rake angle,

$$\alpha_e = \sin^{-1}(\sin I \sin \psi + \cos I \sin \alpha_n \cos \psi)$$

   The above relationships are found geometrically.

2.11.3 Cutting Ratios

In oblique cutting, width of chip is not equal to the width of uncut chip. However, volume of material undergoing plastic deformation remains constant.

$$\frac{L}{L_c} \times \frac{t}{t_c} \times \frac{b}{b_c} = 1$$

where, L = length of uncut chip

t = thickness of uncut chip

b = width of uncut chip

L_c = length of cut chip

t_c = thickness of cut chip

b_c = width of cut chip

The chip ratio,

$$r_t = \frac{t}{t_c} = \frac{b_c L_c}{bL}$$

Chip length ration, $r_L = \dfrac{L}{L_c}$

Chip thickness ratio, $r_c = \dfrac{t}{t_c}$

$$\text{Chip width ratio, } r_b = \frac{b}{b_c}$$

$$\therefore \qquad r_t = \frac{1}{r_b . r_L}$$

2.11.4 Velocity Relationships

The velocity diagram for oblique cutting is shown in Fig. 2.19.

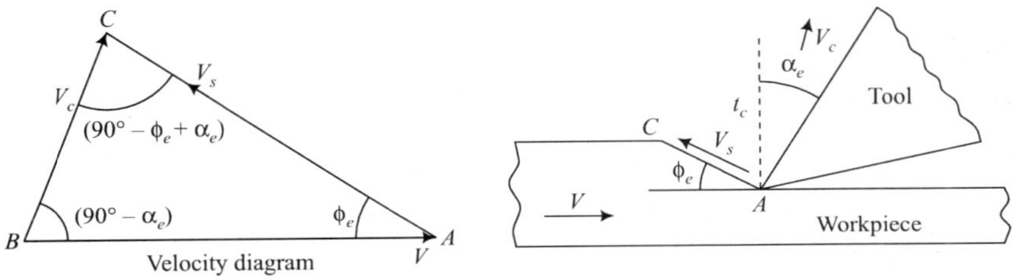

Fig. 2.19 Velocity relationship.

V = uncut chip velocity

V_c = chip velocity

V_s = shear velocity

ϕ_e = effective shear plane angle

α_e = effective rake angle

$$\vec{V_c} = \vec{V} + \vec{V_s}$$

From properties of triangle ABC,

$$\frac{V_c}{V} = \frac{\sin \phi_e}{\cos (\phi_e - \alpha_e)}$$

$$\frac{V_s}{V} = \frac{\cos \phi_e}{\cos (\phi_e - \alpha_e)}$$

Expressing the relationship in terms of chip flow angle ψ, angle of obliquity I and normal rake angle α_n,

$$\frac{V_c}{V} = \frac{\sin \phi_n \cos I}{\cos (\phi_n - \alpha_n) . \cos \psi}$$

and,
$$\frac{V_s}{V} = \frac{\cos I \cos \alpha_n}{\cos \psi \cos (\phi_n - \alpha_n)}.$$

2.11.5 Shear Plane Angle

The angle between shear velocity vector, V_s and the cutting edge is called shear plane angle. Similar to rake angle, shear angle can be measured in different planes. Refer to Fig. 2.19.

$$\tan \phi_e = \frac{V_c/V \cos \phi_c}{1 - V_c/V \sin \alpha_e}$$

Normal shear angle is measured in a plane normal to cutting edge.

$$\tan \phi_n = \frac{\left(\dfrac{t}{t_c}\right) \cos \alpha_n}{1 - \left(\dfrac{t}{t_c}\right) \sin \alpha_n}$$

2.12 MECHANICS OF TURNING PROCESS

A turning tool has complex geometry with the following important angles which affect the chip flow direction.

1. Rake angle (α)
2. Back rake angle (α_b)
3. Side rake angle (α_s)
4. Side cutting edge angle (γ_s)
5. End cutting edge angle (γ_e)
6. Nose radius

The chip flow direction becomes complex due to the presence of nose radius and for mathematical analysis, the presence of nose radius is neglected.

The top view of a turning tool is shown in Fig. 2.20, where

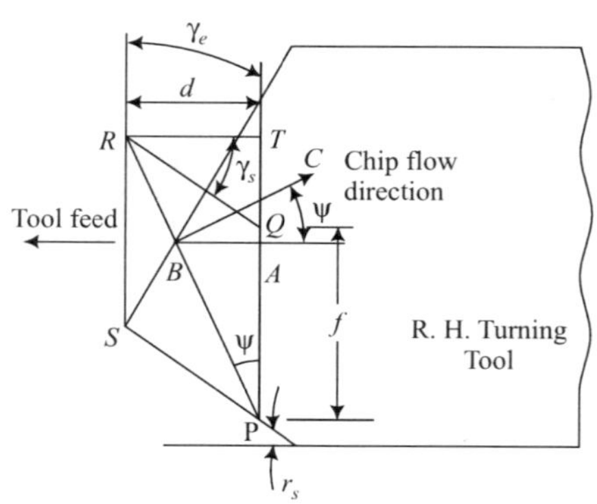

Fig. 2.20 Chip flow direction in turning.

PS = engagement length of side cutting edge during cutting

$PQ = RS$ = feed, f

ψ = chip flow angle

The chip flow direction is normal to the major diagonal of parallelogram $PSRQ$. Draw $RT \perp$ to PQ and $BC \perp$ to PR and $BA \perp$ to PQ.

$$\angle ABC = \angle RPQ = \psi = \text{chip flow angle}$$

From ΔRPQ,

$$\tan \psi = \frac{RT}{PT} = \frac{RT}{(PQ + QT)}$$

$$= \frac{d}{f + d \tan \gamma_s}$$

If the nose radius r_n is small,

$$\tan \psi = \frac{(d - r_n)}{f + (d - r_n) \tan \gamma_s}$$

2.12.1 Effective Rake Angle

Assuming nose radius as zero, the effective rake angle for a turning tool can be calculated from the following relationship.

$$\tan \alpha_e = \tan \alpha_s \sin \psi + \tan \alpha_b \cos \psi$$

where, α_s = side rake angle

ψ = chip flow angle.

α_b = back rake angle

2.12.2 Forces and Power Consumption

The turning process is shown in Fig. 2.21. There are three components of the cutting force

F_r = Radial component

F_t = Tangential component

F_a = Axial component

There is no movement of tool in radial direction and the feed velocity is small, therefore, no work is done by radial and axial components of cutting force.

Power consumption in turning = work done in turning

$$\therefore \quad P = \frac{F_t \times V}{60} \text{ [W]}$$

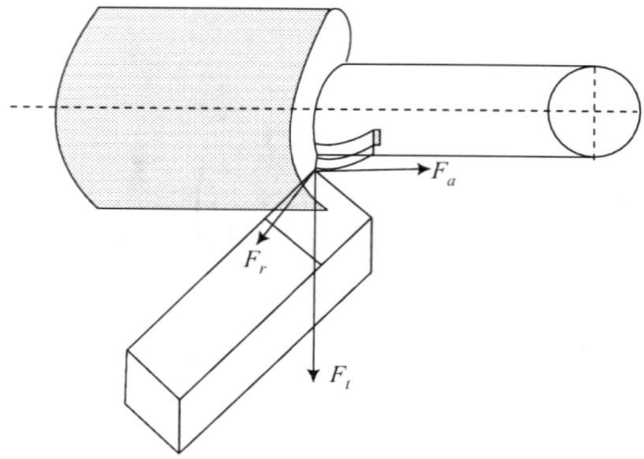

Fig. 2.21 Forces in turning.

where, F_t = tangential force component [N].
V = cutting velocity [m/min].

The cutting force components can be expressed by the following equations.

$$F_t = R_1 f^{m_1} d^{n_1} \text{ [N]}$$

$$F_a = R_2 f^{m_2} d^{n_2} \text{ [N]}$$

$$F_r = R_3 f^{m_3} d^{n_3} \text{ [N]}$$

where, f = feed
d = depth of cut
R_1, R_2, R_3 = material coefficients
m_1, m_2, m_3 and n_1, n_2, n_3 are exponents.

Various values of coefficients and exponents are given in Table 2.1.

Table 2.1 Coefficients and exponents for turning forces

Material	*Coefficients*			*Exponents*					
	R_1	R_2	R_3	m_1	m_2	m_3	n_1	n_2	n_3
0.2% carbon steel	1590	337	397	0.85	0.8	0.67	0.98	1.46	0.47
Brass	1210	220	558	0.81	0.91	0.97	0.96	1.43	0.38
18-8 stainless steel	1930	368	876	0.85	0.48	0.71	0.96	1.26	0.69

2.12.3 Specific Energy

Specific energy is the power consumption per unit volume of material removed.

$$P_s = \frac{F_t}{(f.d)}$$

The metal removal rate,

$$Z_w = f.d.V. \times \frac{1000}{60} \text{ [mm}^3\text{/s]}$$

where, f = feed (mm)

d = depth of cut [mm]

V = cutting speed [m/min]

Power consumption,

$$P = F_t \times \frac{V}{60} \text{ [W]}$$

$$P_s = \frac{F_t}{1000 \times f \times d} \text{ [kW/cm}^3\text{/sec].}$$

Example 2.5 Calculate the specific energy and unit power in a turning process. The machining data are:

Diameter of workpiece = 50 mm

Cutting speed = 40 m/min

Feed = 0.24 mm/sec

Depth of cut = 1.8 mm

Tangential component of force = 800 N

Axial component of force = 290 N

Solution: Metal removal rate,

$$Z_w = fd\frac{V}{60} \text{ [cm}^3\text{/s]}$$

$$= \frac{0.24}{10} \times \frac{1.8}{10} \times \frac{40 \times 100}{60} = 0.288 \text{ cm}^3\text{/sec.}$$

Power consumed,

$$P = \frac{F_t \times V}{60} \text{ [W]}$$

$$= 800 \times \frac{40}{60} \text{ [W].}$$

Assume that power consumped by feed force is negligible.

Unit power, $$P_s = \frac{P}{Z_w} = \frac{800 \times 40}{60 \times 0.288} = 1.5 \text{ kW/cm}^3\text{/sec}$$

2.13 MECHANICS OF MILLING PROCESS

Slab milling and end milling processes are the most versatile milling processes for producing flat and formed surfaces. The cutter can move in both directions and the processes are called up-milling and down-milling. The average chip thickness is higher in down-milling as compound to up-milling. The power consumption is less in down-milling.

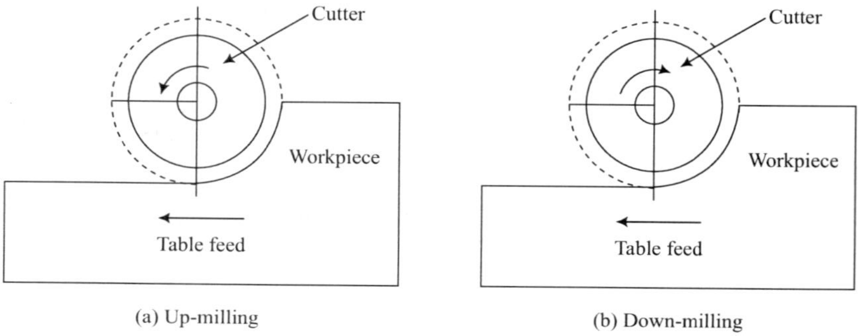

Fig. 2.22 Types of milling processes.

2.13.1 Chip Thickness in Slab Milling

The two positions of milling cutter are shown as C and C' during the distance travelled equal to feed per tooth, f_t.

The uncut chip thickness,

$$t = f_t \sin \theta$$

where, θ = angle of a radius with the vertical.

α = contact angle of cutter with workpiece.

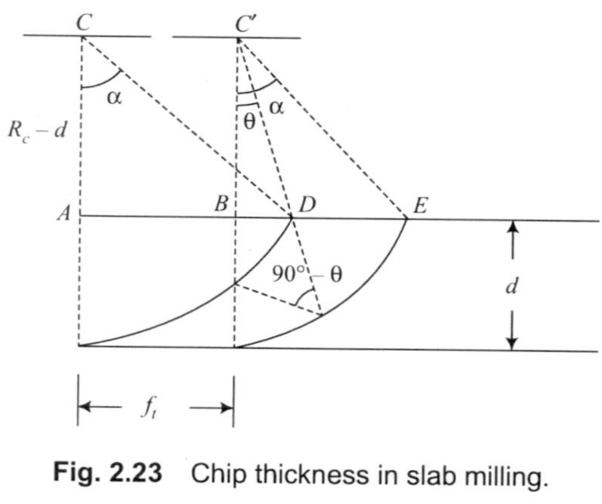

Fig. 2.23 Chip thickness in slab milling.

∴ Average chip thickness,

$$t_a = \frac{1}{\alpha}\int_{\theta=0}^{\theta=\alpha} f_t \sin\theta d\theta$$

$$= \frac{f_t(1-\cos\alpha)}{\alpha} \qquad ...(1)$$

Maximum chip thickness

$$t_m = f_t \sin\alpha \qquad ...(2)$$

But $\cos\alpha = \dfrac{(R_c - d)}{R_c}$

where, R_c = cutter radius

d = depth of cut

$$\sin\alpha = \sqrt{(1-\cos^2\alpha)}$$

$$= \frac{\sqrt{2R_c d - d^2}}{R_c}$$

Substituting into equation (2)

$$t_m = f_t\sqrt{\left[\frac{2d}{R_c} - \left(\frac{d}{R_c}\right)^2\right]} \qquad ...(3)$$

Substituting the value of cos α into equation (1)

$$t_a = f_t \frac{d}{\alpha R_c} = f_t\sqrt{\frac{d}{2R_c}}.$$

2.13.2 Chip Thickness in Face Milling

Face milling process is shown in Fig. 2.24. The positions of milling cutter are O and O' for a feed of f_t (feed per tooth).

The average chip thickness

$$t_a = f_t \sin\theta$$

$$= f_t\left(\frac{\cos\alpha_2 - \cos\alpha_1}{\alpha_2 - \alpha_1}\right).$$

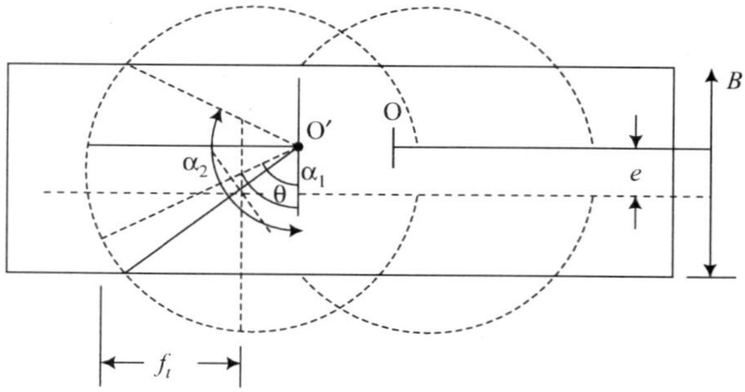

Fig. 2.24 Chip thickness in face milling.

2.14 MECHANICS OF DRILLING PROCESS

There are empirical relations for thrust and torque in drilling process.

The thrust,

$$F = C_1 f^{m_1} \left(B.D + \frac{W}{D} \right)^{n_1}$$

where, D = drill diameter

f = drill feed

W = web thickness

B, C_1 and C_2 = constants for a material

m_1 and n_1 = exponents for a material

(i) For 0.2% carbon steel,

$$F = 242.3 \times 10^3 f^{0.87} \left(\frac{D}{127} + \frac{W}{D} \right)^{2.12} \text{ [N]}$$

(ii) For cast iron,

$$F = 67.4 \times 10^3 f^{0.87} \left(\frac{D}{127} + \frac{W}{D} \right)^{1.9} \text{ [N]}$$

(iii) For aluminium,

$$F = 196.2 f^{1.1} D^{1.2} \text{ [N]}$$

(iv) For brass,

$$F = 174 f^{0.6} D \text{ [N]}$$

The torque,

$$M = C_2 f^{m_2} D^{n_2}$$

(i) For 0.2% carbon steel

$$M = 0.7 f^{0.78} D^{1.8} \text{ (N-m)}$$

(ii) For cast ion,

$$M = 0.292 f^{0.6} D^{1.7} \text{ (N-m)}$$

(iii) For aluminium,

$$M = 0.1265 f^{0.83} D^{1.9} \text{ (N-m)}$$

(iv) For leaded brass,

$$m = 0.167 f^{0.72} D^{1.9} \text{ (N-m)}.$$

Example 2.6 A drill of 12 mm diameter is used to drill holes with a feed of 0.2 mm per rev. and speed of 300 rpm in C.I. Calculate the thrust, torque and power by empirical relations.

Solution: The thrust,

$$F = 67.4 \times 10^3 f^{0.87} \left(\frac{D}{127} + \frac{W}{D} \right)^{1.9} \text{ [N]}$$

$$= 67.4 \times 10^3 (0.2)^{0.87} \left(\frac{12}{127} + 0.153 \right)^{1.9}$$

$$= 5337 \text{ N}$$

$$\text{Torque, } M = 0.292 (0.2)^{0.6} (12)^{1.7} \text{ N-m}$$

$$= 7.6 \text{ N-m}$$

$$\text{Power, } P = MV = 7.6 \times \frac{2\pi \times 300}{60} = 238.76 \text{ W.}$$

QUESTION BANK NO. 2

1. Differentiate between oblique cutting and orthogonal cutting.
2. Explain various rake angles.
3. Discuss the formation process of various types of chips.
4. How can shear plane angle be estimated in orthogonal cutting?
5. What are various components of forces in orthogonal cutting? How can these be estimated?
6. How can optimum value of shear angle be estimated by using theory of Ernst and Merchant?
7. What are the basis of theory of Lee and Shaffer?

8. Establish stress relationships on rake face in idealized orthogonal cutting conditions.
9. Write notes on:
   (a) Ploughing force
   (b) Chip velocity
   (c) Machining with variable uncut chip thickness.
10. How can chip flow direction be established in oblique cutting process?
11. Write notes on the following for oblique cutting.
    (a) Velocity relationships
    (b) Cutting ratios
    (c) Rake angles
    (d) Shear plane angle.
12. How are the following estimated in turning process?
    (a) Chip flow direction
    (b) Effective rake angle
    (c) Specific energy and unit power
13. Explain mechanics of milling and drilling processes.

3

HEAT GENERATION IN METAL CUTTING

Introduction; Heat sources and heat distribution; Shear plane temperature in orthogonal cutting; Measurement of cutting temperatures. Question Bank.

3.1 INTRODUCTION

Heat is generated in the region of tool cutting edge raising the temperature of tool, workpiece and the chip. When the tool deforms the workpiece during cutting the material is first deformed elastically and then plastically. During elastic deformation, the energy required for operation is stored in the material as strain energy, and no heat is generated. However, during plastic deformation, most of the energy used is converted into heat. The elastic deformation forms a very small proportion of the total deformation; therefore, it may be assumed that all the energy is converted into heat.

The flow of chip against friction on the tool rake face is another source of heat. Similarly, the rubbing of the machined surface on the flank face (due to elastic recovery and wear of flank face) is yet another source of heat generation. All these four sources of heat tend to raise the temperature of the chip, tool and workpiece.

The temperatures raised during metal cutting have controlling influence on the tool life, accuracy of machining, chip formation and machining conditions. The tool life decreases with temperature and there is early wear of tool. There can be damage to the dimensional accuracy and surface properties of the machined component. The built-up edge of the chip increases with temperature leading to increased friction.

3.2 HEAT SOURCES AND HEAT DISTRIBUTION

All the mechanical work done in metal cutting is converted into equivalent amount of heat. Out of total energy consumed in the cutting process, this maximum portion (about 80%) is consumed in the primary and secondary zones of deformation. The energy loss in friction between chip and rake face of tool may go upto 18% and the friction loss at flank face may account for about 2%. A small fraction of energy consumed in deformation is stored in the chip and the rest appears as heat.

Conversion of energy into heat occurs in the following zones of plastic deformation as shown in Fig. 3.1.

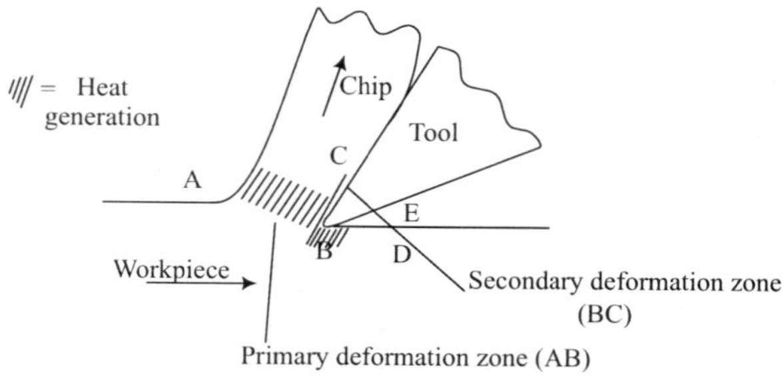

Fig. 3.1 Generation of heat in orthogonal cutting.

1. Shear zone or primary deformation zone, *AB*.
2. Secondary deformation zone, *BC*.
3. Zone of friction between new workpiece surface and tool, BD. This heat source would be small and can be neglected.

   Therefore, equivalent rate of heat generation during machining,

$$P_m = P_s + P_f$$

where, P_s = Rate of heat generation in the primary deformation zone (shear-zone heat rate)

P_f = Rate of heat generation in the secondary deformation zone (frictional heat rate)

$$P_m = F_c V$$

where, F_c = cutting component of resultant tool force

V = cutting speed

$$P_f = F_f V_0$$

where, F_f = Frictional force on the tool face

V_0 = Velocity of chip flow

$$P_s = P_m - P_f.$$

Heat is removed from these zones by the workpiece, chip and the tool. The relative amount of heat transformed to chip, workpiece and tool at different cutting speeds is shown in Fig. 3.2. This diagram is for machining of steel with a

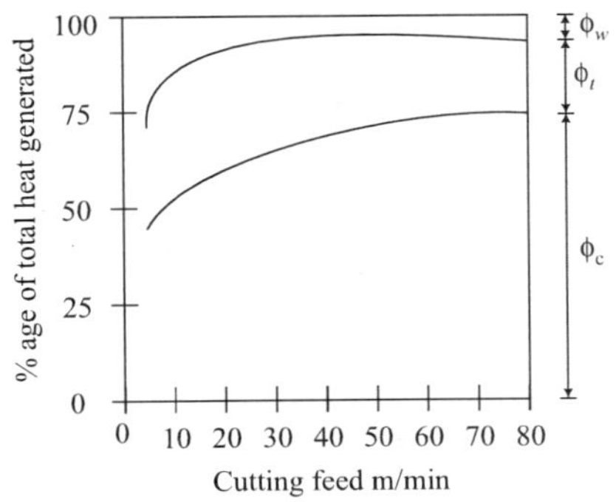

Fig. 3.2 Distribution of heat during metal cutting.

single-point tool having cemented carbide tip. As the cutting speed increases, more heat is carried away by the chip and less heat is transferred to the workpiece and the tool. High speed machining is, therefore, advantageous to the tool life. The total rate of heat generation,

$$P_m = \phi_c + \phi_w + \phi_t$$

where, ϕ_c = rate of heat transportation by the chip,

ϕ_w = rate of heat conduction into the workpiece.

ϕ_t = rate of heat conduction into the tool.

The chip material near the tool face is flowing rapidly. It can remove the heat from the tool fast. Therefore, ϕ_t forms a very small proportion of P_m and may be neglected.

3.3 SHEAR PLANE TEMPERATURE IN ORTHOGONAL CUTTING

The rate of heat generation in the primary deformation zone is P_s, and a fraction of this heat, Γ is conducted into the workpiece, the remainder $[1 - \Gamma]P_s$, is transported with the chip. The average temperature rise θ_s, of the material passing through the shear plane is given by:

$$\theta_s = \frac{(1-\Gamma)O_s}{\rho C V a_c a_w}$$

where, a_w = width of the chip

ρ = density of work material

C = specific heat of work material

a_c = depth of cut or uncut chip thickness

V = cutting velocity [m/min].

Example 3.1 A steel alloy of yield strength 300 N/mm^2 (in shear) is orthogonally cut with a tool of 10° rake angle. Estimate the temperature rise at the shear plane. Take $\Gamma = 0$ and angle of fiction $\beta = 42°$.

Solution: The shear plane angle,

$$\phi = 45° - 1/2(\beta - 10°) \qquad \text{(Merchant's model)}$$

$$= 45° - 1/2(42° - 10°) = 29°$$

$$P_s = \frac{F_s V_s}{a_w a_c V} \text{ [N-m/cm}^3\text{]}$$

where, F_s = Shear force on the shear plane [N]

V_s = Shear velocity [m/min]

$$\therefore \quad P_s = \frac{F_s \cos\alpha}{\sin\phi \cos(\phi - \alpha)}$$

$$= \frac{300 \cos 10°}{\sin 29° \cos 10°} = 704.67 \text{ N.m/cm}^3$$

Temperature rise in shear plane,

$$\theta_s = \frac{P_s(l - \Gamma)}{0.44 \times 7.87} = \frac{704.67 \times 1}{0.44 \times 7.87}$$

$$= 203.5°\text{C}.$$

Example 3.2 Estimate the temperature rise at the shear plane of orthogonal cutting of mild steel from the following data.

1. Density of mild steel, $\rho = 7.87$ gm/cm^3.
2. Specific heat of mild steel, $C = 0.44$ J/gm
3. $\Gamma = 0$.
4. Force component in the direction of cutting velocity = 1600 N
5. Force component normal to machine surface = 500 N
6. Depth of cut, $a_c = 0.3$ mm
7. Width of cut, $a_w = 5$ mm
8. Chip thickness ratio, $r = 0.42$
9. Tool rake angle, $\alpha = 10°$
10. Cutting velocity, $V = 35$ m/min

Solution: The shear plane angle, $\phi = \tan^{-1}\left[\dfrac{r\cos\alpha}{1 - r\sin\alpha}\right]$

$$\phi = \tan^{-1}\left[\frac{0.42 \cos 10^\circ}{1 - 0.42 \sin 10^\circ}\right]$$

$$= 26.6^\circ$$

$$\text{Resultant force, } F_s = \sqrt{(1600^2 + 500^2)} = 1676 \text{ N}$$

$$\tan(\beta - \alpha) = \frac{500}{1600} = 0.3125$$

$$(\beta - \alpha) = 19.28^\circ$$

$$\beta = 19.28^\circ + 10^\circ = 29.28^\circ.$$

$$F_s = R \cos(\phi + \beta - \alpha)$$

$$= 1676 \cos(26.6 + 29.28 - 10)$$

$$= 1676 \cos 44.88^\circ$$

$$= 1259.27 \text{ N.}$$

$$V_s = \frac{V \cos\alpha}{\cos(\phi - \alpha)} = \frac{35 \cos 10^\circ}{\cos(26.6 - 10^\circ)} = 35.78 \text{ m/min}$$

$$P_s = \frac{F_s V_s}{a_c a_w V} = \frac{1259.27 \times 35.78}{(5 \times 0.3 \times 35 \times 1000)}$$

$$= 0.858 \text{ N-m/mm}^3$$

$$\theta_s = \frac{0.858 \times 1000}{0.44 \times 7.87} = 247.7^\circ\text{C}$$

3.4 MEASUREMENT OF CUTTING TEMPERATURES

Following analytical methods have been used to predict temperature distribution in machining:

1. Finite element simulation using Johnson-Cook model
2. Finite element simulation using Zerelli-Armstrong model
3. Finite element simulation using power law model
4. Finite difference calculation.

It has been found that experimental peak temperatures measured are very near to those found by finite difference calculation. A number of methods have been developed for the measurement of temperatures in metal cutting.

1. Tool-workpiece thermcouple
2. Embedded thermocouple
3. Infrared photographic technique

3.4.1 Tool-Work Material Thermocouple

The workpiece is insulated from the chuck and tail stock centre. The external connection is made with the mercury cup through a galvanometer to the tail end. The magnitude of the current depends upon the temperature difference between the hot junction (tool-chip interface) and cold junction (mercury cup). This method gives an average temperature at the tool-chip interface.

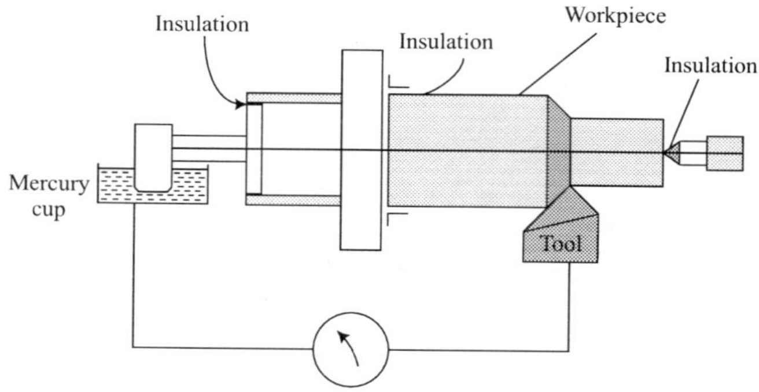

Fig. 3.3 Tool-Workpiece Thermocouple.

3.4.2 Embeded Thermocouple

Thermocouples are implanted in fine holes of 0.45 mm diameter made in HSS tool from the bottom face. A large number of such thermocouples are distributed over the rake face of the tool. Graphs are plotted between temperature and distance from cutting edge along contact lengths for different feeds as shown in Fig. 3.4.

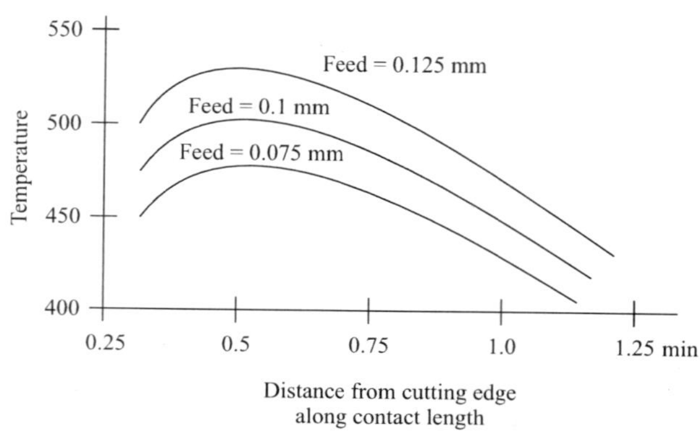

Fig. 3.4 Embedded Thermocouple.

3.4.3 Infrared Photographic Technique

The side face of the tool-chip along with strips of known temperature are photographed during cutting operations. The temperature distribution on the tool, chip and workpiece are estimated by comparison with the intensities of strips of known temperature.

3.4.4 Hardness and Microstructure Changes in Tools

The hardness and microstructure of tool steel change with temperature and time of heating. These changes can be observed using optical and electron microscopes. These changes are effective indicators to determine temperature distribution in the tool during cutting. Measurements of micro-hardness on tools after cutting can be used to plot constant temperature contours in the tool.

QUESTION BANK NO. 3

1. Describe a method to determine tool-chip contact temperature.
2. Discuss the sources of heat generation and its distribution among chip, tool and workpiece.
3. How can the temperature in the shear plane be determined analytically and experimentally?

4

CUTTING FLUIDS

Introduction; Beneficial and harmful effects of cutting fluids; Desirable properties; Categories of cutting fluids; Composition of cutting fluids; The action of cutting fluids; Selection of cutting fluids; Application of cutting fluids; Maintenance and disposal of cutting fluids.

4.1 INTRODUCTION

Cutting fluids also called lubricants or coolants are liquids and gases applied to the tool and workpiece to assist in the cutting operations. The cutting fluids help to produce components of high accuracy and surface finish more economically. The above achievements are as a result of combined contribution of tool, work material, cutting fluid and cutting conditions. The cutting fluid can act as:

1. **Lubricant:** The lubricating action is very important at low cutting speeds.
2. **Coolant:** The cooling effect is very important at high cutting speed due to large heat generation by the chip removal process.

There are many types of cutting fluids used for different machining operations. The most widely used cutting fluids are water based emulsions of oils or synthetic fluids. Neat oils are used for slow speed cutting operations such as screw cutting, broaching and gear cutting when lubricating action of cutting fluids is of most important consideration.

The cost of cutting fluids can be as high as 20% of total manufacturing cost in machining as compared to only 7.5% cost of actual cutting tools. Therefore, special attention should be paid to reducing the quantity and extending the useful life of cutting fluids in addition to their proper selection of grade.

4.2 BENEFICIAL AND HARMFUL EFFECTS OF CUTTING FLUIDS

4.2.1 Beneficial Effects

The Cutting Fluids are Used for the Following Purposes:

1. **Cooling of tool**

Cooling of tool in the region of the tool cutting edge helps in the reduction of temperature and metallurgical damage to the tool material. The cooling action of the cutting fluids is by two mechanisms:

(i) There is decrease of friction at the tool-chip interface and at the tool-workpiece interface. Decreasing friction results in not only less power requirement but also less heat generation leading to increased tool life and good surface finish.

(ii) The cutting fluid carries away the heat developed by plastic deformation of the shear plane and that due to friction.

Hence, a good coolant must possess high specific heat, high thermal conductivily and high film-coefficient for heat transfer. Water has all the above properties and can be a very effective coolant. But it is objectionable due to its corrosion properties and lack of lubricating property to reduce friction and wear.

The tool life depends upon its temperature as discussed in the previous chapter.

2. **Cooling of workpiece**

Cooling of workpiece results in reduction of thermal distortion caused by temperature gradients generated within the workpiece during machining. This factor is very important in grinding process. This cooling effect of workpiece gives improved surface finish and accuracy of size.

3. **Lubrication effects**

The lubrication property of cutting fluids can give the following advantages:

(i) Reduction of cutting forces and energy consumption in removing the metal as a result of reduction of friction.

(ii) Reduction of abrasion and wear on the cutting tool thereby increasing tool life.

(iii) Reduction in heat generation results in operation of tool at lower temperatures thereby enhancement of its life and cutting conditions.

(iv) Assistance in flowing out of chips from flutes of drills, taps, dies, saws, broaches, etc.

(v) Improvement in the cutting operation as the built-up edge is reduced leading to decreased friction at the tool-workpiece interface with cooler tool.

Proper selection of lubricant can ensure optimum cooling effect and lubrication condition in metal cutting.

4. **Surface condition**

The surface finish is improved and the finished surface is protected from corrosion. Corrosion inhibitors such as sodium nitrate or triethanolamine are added to water to protect the finished surface from corrosion.

5. **Handling of workpiece**

It is easier to handle cooled finished workpiece especially in the grinding process.

6. **Chip management**

The cutting fluids can flush away the chips from the cutting area. Properly placed fluid nozzles can help prevent the packing of chips in the flutes of drills or milling cutters. This helps to prevent fouling of the cutting tool with the workpiece.

The cutting fluid also helps to break up chips into small pieces. It is difficult to remove hot and sharp long ribbon of chips from the workpiece.

4.2.2 Harmful Effects

Some cutting fluids may produce harmful effects as toxic vapours produced can adversely affect the human operator. The component surface can also be affected.

1. Some cutting fluids can cause allergic reaction on the skin.
2. Additives like N-Nitroamines and Polychlorinated Biphenyls are cancer generating chemicals and should not be used.
3. Microbial contamination of cutting fluid can be a health hazard.
4. The working environment can become unpleasant due to emission of unpleasant odours in addition to cause of occupational hazards.
5. Some cutting fluids can pose a fire hazard.
6. The work surface may be stained by the chemical action of some cutting fluids.
7. The cutting fluids can corrode slide ways and other parts of machine tools affecting its working and life.

The cutting fluid should lubricate the slide ways of machine tool.

4.3 DESIRABLE PROPERTIES

The cutting fluids should process the following general properties to realize the full benefits.

1. *Good lubricating property* to produce low coefficient of friction.
2. *High heat absorption capacity* to readily remove heat generated during cutting action.
3. *Low viscosity* to ensure free flow of fluid.
4. *Chemically stable and non-corrosive* to the workpiece and machine tool.
5. *High flash point* to eliminate fire risks.
6. *Transparent* to allow complete visibility of cutting action of the tool.
7. Emit *no toxic fumes* when put on hot surface. The fumes, if produced, should be *odourless*.
8. *Chemically neutral* and harmless to the skin of the operators and bearings and other parts of the machine tool.
9. Easily available at low cost to minimize production cost.

4.4 CATEGORIES OF CUTTING FLUIDS

Cutting fluids are classified in seven main groups. The selection of a cutting fluid depends upon work material and characteristics of machining process.

1. **Water:** Water can be used as a coolant in plain form or a salt or alkali solution. No oil or soap is added. However, in most cases, pure water is objectionable due to its corrosive property.
2. **Soluble oils:** Soluble oils are used as emulsions of water, soap and mineral oil. Soap is added as emulsifier which breaks the oil into minute particles and which thus distributed throughout water. It contains more than 80 per cent water which increases the cooling effect. The mineral oil provides the lubricating property and prevents rusting. Cutting fluids can be prepared with various proportions of oil, soap and water to get a wide range of cooling and lubricating properties. These fluids are widely used in industry due to low cost.
3. **Neat oils:** These oils are used undiluted and have both cooling and lubricating properties. These are preferred for light machining operations. These oils may be:
   (a) Mineral oils, kerosene and petroleum products
   (b) Vegetable, animal or synthetic oils.
4. **Mixed oils:** Mineral and fatty oils are mixed to get excellent cooling and lubricating properties. These are used for high precision and good finish of components using automatic screw-machine or other light machining operations.
5. **Sulphurized and chlorinated mineral oil:** Sulphur and chlorine are used to increase both the cooling and lubricating properties of mineral oils. Sulphurized mineral oils are used for machining tough, stringy, low-carbon steels. Chlorinated mineral oil promotes anti-weld properties.
6. **Chemical compounds:** Chemical compounds are solutions of sodium nitrate and other rust inhibitors in water. These are mostly used in grinding and machining where formation of rust has to be avoided.
7. **Solid lubricants:** Stick wax and bar soap can be conveniently applied as a lubricant to cutting tool.

4.5 COMPOSITION OF CUTTING FLUIDS

The cutting fluids are formed by mixing different chemicals to achieve favourable properties and to control the harmful effects. The cutting fluids should be formulated with the following chemicals.

1. Base oil
2. Emulsifiers
3. Corrosion inhibitors for non-ferrous alloys and rust inhibitor for ferrous alloys.

4. Lubricating, anti-wear and high pressure additives.
5. Neutralizing agents.
6. Biocides and fungicides.
7. Foam inhibitors.
8. Stabilizing agents.

The important additives along with their functions are given in Table 4.1.

Table 4.1 Important constituents of cutting fluids.

Sl. No	Constituent	Function
1.	Mineral oils and other hydrocarbons	Base oils
2.	Water soluble oils such as polyglycoether	Emulsifier.
3.	Aliphatic amines (for water soluble oils)	Neutralizing agent.
4.	Aliphatic amines in neutralizing form	Corrosion inhibitors
5.	Sulphonates	Corrosion inhibitors and high pressure additives.
6.	Fatty acid amines	Lubrication property enhancement.
7.	Sulphur/Phosphorus compounds	High pressure additives.
8.	Aldehyde derivatives (for water based coolants)	Biocides.

The formulation of cutting fluids is a highly complex process. Some additives have harmful effects and should not be used. Some agents have potential for cancer generation and some additives have been identified as harmful wastes and their disposal requires high cost processes. The following chemicals should not be used.

1. Chlorinated paraffins
2. Benzopyrene
3. N-Nitroamines and polychlorinated biphenyl.

   The above additives can produce cancer.

4.6 THE ACTION OF CUTTING FLUIDS

The cutting fluids are applied for the following two effects:

1. Cooling effect
2. Lubrication effect.

4.6.1 Cooling Action

The cutting fluid has the following advantages as coolant.

1. Temperature reduction in the cutting region helps to increase tool life.
2. There is decrease in thermal distortion of workpiece caused by temperature gradients.
3. Handling of cooled workpiece is easier.
4. The cutting fluid helps to flush out chips from the cutting area.

4.6.2 Lubrication Action

A lubricant boundary layer is formed which reduces the area of intimate metallic contact between the surfaces. The high pressure additives like sulphur, phosphorus or chlorine form solid lubricant layer by chemical reaction between the cutting fluid and the metal surface. At melting point, the lubricant gives protection during sliding.

1. The action of cutting lubricants results in considerable reduction in mean coefficient of friction on the tool face.
2. The action of cutting lubricants have startling effects on cutting process, reducing power consumption by as much as 60% in many cases.
3. The effect of a lubricant is to eliminate the built-up edge and secondary deformation.
4. There is reduction in friction at the chip-tool interface, reducing the forces required to form the chip.
5. The friction reduction on tool face can cause a reduction in power consumption, an increase in tool life, and an improvement in the surface finish of the machined component by reducing the occurrence of a built-up edge.

The cutting fluid should have the following properties to act as an efficient lubricant.

1. Small molecular size for rapid diffusion and penetration to the chip-tool interface.
2. Should have reactive gradient to form a compound to act as boundary lubricant of low shear strength.
3. It should be unstable and break down under pressure and temperature existing at the chip-tool interface.

Certain tetrachloride (CCl_4), chloroform ($CHCl_3$), tri-chloroethane (CH_2CCl_3) and other chlorinated hydrocarbons have sufficient lubricating properties even when cutting many metals at low speeds.

Pure chemicals are toxic in nature. Agents and additives are used for the formulation of cutting fluids.

4.7 SELECTION OF CUTTING FLUIDS

The selection of a proper cutting fluid for a given application is a difficult problem. The initial cost of the cutting fluid is less important than the resultant overall cost of production. The following guidelines should be followed.

1. Neat oil of high pressure characteristics should be used for low speed machinery. At low speeds, the heat transfer characteristics are not important but lubrication property should be high. For high speed machining, good heat transfer properties are very important.
2. Water soluble oils may be used as cutting fluids for medium and high speed cutting.

The choice of cutting fluid depends upon the following factors:

1. Type of machining operation
2. The rate of metal removal
3. Material of the workpiece
4. Tool material
5. Surface finish required
6. Cost of cutting fluid.

Some of the recommended cutting fluids for common work materials are given below:

1. **Aluminium:** The chips of aluminium have the tendency to weld onto the tool. Kerosene or light mineral oils mixed with kerosene are the suitable cutting fluids for machining of aluminium workpiece.
2. **Copper and copper alloys:** Some copper alloys produce tough and stingy chips. Mineral oil with 5 to 25% lard oil is used for machining of such alloys. For other alloys, water soluble oils in the ratio of 1 : 20 may be used.
3. **Carbon steels:** Carbon steels are high-strength alloys and different cutting fluids are used for different machining operations.

   For broaching, oil with EP additives are used. For machining operations with low cutting speed, high feed and depth of cut, mineral oils with additives are used.
4. **Free machinery steels:** These steels are normally machined dry but use of sulphurized mineral oils are helpful.
5. **Heat resistant steels:** Sulphurized mineral oils or water-soluble oils in the ratio of oil to water 1:20 to 1:40 may be used.
6. **High strength steels:** Sulphurized mineral oils, fatty oils, or water-soluble emulsion with high % age of oil may be used.
7. **Magnesium alloys:** There is danger of chip catching fire. These alloys are machined at low speed. For high-speed machining, copious amount of low viscosity mineral oil may be used. Do not use water soluble fluids.
8. **Stainless steels:** For low speed machining, i.e., broaching, tapping, etc. mineral oils with additives are used. For high-speed machining operations, water-soluble oils may be used.

The application of cutting fluids for different material and machining operations are given in Table 4.2.

4.8 APPLICATION OF CUTTING FLUIDS

For efficient performance, cutting fluid must be applied to the cutting zone. A lubricant film must be formed at the sliding surface. The coolants must be flooded in the cutting area to remove the generated heat effectively. Continuous

Table 4.2 Cutting fluids for different operations.

Sl. No.	*Work material*	*Machining operations*					
		Turning and Boring	*Threading*	*Drilling*	*Reaming*	*Shaping Planning slotting*	*Milling*
1.	Cast iron.	Dry	Dry	Dry	Dry, Tallow, Lard oil	Dry	Dry
2.	Soft steel	Cutting compound, cutting oil, soap water	Cutting compound, cutting oil, soap water	Dry Coolant	Cutting compound	Soap water Soda water	Soluble sulphurized mineral oil
3.	Hard steel	Mineral lard oil	Mineral lard oil	Kerosene, strong soda water	mineral lard oil	Mineral lard oil	Mineral lard oil or soluble sulphurized mineral oil
4.	Brass	Dry	Dry, Kerosene, turpentine	Dry	Dry, kerosene, turpentine	Dry	Dry
5.	Bronze	Mineral lard oil	Mineral lard oil	Dry, any coolant	Mineral lard oil	Dry	Soluble sulphurized mineral oil or kerosene
6.	Aluminium	Kerosene	Kerosene	Dry, kerosene	Kerosene with 25% soluble cutting oil	Kerosene	Soluble sulphurized mineral oil or kerosene
7.	Copper	Mixture of lard oil and turpentine	Dry or mixture of lard oil and turpentine	Dry, cooling compound lard oil and turpentine	Dry or any coolant	Dry	Soluble sulphurized mineral lard oil

application of fluid is better than intermittent application. The following methods are used for application of cutting fluid.

1. Manual application
2. Flood application
3. Jet application
4. Mist application
5. Through the tool.

4.8.1 Manual Application

For small batch and jobbing shop production, the cutting fluids and lubricants are applied manually with the help of a brush. This is the easiest and least expensive method.

The main disadvantages are:

(i) intermittent fluid application
(ii) poor chip removal
(iii) limited access of fluid to the cutting zone.

4.8.2 Flood Application

A low-pressure fluid is delivered to flood the tool, workpiece, and cutting zone with the help of a directed nozzle. The cutting fluid is collected in a sump, filtered and pumped back to the delivery nozzle. The nozzle is carefully oriented to avoid deflection of fluid away by rotating tool or workpiece.

Flood application ensures a continuous and sufficient fluid flow to cutting zone. The heat and chips are removed effectively.

4.8.3 Jet Application

A high pressure and high velocity jet of fluid is directed at the cutting zone. The fluid is collected in the sump, filtered and pumped back to the delivery nozzle. This form of application is suitable for grinding and high speed machining. For details see chapter 15.

4.8.4 Mist Application

An air-carried mist of cutting fluid can be applied where flooding is impracticable (inaccessible areas) and for processes where cutting speeds are high and uncut chip thickness is small as in end milling operation. Very good ventilation is required to avoid health risk due to inhalation of mist by the operator.

4.8.5 Through-the-tool Application

The cutting fluids and lubricants are applied through the shank and cutting edge of drills, spindles and tools for end milling operation and other operations where it is difficult to apply cutting fluid directly to the cutting zone. This results in improved tool life and effective flushing of chips from cutting zone.

4.9 MAINTENANCE AND DISPOSAL OF CUTTING FLUIDS

Good maintenance practice can extend the useful life of costly cutting fluids, reduce replacement cost and frequent disposal cost.

Machining coolants do not wear out with time but can get contaminated with the following substances.

1. Tramp oils from hydraulic and lubrication systems of machine tool.
2. Oxidation products due to heat generated during machining.
3. Solid particles of work material
4. Separated emulsion oil
5. Bacteria and fungi in the fatty compounds of cutting fluids.

In addition, the coolant concentration changes due to evaporation. Therefore, the concentration of cutting fluid should be regularly restored and contamination removed as follows:

1. **Removal of contamination:** The cutting fluid is collected in a sump, the light contamination is removed by skimmers, barrier filters and centrifugal separators are used to remove the residual substances.
2. **Bacterial growth:** The bacterial growth in the fatty acid and animal fat additives can be cause of health hazard to operator in the form of skin rashes. Addition of toxic biocides can control the bacterial growth.
3. **Periodic testing of cutting fluids:** The following tests are periodically carried out to control the cleanliness and concentration of cutting fluids:
   (i) pH value is regularly measured with a dip strip or pH meter.
   (ii) Refractometers are used to determine the oil concentration.
   (iii) Solid particle concentration is found by allowing samples of fluid by sedimentation.

The troubleshooting guide for cutting fluids is given in Table 4.3.

Table 4.3 Troubleshooting guide for cutting fluids.

Sl. No.	Problem	Cause	Solution
1.	Foaming	– Concentration high – High tramp oil content – Soft water – Machine cleaner in sump – Mechanical defect of pump and hose.	– Check pH, adjust concentration – Skim off oil – Sample water and treat – Run machine, remove cleaner – Check machinery and repair.
2.	Rusting	– Concentration high – Poor mixing of soluble oil – High tramp oil concentration.	– Adjust concentration – Add concentrate to water – Skim off oil.
3.	Poor tool life	– Low concentration – Wrong cutting fluid – Large biocide amount – High tramp oil concentration.	– Adjust concentration – Consult fluid supplier – Consult fluid supplier – Skim off oil.
4.	Odour	– Low concentration – Low pH – High tramp oil content.	– Adjust concentration – Check pH or consult supplier – Skin off oil, check leaks,
5.	Skin irritation	– High concentration – High pH value – High tramp oil content – Dirty shop clothes – Allergies – Out-of-shop influences	– Adjust concentration – Check pH and consult supplier – Skim off oil, check leaks, – Use clean clothes only – Check the operator for allergies – Contact physician.
6.	Residue in machine	– High concentration – High tramp oil concentration – Incorrect mixing – High misting operations	– Adjust concentration – Skim off oil, check leaks – Consult supplier – Adjust coolant nozzle, check ventilation system

4.9.1 Disposal of Cutting Fluids

The cutting fluid must be treated before disposal. The treatment system must take care of the following:

1. National, state and local pollution control regulations must be followed.
2. The method of treatment should be suitable for the type of cutting fluid.

The treatment plant includes the following stages:

1. Cleaning, filtering and recycling of neat oils.
2. Removal of water and disposal of solids separated in water-diluted coolants.
3. *Primary stage.* Skimming of free oil and filtration and settling of suspended solids.
4. Ultra filtration and chemical treatment to separate fluids.
5. *Final stage.* Final purification of water by carbon filtration or reverse osmosis and reuse of water.

QUESTION BANK NO. 4

1. Discuss the role of cutting fluids in machining operation.
2. What are the beneficial and harmful effects of cutting fluids.
3. List the desirable properties of cutting fluids.
4. What are the main categories of cutting fluids?
5. What are main constituents of cutting fluids? What are the roles of various additives?
6. Discuss the theory of action of cutting fluids as coolant and lubricant.
7. What factors decide the selection of cutting fluid for various common materials and machining operations?
8. What are the merits and demerits of application method of cutting fluids?
9. Write a brief note on maintenance and disposal of cutting fluids.
10. Draw a troubleshooting guide for cutting fluids.

5

SURFACE QUALITY AND SURFACE FINISH

Introduction; Surface quality; Importance of surface quality; Control of surface quality; Surface roughness; Measurement of surface roughness; Surface finish and machining cost, Question Bank.

5.1 INTRODUCTION

Machining operations are performed to remove unwanted material from the surface of a workpiece and to achieve accurate and desired shape and size. The machining generates new surfaces by plastic deformation and crack propagation. The workpiece is subjected to high mechanical stresses and heating. There are marks left by the tool on the machined surface.

The machine, cutting tool, and workpiece form a structural system having complicated dynamic characteristics. Under certain conditions, vibration of the structure system may occur. The vibration affects the distance between the tool and workpiece, continuously damaging the surface quality and dimensions of the component.

There are two types of surface damage.

1. **Surface quality:** An altered material zone (*AMZ*) is formed just below the machined surface due to plastic deformation and local heating, the depth and nature of which depend upon machining parameters and use of cutting fluids. The surface alteration can be in the form of feed marks, residual stresses, work hardening, phase transformation, micro and macro-cracks depending upon machining process. Most of the machining operations can damage the surface quality or integrity causing premature failure of the component. However, the quality of the surface can be improved by certain operations like cold rolling and polishing of the surface.
2. **Surface finish:** The surface finish or roughness is characterized by the topography of the surface. It describes the micro-geometrical properties or

texture of the machined surface in terms of roughness, waviness, lay and other surface irregularities. Good surface finish is required on bearings and locating surfaces and surfaces requiring low friction for flow of liquids and gases and reflecting surfaces for light rays.

The surface finish depends upon tool geometry, workpiece geometry, condition of machine, cutting conditions and cutting fluid.

5.2 SURFACE QUALITY

The surface quality or integrity is the soundness of machined surface and following types of damages can be caused.

1. Plastic deformation.
2. Micro-structural changes including phase change, inter-granular attack, recrystallization of grain size, etc.
3. Porosity
4. Micro-cracks
5. Residual stress

5.2.1 Plastic Deformation

The surface layer undergoes plastic deformation due to frictional force at the chip tool interface. The ductility of the surface decreases and the component may fail even under stresses and deformation well within design limits.

The amount of plastic deformation depends upon process parameters, i.e., cutting speed, size of the cut (feed and depth of cut), tool geometry work material and the condition of the machine tool.

The main reasons of plastic deformation are as follows:

1. The deformation and depth of affected surface layer depends upon the shear plane angle.
2. Rubbing between machined surface and tool flank.
3. The hardened chips or debris produced by built-up edge, grits torn out of grinding wheels, etc. may get embedded in the surface layer.

5.2.2 Microstructural Changes

There is intense heating of surface layer during machining especially at high speeds resulting in high temperatures which cause microstructural changes.

1. **Phase transformation:** The surface gets sufficiently heated during machining to produce untempered martensite with hardness *RC* 61. The layer between base material and hardened surface layer is overtempered martensite. Both untempered and overtempered martensite layers have low fatigue life.
2. **Recrystallization:** When a component is heated, it undergoes recry-stallization accompanied with grain growth affecting the ductility of the

material. The coarse grain structure is responsible for failure of component well below design limits.

3. **Intergranular Attack:** During electrochemical or chemical machining, the grain boundaries are weakened due to loss of material. This is due to different rates of reaction at the grains and grain boundaries. The etching of the grain boundaries is faster than that of grains. It is called selective etching.

5.2.3 Porosity

Metal is removed through vaporization in some thermal machining processes like electro-discharge machining (*EDM*) and laser beam machining (*LBM*). The vaporized material may get redeposited after cooling on the work surface in the form of a thin and porous layer. This cracked layer is very harmful as the component has low fatigue life.

5.2.4 Microcracks

Microcracks are frequently formed in the following cases:

1. Machining of brittle materials due to intense plastic deformation and thermal stresses.
2. Embedding of built-up edge fragments as microcracks are formed around them.
3. Heavy grinding of some material, i.e., cast nickel alloys and cobalt-base alloys result in cracking.

   Microcracks at the surface results in low fatigue life and surface corrosion and rusting.

5.2.5 Residual Stress

When a machining load is removed from a plastic deformed surface or after cooling of heated machined surface, residual stresses are produced. The residual stress may be tensile or compressive, high or low, deep or shallow. The distribution of residual stress depends upon cutting load, variation of mechanical and thermal stresses during chip formation.

1. **Grinding:** The residual stress distribution in grinding of a steel with different grinding conditions is shown in Fig. 5.1.
   - (i) Gentle grinding produces compressive residual stress in a shallow layer of 0.05 mm thickness.
   - (ii) Normal grinding (feed = 0.025 mm per pass) produces tensile and compressive residual stressless than heavy grinding.

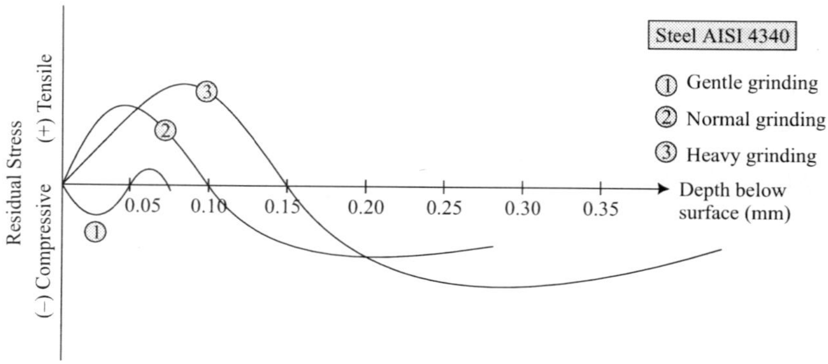

Fig. 5.1 Residual stress distribution in grinding.

(iii) Heavy grinding (feed = 0.5 mm per pass) produces high residual stress. Tensile stress peaking at 0.12 mm below surface. Compressive stress peaking at 0.035 mm below surface.

2. **Milling:** A compressive residual stress is produced below the surface layer when using a sharp cutter during face milling. Both tensile and compressive residual stresses are produced in the surface layer when a worn mill is used.
3. **Unconventional machining process:** There is no residual stress produced in electro-chemical machining (*ECM*). High tensile residual stresses are produced in the surface layer in electro-discharge machining (*EDM*).

5.3 IMPORTANCE OF SURFACE QUALITY

It is very important to control the quality of surface of the component. The presence of microcracks affects the fatigue strength of the component and the residual stresses produce distortion especially if the component is thin.

5.3.1 Fatigue Strength

The endurance limit in terms of failure of a component at relatively low number of cycles is reduced if the part is subjected to alternating stresses. The microcracks at the machined surface also act as stress concentrators affecting the fatigue strength.

Adverse conditions of machining can cause plastic deformation and phase transformation in the surface layer of a part causing microcracks and residual stresses. The finish grinding can be carried out under gentle grinding conditions by the use of a soft wheel, low wheel speed, small down feed and an effective use of cutting fluid. The finish grinding can also be very rough by the use of hard wheel, high wheel speed, large down feed and no cutting fluid. The effect of grinding condition on the endurance of a steel is shown in Fig. 5.2.

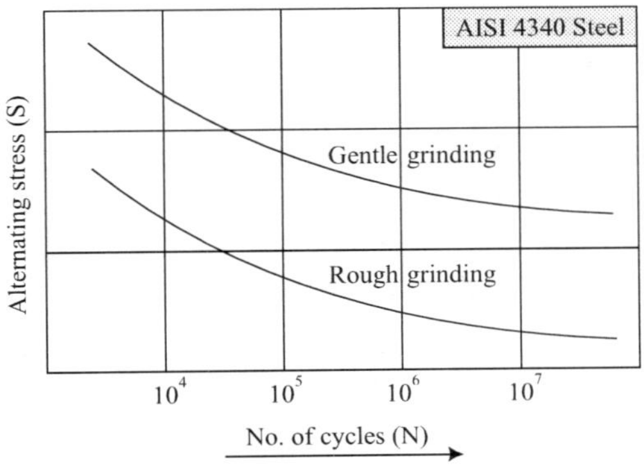

Fig. 5.2 S-N curve.

The endurance limit may decrease even by 30% under adverse grinding conditions.

The fatigue strength decreases in electrical-discharge machining (*EDM*) as well as electro-chemical machining (*ECM*).

5.3.2 Distortion of Component

There is distortion of component due to residual stress induced during machining. The machining process can also remove pre-existing residual stress in the work-piece thereby causing distortion.

The amount of distortion can be noticeable and measurable. It depends upon the amount of residual stress produced during machining and the latter depends upon size of the workpiece, its modulus of elasticity and the size of the cut. A titanium component distorts more than steel because of its low modulus of elasticity. The effect of size of cut in distortion is shown in Fig. 5.3.

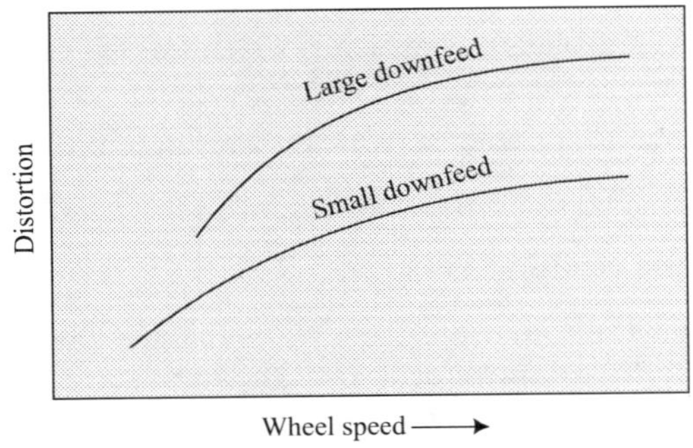

Fig. 5.3 Distortion due to size of cut.

At high speeds of grinding wheel and large downfeed per pass in surface grinding, the amount of distortion will be large.

5.4 CONTROL OF SURFACE QUALITY

There are a number of precautions to be observed during machining to control the surface quality of a component.

1. **Low vibration and chatter:** The machine tool-work system should be dynamically rigid to produce minimum vibrations and chatter.
2. **Sharp cutting tool:** A sharp cutting tool with minimum wear land should be used for machining. A worn milling cutter produces tensile residual stress.
3. **Light cuts:** The cutting speed, feed and depth of cut used should be minimum optimal. There should be no formation of built-up edge, chip formation should be continuous, cutting forces should be small and temperature in cutting zone should be low.
4. **Cutting fluid:** The selection of cutting fluid and application method should be proper. The effective cooling and lubrication action of cutting fluid supplied in abundance can control the temperature in the cutting zone and cutting forces.
5. **Post-machining treatment:** After-machining treatment consists of cold rolling, shot peening, burnishing, tumbling, polishing, etc. These processes can induce compressive residual stress in the surface layer to improve fatigue life.

5.5 SURFACE ROUGHNESS

The action of a cutting tool in machining never results in absolutely smooth and even surface of components. There are always some traces of unevenness, roughness, notches, scores, lines, etc. There defects are both in the direction of primary motion as well as in the direction of feed. Even after scrapping, lapping, etc. there are minor unevenness.

The final roughness is the sum of two independent effects.

1. The *ideal* surface roughness, which is a result of the geometry of the tool and the feed or feed speed.
2. The *natural* surface roughness, which is a result of the irregularities in the cutting operation.

5.5.1 Ideal Surface Roughness

The ideal conditions can be created by the elimination of built-up edge, chatter inaccuracies in machine tool movement, etc. The ideal surface roughness is the best possible finish that can be obtained by a sharp-cornered tool and feed under ideal conditions. An idealized module of surface roughness for a cutting tool with a sharp corner is shown in Fig. 5.4.

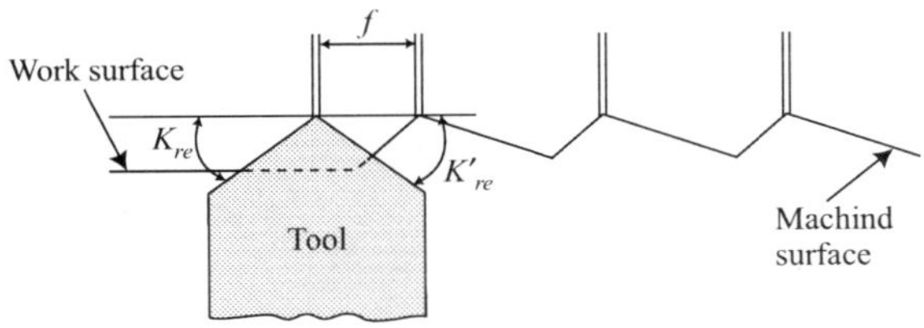

Fig. 5.4 Ideal Surface produced.

f = feed

K_{re} = working major cutting edge angle

K'_{re} = working minor cutting edge angle.

A cross-section of the machined surface is shown in Fig. 5.5. A mean line is parallel to general surface direction and divides, the surface so that sum of areas formed above the line is equal to sum of the areas formed below the line.

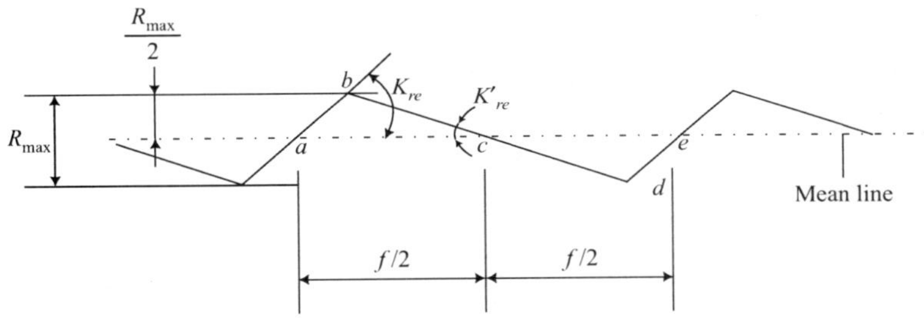

Fig. 5.5 Cross-section of machined surface.

The surface roughness, R_a, is the sum of the absolute values of all the areas above and below the mean line divided by the sampling length.

The surface roughness,

$$R_a = \frac{|\text{area } abc| + |\text{area } cde|}{f}$$

Area abc = Area cde

$$\therefore \quad R_a = \frac{2}{f}(\text{area } abc) = \frac{R_{max}}{4} \quad ...(1)$$

where, $\frac{R_{max}}{2}$ = Height of triangle abc

By geometry,

$$R_{max} = \frac{f}{\cot k_{re} + \cot k'_{re}} \quad ...(2)$$

∴ From equation (1)

$$R_a = \frac{f}{4(\cot k_{re} + \cot k'_{re})} \quad ...(3)$$

The arithmetic mean value of roughness is directly proportional to the feed, working major-cutting-edge angle and minor-cutting-edge angle.

In practice, cutting tools have rounded corner as shown is Fig. 5.6. The surface roughness,

$$R_a = \frac{0.0321 f^2}{r_e}$$

where, r_e = corner radius of tool.

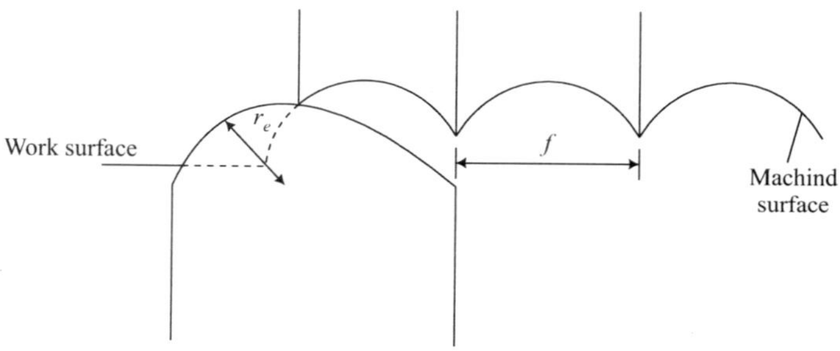

Fig. 5.6 Tool with rounded corner.

Example 5.1 A turning tool has a sharp nose, major cutting edge angle of 60° and minor cutting-edge angle of 15°. What should be the feed rate in order to restrict ideal centre-line average roughness of 5 μm?

Solution: The data given

$K_{re} = 60°$

$K'_{re} = 15°$

$R_a = 5$ μm

$$R_a = \frac{f}{4(\cot k_{re} + \cot k'_{re})}$$

$$\therefore \quad f = 4\,(5 \times 10^{-3})(\cot 60° + \cot 15°)$$

$$= 0.11 \text{ mm/rev.}$$

5.5.2 Natural Surface Roughness

It is not possible to achieve ideal conditions for machining. Natural surface roughness forms a large proportion of the actual roughness. Main factors that commonly contribute to the natural surface roughness in practice are:

1. Occurrence of a built-up edge.
2. Occurrence of chatter or vibrations of machine tool.
3. Inaccuracies in the movements of machined tool.
   Example. Movement of the saddle on a lathe.
4. Irregularities in the feed mechanisms.
5. Defects in the structure of the work material.
6. Discontinuous chip formation when machining brittle materials.
7. Tearing of the work material when ductile metals are cut at low speeds.
8. Surface damage caused by chip flow.

5.5.2.1 Built-up edge

The built-up edge may be formed continuously which may break down. The fractured particles are carried into the under surface of the chip and new work piece surface. The larger the built-up edge, rougher is the surface product. In order to get improved surface finish, the built-up edge should be reduced by:

(i) increase in cutting speed
(ii) use of cemented carbide tool in place of HSS tool
(iii) introduction of free machining materials such as leaded or resulphurized steel
(iv) application of correct cutting lubricant at low cutting speeds.

5.5.2.2 Effect of cutting speed

The effect of cutting speed on surface roughness in turning is shown in Fig.5.7.

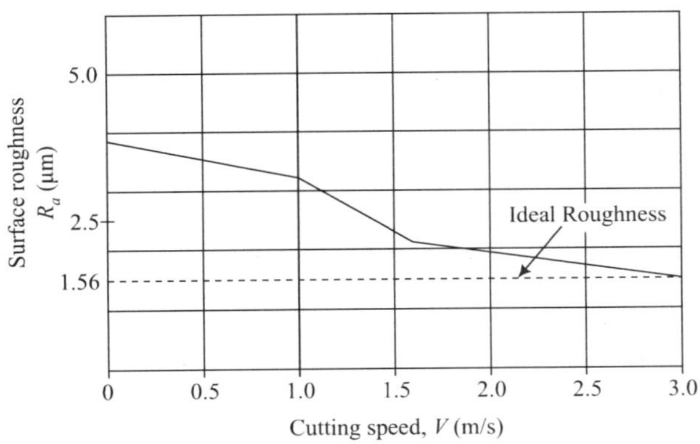

Fig. 5.7 Effect of cutting speed on surface roughness.

The actual surface roughness for a turned component is large at low cutting speeds and becomes reduced as the cutting speed is increased. At very high speeds, the actual roughness approaches the ideal surface rougheness.

5.5.2.3 *Effect of multiunit tool*

There will be inaccuracy in the manufacture of different cutting points. Also there can be inaccuracy in the primary motions of machine tool. Therefore, one tooth of the cutter plays a dominant role in generating the machined surface.

The surface roughness in multipoint machining

$$R_a = \frac{0.0642}{d_t}\left(\frac{v_f}{n_t}\right)^2$$

where, v_f = feed speed

d_t = diameter of cutter

n_t = rotational frequency of cutter

5.5.2.4 *Values of surface roughness*

As a general guide, the surface roughness values by different machining operations are given below

Grinding = 0.05 to 2.5 μm

Diamond turning = 0.10 to 1.0 μm

Milling and turning > 1 μm

5.6 MEASUREMENT OF SURFACE ROUGHNESS

The main derivations from a perfectly smooth surface are roughness, waviness and surface flaws. The surface flaws can be scratches, cracks and other irregularities which occur at random over the surface.

5.6.1 Surface Texture

The surface texture can be assured by the use of following irregularities:

1. **Roughness:** It consists of surface irregularities as a result of machining. The combination of these irregularities form the surface texture. The roughness is measured as follows.
   (i) *Roughness height:* It is also called height of uneverness and is measured with respect to a reference line.
   (ii) *Roughness width*: It is the distance between successive peaks or ridges parallel to normal surface.
   (iii) *Roughness width cut off:* It is the greatest spacing or sample length for measurement of roughness.

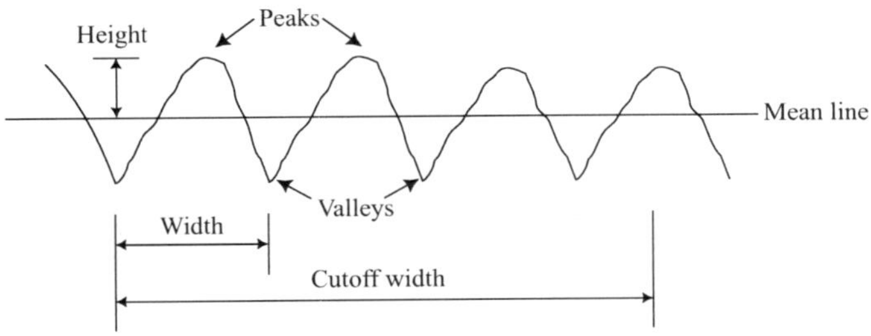

Fig. 5.8 Elements of surface roughness.

2. **Waviness:** Roughness is a fine irregularity. Waviness includes all surface irregularities with spacing more than roughness cut off width. This is caused by vibration, chatter, tool and workpiece deflection, run out, etc.
3. **Lay:** It is the direction of predominant surface pattern. It depends upon the machining method used and the orientations of workpiece and the cutting tool relative to machine. It also depends upon the relative motion between the workpiece and the tool. Measurements of surface are made at right angles to the lay.
4. **Surface flaws:** There are inherent defects such as inclusions, cracks, blow-holes, etc. in the workpiece which are exposed during machining.

5.6.2 Surface Profile

The surface irregularities are enlarged by instruments and surface profile is obtained by tracing the irregularities. The vertical motion of a stylus is amplified as it is drawn slowly across the profile. The profile is continuously recorded.

In modern instruments, the surface profile is digitally sampled and the arithmetical-mean surface-roughness is continuously calculated.

5.6.3 Roughness Grade Numbers

The *ISO* has recommended a series of preferred roughness values and corresponding roughness grade numbers as shown in table 5.1. These grade numbers are used to specify surface roughness on drawings.

5.6.4 Symbols on Drawings

The surface roughness on drawing is represented by triangles. These are classified into four groups.

Group	R_a *(μm)*
∇	8 – 25
∇∇	1.6 – 8
∇ ∇ ∇	0.025 – 1.6
∇ ∇ ∇ ∇	< 0.025

The following symbols are used to fully specify the surface texture on the drawing

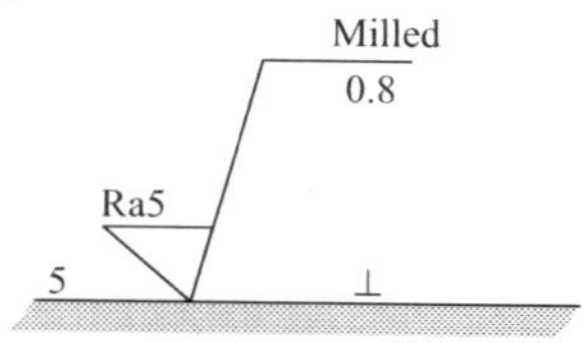

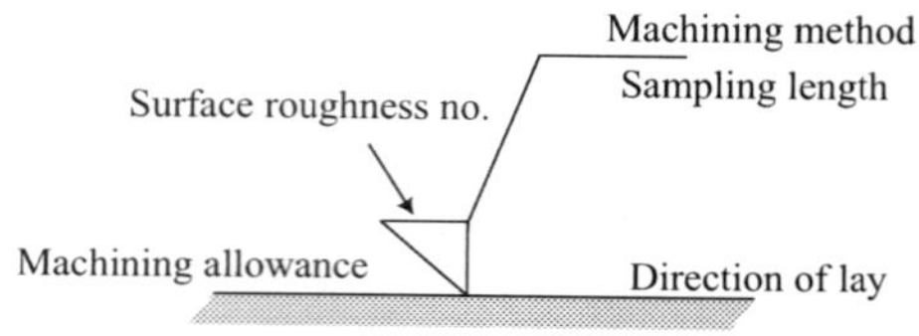

Table 5.1 Roughness Grade Numbers.

Roughness Grade Number	*Roughness value R_a (μm)*
N12	50
N11	25
N10	12.5
N9	6.3
N8	3.2
N7	1.6
N6	0.8
N5	0.4
N4	0.2
N3	0.1
N2	0.05
N1	0.025

5.7 SURFACE FINISH AND MACHINING COST

The machining cost increases with the requirement of surface finish, i.e., smoother the surface required, the greater the machining cost. The increase in cost is due to following reasons.

1. Use of finer feed and depth of cut.
2. Lower permissible wear of cutting edge.
3. More frequent tool changes.
4. Need for more rigid machine tool.
5. Need for more effective clamping methods for tools and workpiece.
6. Need for special tool inserts.
7. Need for more frequent and closer inspection.

The change of machining cost with surface finish is shown in Fig. 5.9.

The machining cost is less for rough machining and increases linearly with surface finish. The cost of machining increases exponentially with the requirement of finer finish and ultra finishing operators such as grinding, honing, lapping, etc.

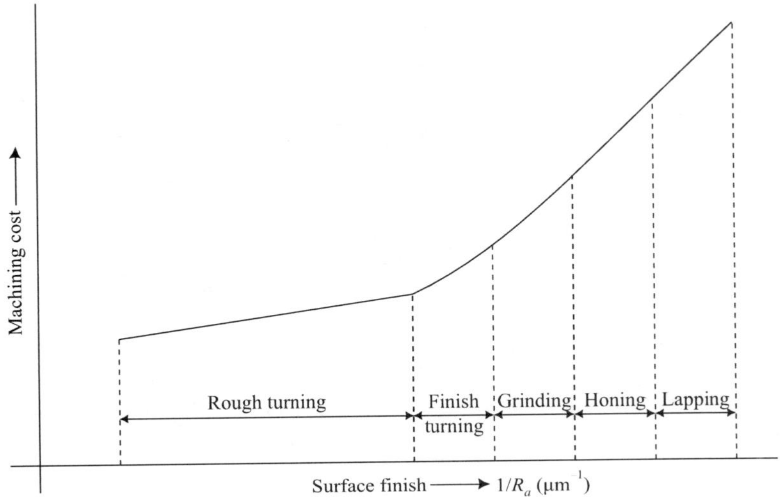

Fig. 5.9 Change of machinery cost with surface finish.

QUESTION BANK NO. 5

1. Distinguish between surface quality and surface finish.
2. What are different parameters of surface quality? How are these related to machining parameters?
3. Why is it important to control surface quality? Discuss the role of surface quality on fatigue strengths and distortion of machined component.
4. Outline different methods of improving surface quality of a machined component.
5. What are ideal surface roughness and natural surface roughness? List the effect of machining parameters on surface finish.
6. Discuss the measurement of surface roughness detailing surface texture, surface profile, roughness grade number and symbols used on drawings.
7. Discuss the variation of machining cost with surface finish requirement.

6

CHIP CONTROL

Introduction: Chip breakers; Radius of chip curvature; Tool wear during chip breaking; Question bank.

6.1 INTRODUCTION

In conventional machining, the continuous chips produced at low cutting speeds have a natural curl and are brittle. There is no serious problem encountered in machining. However, in high speed turning with a carbide tool, a continuous chip is produced which may create several problems.

1. The chip may get entangled with the tool or workpiece or machine tool elements.
2. This can be hazardous to the operator.
3. There can be mechanical chipping of the cutting edge.
4. There can be interference to the flow of coolant leading to alternate heating and cooling of cutting edge.
5. The thermal stress so produced can reduce the life of cemented carbide tools.

There will be a serious economic problem in handling continuous chips in bulk. The handling problem of chips may be expressed by the bulk ratio which is defined as the ratio of total volume occupied by the chips to the volume of solid chip material. Some examples of bulk ratio are:

		Bulk ratio
1.	Unbroken, continuous chips	50
2.	Continuous chips tightly wound	15
3.	Well-broken chips	3

It is clear from above example that handling and disposal of well-broken chip is very easy as these chips occupy only 1/17th volume of unbroken chips. Therefore, it is very important to study and develop methods to break the chips into small coils or fragments when they are produced. This is called chip control.

6.2 CHIP BREAKERS

Chip breakers are important element of tool design. Ductile materials such as low carbon steel, copper, aluminium, zinc, brass, etc. when cut, produce long continuous chips which are difficult to handle and dispose of as they occupy large volumes. The chip breaking and chip curl have assumed great importance.

1. **Free machining metals:** The composition of metal used for large volume of products can be changed by the addition of alloying elements which impart chip breaking properties. The free machining steel and free machining brass produce small chips which are easy for handling and disposal.
2. **Chip breakers:** Chip breakers are obstacles provided in the cutting tools which reduce the swarf into small pieces as they are formed. The metal is plastically deformed during cutting resulting in work hardening. The chip breaker performs effectively with work hardened material. There are basically two types of chip breakers.

   (i) *Groove type:* A small groove is ground behind the leading cutting edge on the rake face. The geometry of the groove is shown in Fig. 6.1.

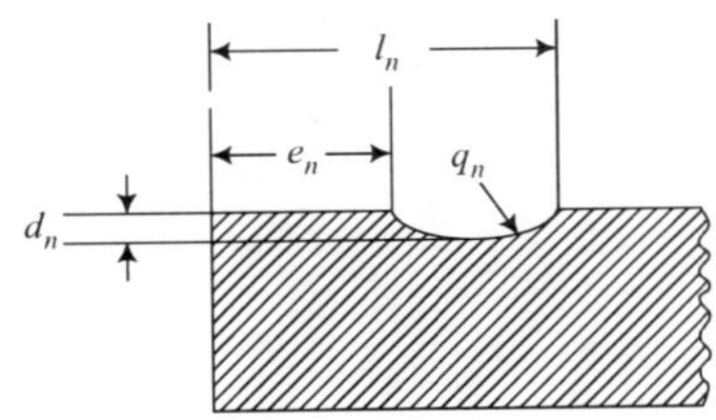

Fig. 6.1 Groove-type chip breaker.

d_n = chip breaker-groove depth
e_n = chip breaker land width
l_n = chip breaker distance
q_n = chip breaker groove radius.

The geometry of the groove determines the radius of chip curvature and its chip breaking performance.

(ii) *Obstruction type:* The obstruction can be integral or attached to the tool as shown in Fig. 6.2.

h = chip breaker height
l_n = chip breaker distance
σ = chip breaker wedge angle.

The obstruction can be made adjustable to cater to different cutting conditions. The bending radius of chip depends upon height (h) and distance (ln) of chip breaker from cutting edge.

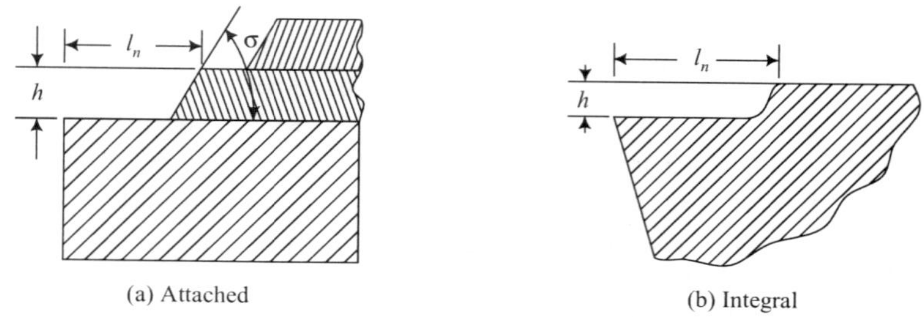

Fig. 6.2 Obstruction-type chip breaker.

3. **Tool Geometry:** Effective control of chip as it moves across the face of the tool may be achieved by proper selection of tool angle. A large positive front rake angle gives rise to a looser chip formation, which flows down the face of the tool, and away from the workpiece, leaving the newly machined surface unscratched. A small positive or negative side rake has the effect of decreasing the radius at which the chip coils which has the tendency to produce short, easily managed chip.
4. **Machining parameters:** The chip breaking characteristic depends upon feed rate and depth of cut. Increased feed gives a thicker chip which breaks more quickly. A small depth of cut with a fine feed allows the chip to form into comparatively small pieces or direct it into the swarf tray. The use of a good stream of coolant that acts as a quenching medium causes the hot chip to become harder and break into small pieces.

6.2.1 Chip Breaking Chart

A chip breaking chart is shown in Fig. 6.3. There are critical feeds and depths of cut below which chip breaking does not take place. These values define the chip breaking limit curve. This curve can be divided into three parts.

1. *Region A.B.* Curl broken chips are formed above the limit curve.
2. *Region C.D.* Side curl spiral continuous chips are formed on the left side of limit curve.
3. *Region B.C.* Three-dimensional chip cutting occurs.

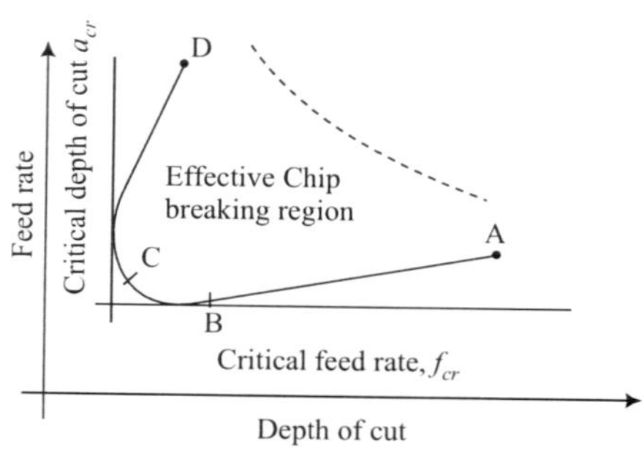

Fig. 6.3 Chip breaking chart.

6.3 RADIUS OF CHIP CURVATURE

The chip breaking depends upon radius of curvature of the chip. A tightly curled chip will usually break. An integral obstruction type chip breaker is shown in Fig. 6.4.

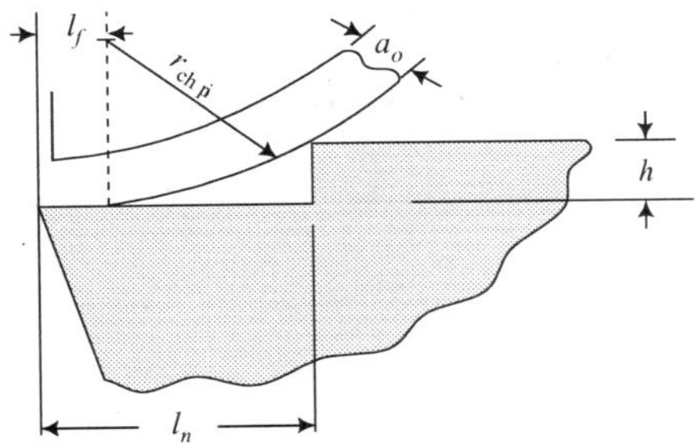

Fig. 6.4 Radius of chip curvature.

The radius of chip curvature at a given point,

$$r_{\text{chip}} = \frac{(l_n - l_f)^2}{2h} + \frac{h}{2}$$

where, l_n = chip breaker distance

l_f = length of chip-tool contact

h = chip breaker height.

The chip-tool contact and chip thickness bear a constant ratio.

$$\frac{l_f}{a_0} = k$$

where, a_0 = chip thickness

k = constant for a given material

$$\therefore \quad r_{\text{chip}} = \frac{(l_n - ka_0)^2}{2h} + \frac{h}{2}$$

with reasonable accuracy k is taken unity.

$$\therefore \quad r_{\text{chip}} = \frac{(l_n - a_0)^2}{2h} + \frac{h}{2}$$

For a groove-type chip breaker, it is found that

$$r_{\text{chip}} = q_n$$

where, q_n = chip breaker groove radius.

Both r_{chip} and a_0 have controlling influence on the efficiency of chip breaking.

6.4 TOOL WEAR DURING CHIP BREAKING

A crater is formed on the tool face. The shape of the crater conforms to the shape of the curled chip. The presence of chip breaker affects the tool wear process. A crater formed on a plain tool can act as chip breaker.

It has been experimentally established that the obstruction type chip breaker causes a reduced rate of wear on the tool face.

QUESTION BANK NO. 6

1. List various types of chip breakers. Why are they used?
2. Write short notes on:
   (i) Chip control
   (ii) Chip breaking chart
   (iii) Radius of chip curvature.
3. Discuss the following effects:
   (i) Tool wear during chip breaking.
   (ii) Effect of tool geometry on chip curl.
   (iii) Effect of machining parameters on chip breaking.

7

GRINDING PROCESS

Introduction; Grinding mechanism; classification of grinding processes; surface grinding; cylindrical grinding; Honing process; Lapping process; High speed grinding; Creep feed grinding; Low stress grinding; Thermal effects in grinding; Cutting fluids; Mechanics of wheel wear; Effect of grinding conditions on wheel behaviour. Question Bank.

7.1 INTRODUCTION

Grinding is a metal cutting operation in which a rotating abrasive wheel is used as a tool. Generally, grinding or abrasive machining process is used as a finishing process to get high surface quality, correct size and accurate shape of the product. The metal removing rate is only 0.25 to 0.50 mm in most operations and dimensional accuracy is of the order of 0.000025 mm. Grinding is also used for machining of hard materials which are difficult to be machined by cutting tools. The development of new grinding wheels has led to bulk removal of metal economically. High speed grinding and creep feed grinding are the two new processes for bulk material removal.

However, grinding will never replace conventional metal cutting as the principal method of machining components. Hard machining has the ability to effectively machine hardened steels and other difficult-to-cut materials and now competes with grinding for many applications.

7.2 GRINDING MECHANISM

A grinding wheel consists of a large number of abrasive particles called grains held together by a suitable material called bond. The active grains are situated at the surface of the wheel and perform the cutting action like a multi-point tool (say, a milling cutter). The cutting points are irregularly shaped and randomly distributed over the active face of the wheel. Each active grain removes a short chip of gradually increasing thickness. The action is similar to that of a slab-milling cutter.

The structure of a grinding wheel is shown in Fig. 7.1 and the cutting action is shown in Fig. 7.2.

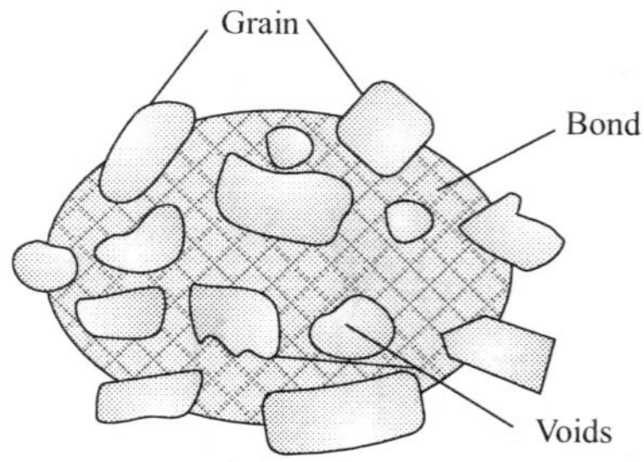

Fig. 7.1 Structure of a grinding wheel.

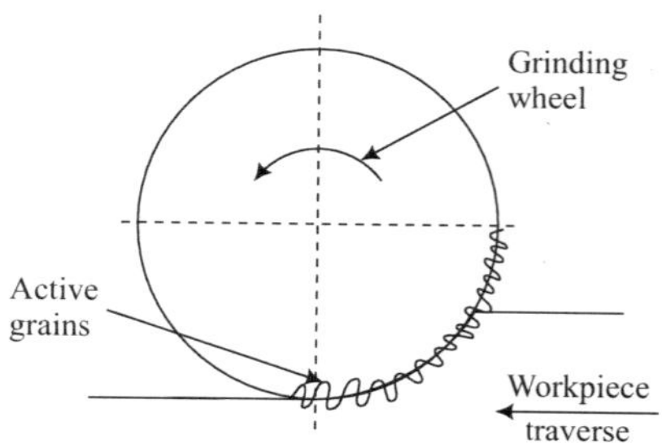

Fig. 7.2 Cutting action of a grinding wheel.

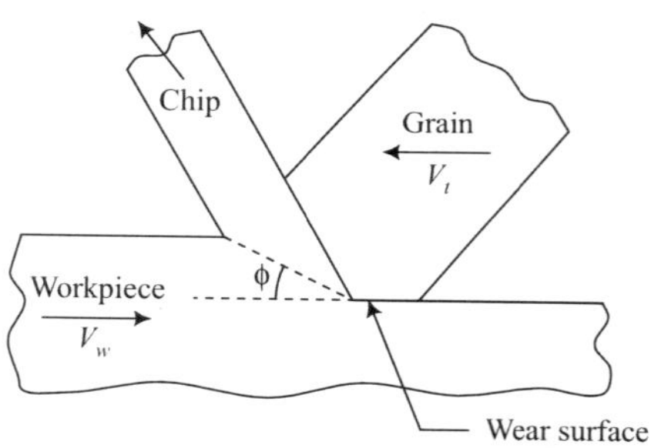

Fig. 7.3 Chip formation by an abrasive grain.

Chip formation

The active grain is a small cutting tool with a large negative rake angle (60° or more). The equivalent shear angle during removal of each chip is very small.

The ploughing action results in progressive wear and formation of worn areas on the active grains. With the progress of grinding, the size and number of worn areas increase. This results in increase of friction and force acting on the grain. When the force becomes very large, the worn grain is fractured and a new cutting edge is exposed. This self-sharpening property of the grinding wheel is very important for its performance.

Redressing

If the wheel is made of tough grains bonded together strongly, the self-sharpening characteristic is poor and a glazed appearance is seen during grinding. The grinding wheel at this stage requires dressing.

Dressing with a diamond-tipped tool removes the worn grains from the wheel periphery and a new sharpened cutting surface is generated.

In case of soft wheel, the volume of wheel removed will be significant as compared to metal removed. Therefore, more frequent dressing will be required. The selection of a grinding wheel is a compromise between two extreme cases of hard and soft wheels.

7.3 CLASSIFICATION OF GRINDING PROCESSES

Grinding processes can be classified according to the shape of the component and type of grinding machine. Many different types of grinding machines have been developed for various types of work. Grinding is done on surfaces of almost all conceivable shapes and materials. The grinding processes have been discussed in chapter 1. The main groups of grinding processes can be as follows.

1. **Precision grinding:** The machined components are subjected to precision grinding to achieve good surface finish and high degree of accuracy. The various types of grinding are :
   - (i) Surface grinding to produce flat surfaces.
   - (ii) External cylindrical grinding to produce external cylindrical or tapered surfaces.
   - (iii) Internal cylindrical grinding to produce internal cylindrical and taper bores.
   - (iv) Form grinding to produce formed surfaces such as gears, threads, splines, holes, spheres, etc.
   - (v) Special grinding processes to grind crankshaft, camshaft, grooves, rollers, etc.
2. **Super finishing:** Honing and lapping processes are carried out to obtain ultra-high accuracy in size and shape and degree of surface finish after precision grinding or machining.
3. **Non-conventional grinding operations:** The grinding processes can be used for bulk removal of metal economically. The following non-conventional grinding operations have been developed.
   - (i) High speed grinding
   - (ii) Creep feed grinding
   - (iii) Low-stress grinding

There are variations in some of the above processes depending upon the type of product to be ground and types of controls provided in the grinding machine. Following types of machines are available, for example, for cylindrical grinding.

1. Plain cylindrical grinding machines.
2. Heavy plain cylindrical or roll grinding machines.
3. Universal cylindrical grinding machines.
4. CNC cylindrical grinders.
5. Contour grinding machine, i.e., computer controlled cam grinder.
6. Crankshaft grinding machine.

In CNC grinders, facilities are available for automatic control of dimensions of the component as well as facility for automatic dressing.

7.4 SURFACE GRINDING

Surface grinding is used to finish flat surfaces. It is also used for grinding irregular, curved, convex and concave surfaces. The work tables can be rotary or reciprocating type. Surface grinding can be horizontal or vertical. The geometry and cutting operation parameters are discussed in chapter 1.

The surface speed of the wheel,

$$V_w = \frac{\pi D_w N}{1000} \text{ (m/min)} \quad ...(1)$$

where, D_w = Diameter of grinding wheel (m)

N = speed of wheel rotation (rpm)

The workpiece speed is same as table speed.

The feed per grain of grinding wheel,

$$f = \frac{V_t}{nV_w} \text{ (mm/grain)}$$

$$= \frac{1000\, V_t}{n\pi D_w N} \quad ...(2)$$

Where, V_t = Speed of workpiece

n = No. of grains per mm length of wheel.

f is the distance moved by workpiece when wheel rotates by one grain dimension. The chip thickness is very small.

The uncut chip thickness at the position of the grain.

$$t_0 = f \sin \theta$$

The maximum uncut chip thickness $t_{max} = t_0$ where $\theta = \alpha$ as shown in Fig. 7.4.

where, α = angle of contact of wheel with workpiece.

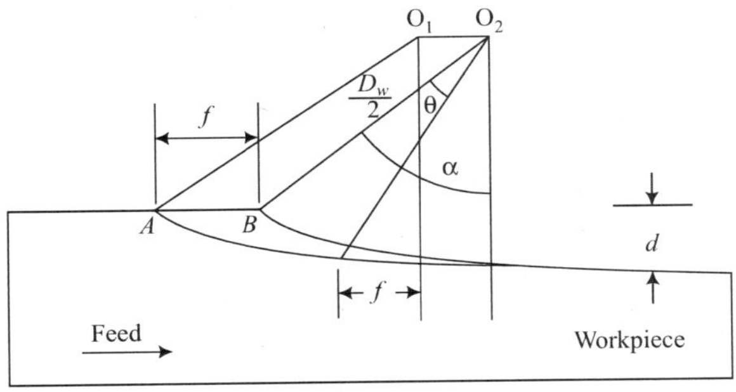

Fig. 7.4 Uncut chip thickness in surface grinding.

Average uncut chip thickness,

$$t_{av} = \frac{1}{\alpha}\int_0^{\alpha} f \sin\theta \, d\theta$$

$$= f\left(\frac{1-\cos\alpha}{\alpha}\right)$$

α is very small.

$$\therefore \qquad t_{av} = f\left(\frac{2\sin^2 \alpha/2}{\alpha}\right)$$

$$\approx \frac{f\alpha}{2} = f\sqrt{\frac{d}{D_w}}$$

But $$f = \frac{1000V_t}{n\pi D_w N}$$

$$\therefore \qquad t_{av} = \frac{1000V_t}{n\pi D_w N}\sqrt{\frac{d}{D_w}}.$$

Example 7.1 The machining parameters for a surface grinding operation are as follows:

Grinding wheel diameter = 150 mm
Wheel speed = 200 rpm
Traverse feed = 5 mm/stroke.
Depth of cut = 0.05 mm
Number of grains/mm = 2
Average table speed = 20 m/min
Specific energy = 15 J/mm^3

Calculate: (i) Average chip thickness
(ii) Tangential grinding force
(iii) Spindle power.

Solution:

(i) Average chip thickness

$$t_{av} = \frac{1000V_t}{n\pi D_w N}\sqrt{\frac{d}{D_w}}$$

Average table speed,

$$V_t = 20 \text{ m/min}$$
$$n = 2$$
$$D_w = 150 \text{ mm}$$
$$N = 2000 \text{ rpm}$$
$$d = 0.05 \text{ mm.}$$

$$\therefore \quad t_{av} = \frac{1000 \times 20}{2 \times \pi \times 150 \times 2000}\sqrt{\frac{0.05}{150}}$$
$$= \mathbf{0.000194 \text{ mm.}} \quad \textbf{Ans.}$$

Material removal rate,

$$Z_w = V_t \times 1000 \times f \times d \text{ (mm}^3\text{/min)}$$
$$= 20 \times 1000 \times 5 \times 0.05$$
$$= 5000 \text{ mm}^3\text{/min.}$$

(ii) Spindle power $P = Z_w p_s$

$$= 5000 \times \frac{15}{60} \times \frac{1}{1000}$$
$$= \mathbf{1.25 \text{ kw.}} \quad \textbf{Ans.}$$

(iii) Spindle power $P = F_t \times \pi\left(\frac{D_w}{1000}\right)\frac{N}{60}$

where, F_t = tangential force on the wheel

$$\therefore \quad F_t = \frac{1.25 \times 1000}{\pi \times \left(\frac{150}{1000}\right) \times \left(\frac{2000}{60}\right)} = \mathbf{79.5 \text{ N}} \quad \textbf{Ans.}$$

Also $$p = \frac{2\pi NT}{60 \times 1000}$$

$$\therefore \qquad \text{Torque, } T = \frac{1.25 \times 1000}{2\pi \times \dfrac{2000}{60}}$$

$$= \mathbf{5.96\ N\text{-}m} \quad \textbf{Ans.}$$

7.5 CYLINDRICAL GRINDING

There are centre-type cylindrical grinding as well as centreless grinding. These are used for grinding plain cylinders, contoured cylinders, fillets, cams, camshafts. Internal grinding is used to finish plain, tapered or formed cylindrical bores. The geometry and machining parameters are discussed in chapter 1. A basic scheme of cylindrical grinding is shown in Fig. 7.5. The workpiece is held between dead centres and rotated by a dog clutch or a clutch. The rotating grinding wheel is automatically fed towards the workpiece for successive cuts.

The relative positions of the workpiece and grinding wheel are shown in Fig. 7.6 over a small interval of time Δt.

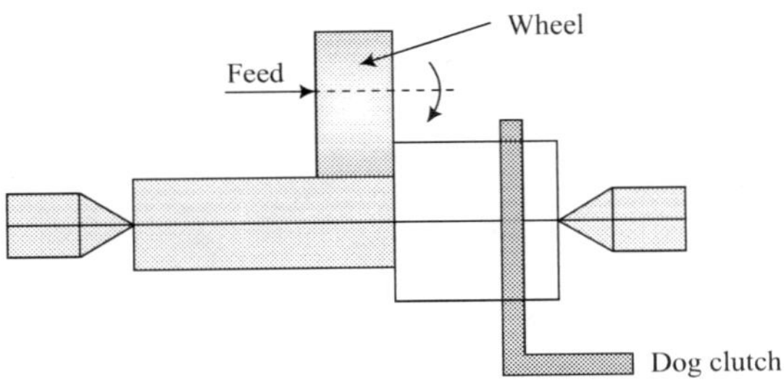

Fig. 7.5 Cylindrical grinding.

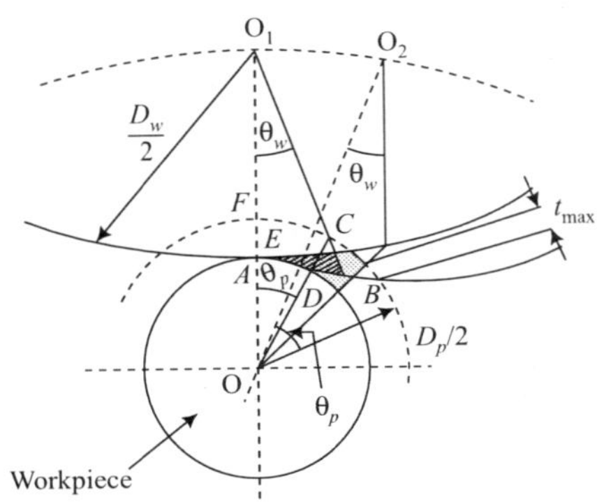

Fig. 7.6 Uncut chip thickness in cylindrical grinding.

The workpiece has centre O and rotates through a small angle θ_p in time Δt. In the same time, the centre of grinding wheel moves for O_1 to O_2.

The workpiece material removed during time Δt = Area $ADBCA$ (shaded)

The maximum uncut chip thickness $t_{max} = CD$

Fig. 7.6 represents the following dimensions.

Radial depth of cut $= d$

Diameter of grinding wheel $= D_w$

Diameter of workpiece $= D_p$

Angle subtended by length EC on wheel centre $= \theta_w$

Angle subtended by length EC on workpiece centre $= \theta_p$

Assume that in time Δt, the wheel moves by a distance equal to linear dimension of grit.

Now $$BC = \frac{V_p}{nV_w}$$

Where, V_p = velocity of workpiece

n = no. of grains/mm

V_w = wheel velocity

In ΔBCD,

$$\angle DBC = \theta_w + \theta_p$$

Assume $\angle CDB = 90°$

The maximum uncut thickness,

$$t_{max} = CD = BC \sin(\theta_w + \theta_p)$$

$$= \frac{V_p}{nV_w} \sin(\theta_w + \theta_p)$$

Consider Δs O_1EC and OEC,

$$\frac{D_w}{2} \sin\theta_w = \frac{D_p}{2} \sin\theta_p$$

$$\therefore \quad \frac{\sin\theta_w}{\sin\theta_p} = \frac{D_w}{D_p}.$$

Angles θ_w and θ_p are very small

$$\therefore \quad \frac{D_w}{D_p} = \frac{\theta_w}{\theta_1} \quad ...(1)$$

Depth of cut,

$$d = AE + EF$$

$$= \frac{D_p}{2}(1 - \cos\theta_p) + \frac{D_w}{2}(1 - \cos\theta_w)$$

$$\approx \frac{D_p}{2}\left(2\sin^2\frac{\theta_p}{2}\right) + \frac{D_w}{2}\left(2\sin^2\frac{\theta_w}{2}\right) \quad ...(2)$$

From equations (1) and (2)

$$d = \frac{1}{4}\left(D_p\theta_p^2 + D_p^2\frac{\theta_p^2}{D_w}\right)$$

$$\therefore \quad \theta_p = \sqrt{\frac{4d\,D_w}{D_p(D_w + D_p)}}$$

$$\text{and,} \quad \theta_w = \sqrt{\frac{4d\,D_P}{D_w(D_w + D_p)}}$$

$$\therefore \quad t_{max} \approx \frac{V_p}{nV_w}(\theta_w + \theta_p)$$

$$\approx \frac{V_p}{nV_w}\sqrt{\frac{4d}{D_w + D_p}}\left\{\sqrt{\frac{D_p}{D_w}} + \sqrt{\frac{D_w}{D_p}}\right\}$$

$$= \frac{V_p}{nV_w}\sqrt{\frac{4d(D_p + D_w)}{D_pD_w}}.$$

Conclusions

1. The maximum uncut chip thickness depends upon surface speed of work material (V_p) surface speed of wheel (V_w), depth of cut (d), diameter of grinding wheel (D_w) and diameter of workpiece (D_p), grit size (n).
2. t_{max} decreases with wheel speed (V_w). The force on the grit also decreases. The wheel will act as a hard wheel.
3. t_{max} increases with workpiece speed (V_p). The force on the grit also increases and the wheel will act as a soft wheel.
4. The behaviour of grinding wheels changes under different operating conditions.

Example 7.2 The machining data for a cylindrical grinding process is as follows.

Diameter of workpiece, $D_p = 100$ mm

Diameter of wheel, $D_w = 125$ mm

No. of grits per mm, $n = 2$

Speed of workpiece, $N_p = 80$ rpm

Speed of grinding wheel, $N_w = 2300$ rpm

Depth of cut, $d = 0.05$ mm

Specific energy for grinding, $p_s = 16$ Nm/mm^3

Transverse feed, $f = 5$ mm/sec. of workpiece

Calculate: (i) Maximum uncut chip thickness per grit.

(ii) Spindle power

(iii) Spindle torque.

Solution: Surface speed of grinding wheel,

$$N_w = \pi D_w N_w$$

$$= \pi \times \frac{250}{1000} \times 2300$$

$$= 1806.4 \text{ m/min.}$$

Surface speed of workpiece,

$$N_p = \pi D_p N_p$$

$$= \pi \times \frac{100}{1000} \times 80$$

$$= 25.132 \text{ m/min.}$$

(i) Max. uncut thickness of chip,

$$t_{\max} = \frac{V_p}{2V_w} \times \sqrt{\frac{2d\left(\frac{D_w}{2} + \frac{D_p}{2}\right)}{\frac{D_w}{2} \times \frac{D_p}{2}}}$$

$$= \frac{25.132}{2 \times 1806.4} \times \sqrt{\frac{2 \times 0.05(125 + 50)}{50 + 125}}$$

$$= \mathbf{0.000368 \text{ mm.}} \quad \textbf{Ans.}$$

Metal removal rate

$$Z_w = \pi D_p N_p f d \text{ mm}^3/\text{min}$$

$$= \pi \times 100 \times 80 \times 5 \times 0.05 = 6283.18 \text{ mm}^3/\text{min.}$$

(ii) Power required at the spindle

$$P_w = p_s Z_w = \frac{16 \times 6283.18}{60} = \mathbf{1.675\ kW} \quad \textbf{Ans.}$$

Also, $$P_w = \frac{2\pi NT}{60 \times 1000} \text{ kW}$$

(iii) Spindle torque,

$$T = \frac{P_w \times 1000 \times 60}{2\pi N_w} = \frac{1.675 \times 1000 \times 60}{2\pi \times 2300}$$

$$= \mathbf{6.954\ N\text{-}m} \quad \textbf{Ans.}$$

7.6 HONING PROCESS

Honing is a surface finishing process where abrasive sticks are slowly moved both by rotation and reciprocation. The abrasive sticks are kept pressed against the work surface. It is very frequently used for finishing cylindrical surfaces both internal and external. Examples are automobile cylinders, hydraulic press cylinders, etc. The rotational speed of honing tool varies between 15 to 40 m/min and reciprocating speed between 1.5 to 20 m/min.

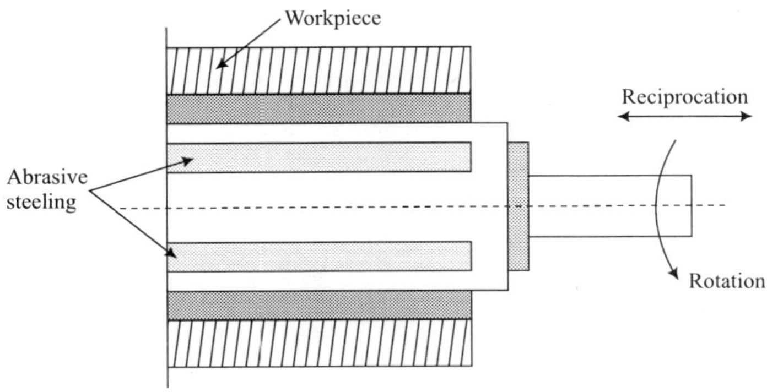

Fig. 7.7 Honing process.

The stroke length of tool ensures sufficient projection on either side of the bore. The surface finish depends upon the relative surface speed of honing sticks and the grit size as given in Table 7.1.

Table 7.1 Grit size and surface finish.

Grit size	*Surface roughness (μm)*
150	0.50
280	0.30
320	0.25
400	0.125
600	0.050

The main advantages of honing are:

1. Good surface finish
2. Straightness of bores or surfaces
3. Accuracy of size = 0.005 mm in a diameter of 50 mm
4. Circularity of surface within 1.25 μm

7.7 LAPPING PROCESS

A lap made of softer material than workpiece rubs abrasive particles lightly on work surface. The process can be used for flat as well as curved surfaces. The abrasive particles are supplied in the form of a paste or slurry between the rubbing surfaces. Water, oil or kerosene is used as carrier of grains. There are three elements used in lapping processes.

1. A lap produces relative motion on work surface.
2. Abrasive particles and carrier fluid.
3. A pressing force on the abrasive particles during sliding.

Lap plate made of castiron rotates at constant speed. The carrier plate can rotate as well as oscillate so that workpieces rub the lap plate on all points. The pressure can be applied hydraulically or pneumatically.

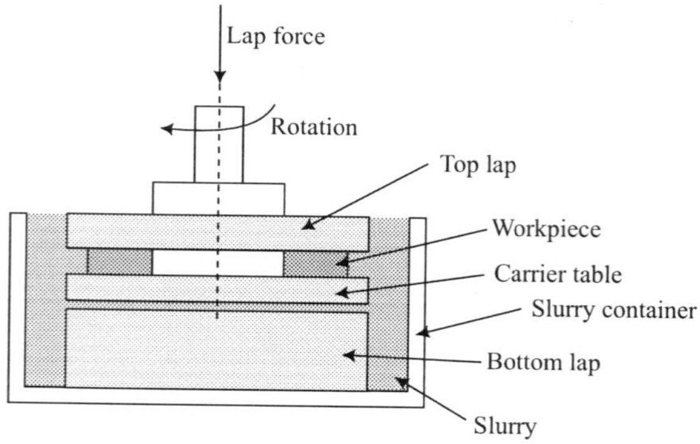

Fig. 7.8 Lapping process.

The metal removal rate and surface roughness depend upon grain size, grain hardness and pressure of laps as given in Table 7.2.

Table 7.2 Grit size and surface finish.

Grit size	*Abrasive*	*Surface roughness (μm)*
320	SiC	0.6-0.8
400	SiC	0.5-0.6
500	SiC	0.4-0.5
800	SiC	0.1-0.25
800	Al_2O_3	0.05-0.08
900	Al_2O_3	0.02-0.05

7.8 HIGH SPEED GRINDING

High speed grinding should be developed for increased metal removal rates. This process has the potential to compete with other machining operations for applications in the area of large stock removal. It has been found that grinding speeds of 180 m/s, well above speeds of conventional grinding, can result in increased metal removal rates. At the saver time, there are no significant changes in grinding forces, surface finish and specific cutting energies at such high speed grinding.

There are particular safety problems in high speed grinding. There can be bursting of highly stressed wheels. Special machines have to be designed to collect broken wheel fragments. The segmental wheels have to be specially designed.

The high speed grinding has not found many production applications due to safety considerations. The process is mostly restricted to groove grinding and grinding of difficult to machine materials

7.9 CREEP FEED GRINDING

Creep-feed grinding is an innovation of conventional grinding for machining of difficult to machine materials. In this process, the depth of cut is many times that in normal grinding process. However, the workpiece feed is reduced. The process is similar to slab milling. The relatively slow rotating milling cutter is replaced by a high speed rotating grinding wheel. Different machining parameters such as length of stroke, depth of cut, speed of workpiece and infeed per revolution of creep-feed grinding are compared with conventional surface grinding.

The creep-feed grinding has the following characteristics.

1. Increased total grinding forces.
2. Reduced average forces on the individual grains.
3. Increased temperatures in the wheel-work contact zone.
4. Reduced temperatures of the newly generated work surface.
5. Large arc of contact between wheel and work.

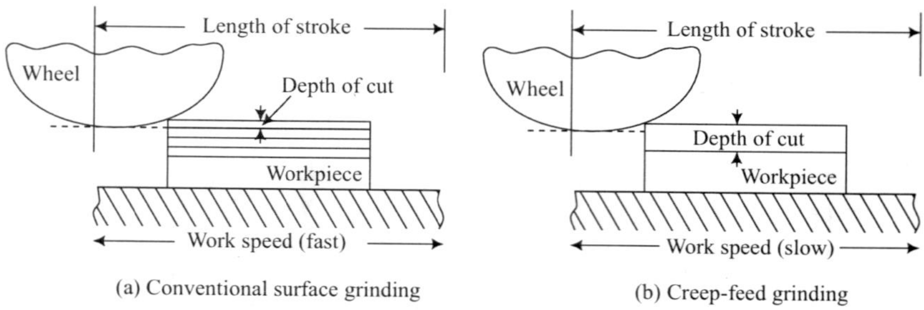

Fig. 7.9 Comparison of feed, length of stroke and depth of cut.

The results of creep-feed grinding of hard and soft steel workpieces are shown in Fig. 7.10. The specific energy per unit volume removed is higher for harder materials than softer materials.

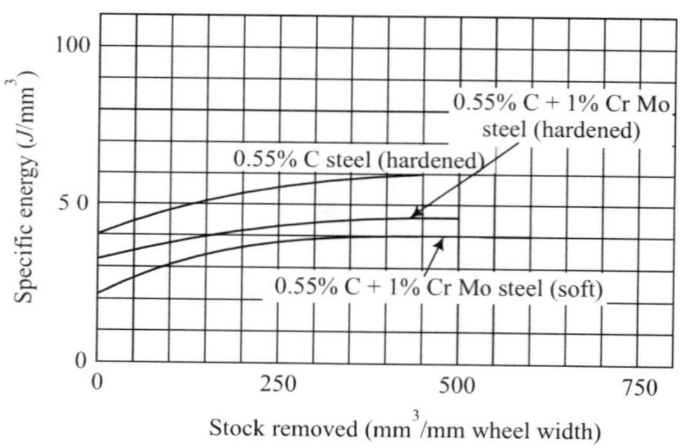

Fig. 7.10 Stock removal rate in creep-feed grinding.

7.9.1 Grinding Process

A large amount of material is removed in one pass of creep-feed grinding. However, a final finishing pass is given to improve the flatness and accuracy of the surface produced. The wheel is redressed before finishing pass and hence the wheel is very sharp.

A typical machining sequence is as follows:

1. A freshly dressed wheel is used for a skin cut of 0.05 mm on the previously cut workpiece.
2. The wheel is then used for the creep-feed rough grinding of the next workpiece.
3. The wheel is redressed and final finishing skin cut is taken as in step 1.

The dressing of the wheel is done by diamond roll dressing. However, conventional crush dressing is also used.

7.9.2 Applications

The creep-feed grinding machines are very rigid with high-efficiency grinding fluid systems. The wheel profile stability is good. Typical applications of creep-feed grinding are:

1. Grinding of deep slots with high depth-to-width ratios.
2. Grinding of profile on difficult-to-grind materials.
3. Form grinding of turbine blades, root profiles of high temperature super alloys.
4. **Deep grinding:** It is cylindrical grinding process. It involves plunge grinding of stationary workpiece. The workpiece is then slowly rotated relative to the wheel. The complete stock removal is then completed in one revolution of workpiece.

7.10 LOW-STRESS GRINDING

Tensile residual stress can be greatly reduced near the surface of workpiece by proper selection of grinding parameters. Low stress grinding is very beneficial on workpiece which will be subjected to high stresses in corrosion environments.

The following machining parameters are used for low stress grinding:

1. Soft and open wheels.
2. Low cutting speeds of the order of 18 m/s. For conventional grinding a speed of 30 m/s is used.
3. Oil based grinding fluids are used.
4. Low infeed rates of the order of 0.005 mm per pass.
5. The machining parameters are selected to minimize friction effects at the work surface.

The residual stresses for low-stress grinding are compared with conventional grinding and high stress grinding in Fig. 7.11. The high-stress grinding is achieved under dry grinding conditions with hard wheel. In order to improve the fatigue resistance of workpiece, a small compressive residual stress can be induced near the work surface by low-stress grinding process.

7.11 THERMAL EFFECTS IN GRINDING

The specific cutting energy is higher in grinding than for other cutting operations due to high cutting speeds. The temperatures at grain tip may reach 1700°C and some chips may melt during formation. Sixty to ninety percent of energy consumed in grinding flows instantaneously into the workpiece.

The proportion of total energy flowing into workpiece,

$$E = (0.6 + 0.05A_g) \times 100\ (\%)$$

where, A_g = Relative contact area of grains (%) = 0 – 5%.

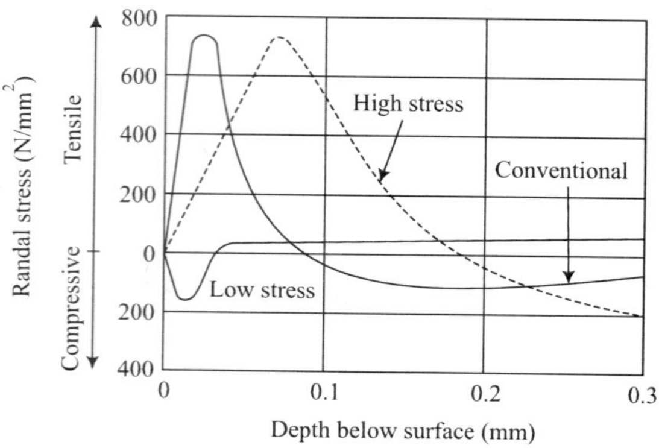

Fig. 7.11 Residual stress.

The value of Ag increases as the wheel becomes dull. There can be following types of thermal damages in grinding.

1. **Dimensional Errors:** There is expansion of workpiece due to heating. There will be dimensional errors on cooling after the grinding process.
2. **Structural changes:** Metallurgical changes like softening and tempering of hardened surface layers of steel may take place.

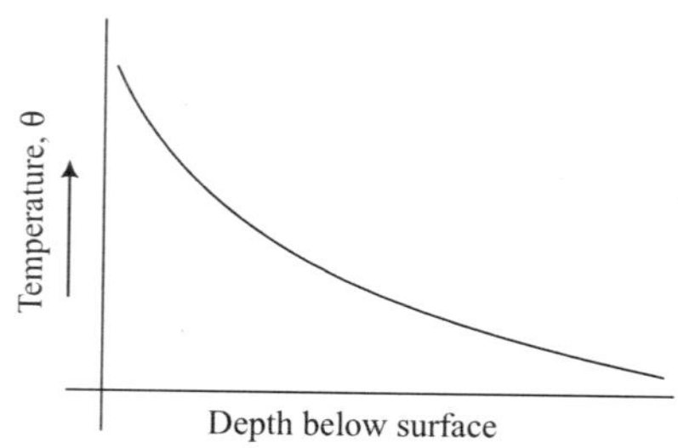

Fig. 7.12 Thermal effect of grinding.

3. **Burning and surface cracking:** Surface oxidation and burning may take place due to high temperatures. There will be microscopic cracks reducing fatigue and stress corrosion resistance of workpiece. In severe cases, there may be visible surface cracks.
4. **Residual stress:** Residual stresses are caused due to temperature gradients and plastic deformations due to temperature changes. Tensile residual stress decreases the fatigue strength of workpiece. These effects can be reduced by the use of soft wheels and low grinding speeds.

   The surface temperature depends upon:

   (i) wheel speed

(ii) work speed

(iii) force intensity

(iv) wheel sharpness

The wheel sharpness and induced force intensity can be monitored and controlled during grinding cycle.

7.12 CUTTING FLUIDS

Cutting fluids for machining operations have been discussed in Chapter 4. These fluids have same functions in grinding also.

1. Lubrication of chip-grit and grit-workpiece interfaces
2. Reduction in power consumption.
3. Reduction in heat generation
4. Cooling of workpiece and reduction of possible thermal damage.

In general, oil and water emulsions are used in grinding. The lubricating property is controlled by proportion of oil and water in the mixture. Neat oils are used for difficult-to-grind materials and form grinding. Neat oils have limited cooling capacity, high costs, potential fire hazard and safety risks due to release of toxic fumes.

7.12.1 Action of Cutting Fluid

The grinding wheels operate at high speeds. A thin air film is formed around the rotating wheel which resists the entry of the cutting fluid into the grinding zone. The cutting fluid should be supplied at the same velocity as the wheel velocity. A laminar stream of fluid follows the contour of the wheel without turbulence. At the end, it leaves the wheel surface due to centrifugal force. Thus, grinding fluid provides the necessary lubrication between the wheel and workpiece.

7.12.2 Application of Grinding Fluids

An adequate supply of fluid should be ensured both in external and internal grinding. Following methods are used for application of grinding fluid.

1. **Flooding:** A large volume of fluid is applied at low pressure in surface grinding.
2. **Fluid jet:** A high pressure fluid jet is applied through near wheel and workpiece in the direction of wheel rotation to overcome the resistance of air cushion. The fluid velocity approaches wheel speed to form a laminar stream.
3. **Grinding wheels:** Radial passages are formed on the side of the wheel which carry the fluid to grinding zone. The neat oil can be applied by flooding and oil-water emulsion applied externally.

4. **Chamber or air deflector nozzle:** The air-deflector nozzle is shown in Fig. 7.13. The air deflector removes the air cushion around grinding wheel and low pressure fluid is carried into grinding zone.

7.13 MECHANICS OF WHEEL WEAR

The wear of grinding wheel follows a curve as shown in Fig. 7.14. It has three zones.

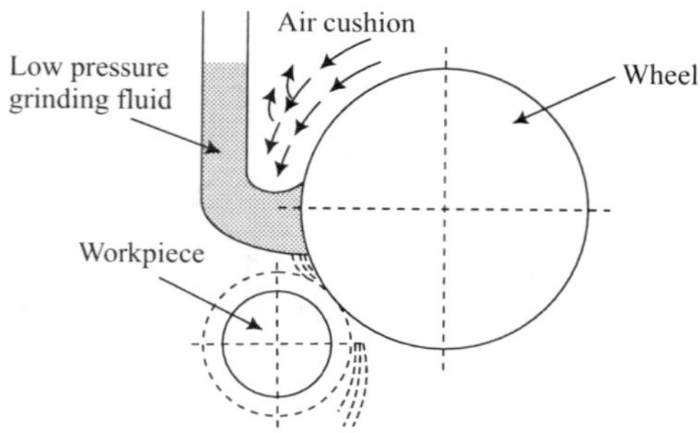

Fig. 7.13 Air deflector nozyle.

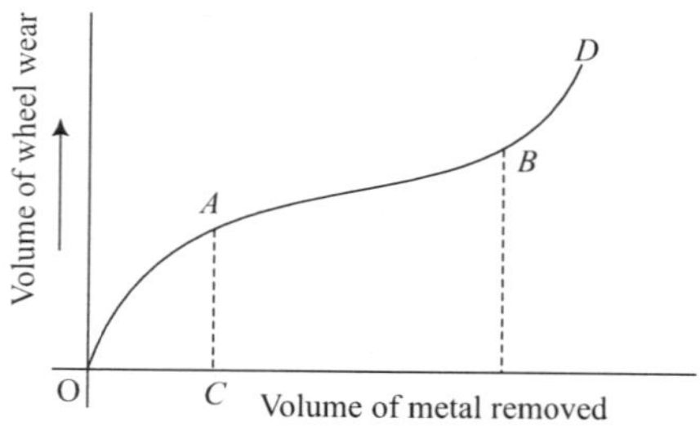

Fig. 7.14 Wear of grinding wheel.

1. **OA** : Initial rapid wear zone. Low grinding ratio.
2. **AB** : Zone of constant wear rate. High grinding ratio.
3. **BD** : Increased wheel wear zone. Low grinding ratio.

At point *B*, the wheel requires redressing.

The wear performance of the wheel is specified by grinding ratio G_r which is defined as the ratio of volume of metal removed to the volume of wheel worn away.

$$G_r = \frac{A_w}{A_t}$$

The grinding wheel wear is of three types:

1. Attrition wear of abrasive grains
2. Grain fracture
3. Bond fracture

7.13.1 Attrition Wear

It is similar to flank wear of a cutting tool. The wheel surface is projected on screen and flat areas on the grain are measured. The attrition wear is measured as the ratio of total flat area on the grain and total wheel area taking part in grinding. It is about 4% of total wheel wear. The horizontal and vertical components of grinding forces (F_h and F_v) increase linearly with flat area. At certain value of flat area, there is sudden increase in grinding forces as shown in Fig. 7.15.

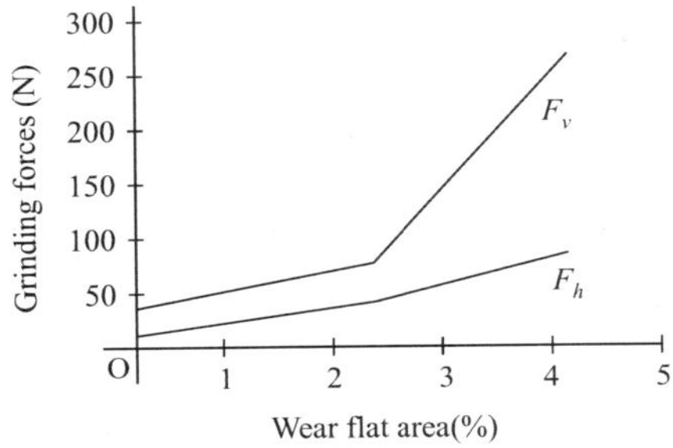

Fig. 7.15 Attrition wear.

7.13.2 Grain fracture

Gain fracture accounts for 10 to 15% of total wheel wear. The forces acting on individual grains are very high. This can be minimized by:

- controlled force grinding, and
- high speed grinding.

Grinding ratios of 4000 can be achieved in high speed grinding, where the force between wheel and surface decreases. However, rotational stresses increase on the wheels at high speed. In ordinary grinding, ratio is about 100 to 200.

7.13.3 Bond Fracture

The main reason for abnormally high wear rate of grinding wheels as compared to cutting tools is due to weaker bond of the wheels. This is particularly true for soft wheels where maximum wear is due to bond fracture.

7.14 EFFECT OF GRINDING CONDITIONS ON WHEEL BEHAVIOUR

The wheel behaviour or performance of wheel is the useful life between dressings. It is measured by grinding ratio G_r, surface finish and surface integrity (quality of surface). The performance of wheel depends upon the following grinding conditions.

1. Wheel speed
2. Workpiece speed
3. Infeed

4. Transvere feed
5. Severity of dressing
6. Application of cutting fluids

7.14.1 Grinding Wheel Speed

The grinding wheel speed has the following effects:

(i) The uncut chip thickness decreases and hence the forces acting on the grit decreases.

(ii) The grinding ratio increases with the decrease of forces acting on the grit. Grinding ratio as high as 4000 have been achieved. Normally this value lies between 100 and 200.

(iii) There is increase in centrifugal force on the wheel. The wheels have to be designed with decreasing thickness towards periphery.

(iv) Special design of coolant application have to be used to penetrate it into grinding zone against air cushion developed around the wheel at high speed.

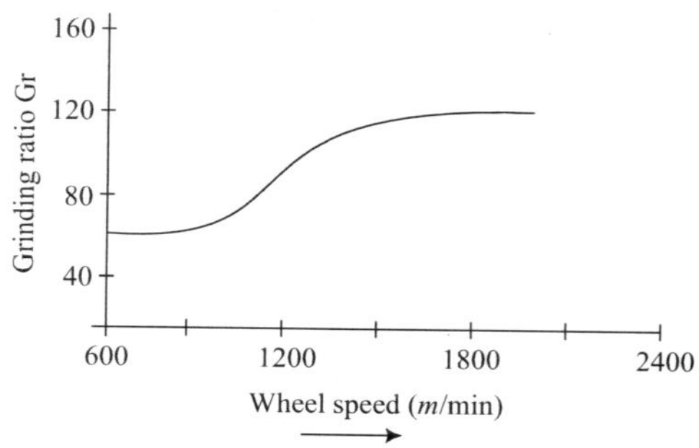

Fig. 7.16 Performance of grinding wheel.

7.14.2 Workpiece Speed

The effect of workpiece speed is opposite to that of wheel speed. The uncut chip thickness increases with the increase of workpiece speed. Therefore, there is increase of force acting on the grit and wheel will act as soft.

7.14.3 Infeed

The effect of infeed are shown in Fig. 7.17.

(i) Surface roughness increases linearly with increase of infeed.

(ii) Power consumption increases linearly with increase of infeed.

(iii) Grinding ratio initially decreases, has a minima and then increases with infeed.

7.14.4 Transverse Feed

The effects of transverse feed on wheel performance are shown in Fig. 7.18.

(i) The grinding ratio decreases with increase of transverse feed.

(ii) The surface roughness increases.

(iii) Power consumption increases.

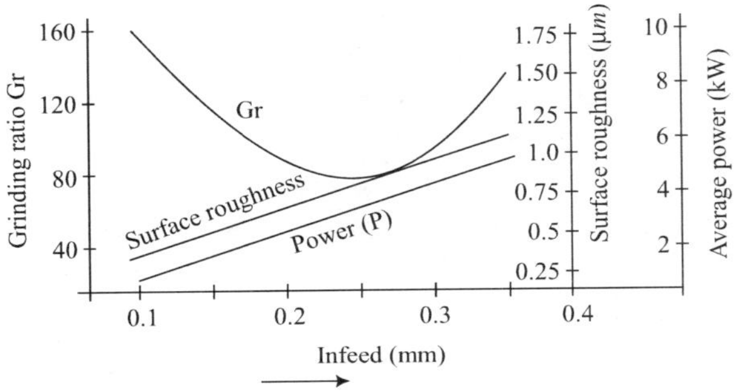

Fig. 7.17 In feed and wheel performance.

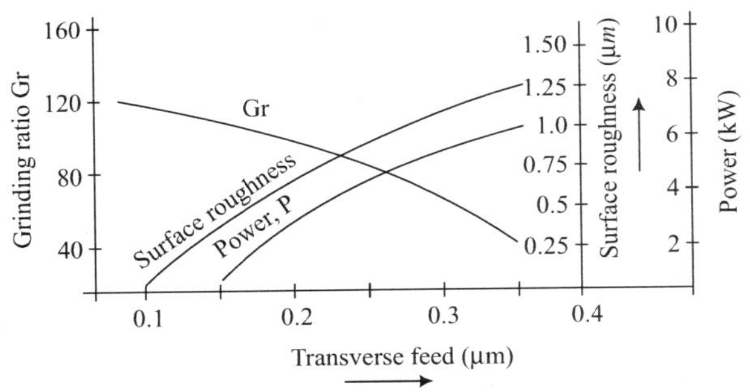

Fig. 7.18 Transverse feed and wheel performance.

7.14.5 Dressing

The performance of grinding wheel depends upon the intensity of dressing and proper dressing procedure is very important to enhance its life.

(i) Heavy dressing results premature pull-out of grits decreasing its working life.

(ii) Soft dressing leaves flat surface on many grains. There is increase of temperature and rapid wear as a result.

7.14.6 Cutting Fluids

Cutting fluids influence cooling effect, surface finish, wheel surface temperature and surface integrity of workpiece.

The cutting fluid should remove the heat generated effectively. It should cool the wheel to reduce its wear and thermal damage of workpiece.

QUESTION BANK NO. 7

1. Explain the basis of the following:
   (i) Centreless grinding operation
   (ii) Honing
2. Derive an expression, from first principles, for the maximum chip thickness in cylindrical grinding operation and interpret it to show that the behaviour of grinding wheels changes under different operating conditions.
3. Discuss the wear mechanism of grinding wheel.
4. Discuss the parameters and potential for the following conditions of grinding.
   (i) High-speed grinding
   (ii) Creep-feed grinding
   (iii) Low-stress grinding.
5. Explain the technological fundamentals and application of deep and creep feed grinding.
6. Explain the basic mechanism of grinding.
7. Give a brief classification of grinding process.
8. Derive an expression for uncut chip thickness in surface grinding.
9. Compare the following:
   (i) Honing and lapping
   (ii) Surface and cylindrical grinding.
10. Comment on thermal effect in grinding.
11. Explain the role and application of grinding fluids.
12. Briefly explain the effect of grinding conditions on wheel performance.

8

ECONOMICS OF METAL CUTTING OPERATIONS

Introduction; Feed selection; Selection of cutting speed; Costs for single pass turning operations; Optimum cutting speed in turning for minimum cost; Optimum cutting speed in turning for maximum production rate; Machining at maximum efficiency, Question Bank.

8.1 INTRODUCTION

In any manufacturing operation, production costs and production rates are of vital importance. Normally, a high production rate should result in low average production costs but the manufacturing conditions for maximum production rate may not be identical to those conditions giving minimum production costs.

Empirical rules have been developed over the years for the selection of optimum parameters including cutting speed, feed and depth of cut for different machining operations. The cutting parameters to be selected are cutting speed and feed if proper cutting tool and cutting fluid have been selected.

Feed is the distance moved by the tool relative to the workpiece in the direction of feed for each revolution of the tool or workpiece or each stroke of the tool or workpiece. In case of machining operations by multipoint tool such as milling, the feed is the relative speed between the tool axis and the workpiece in the direction of feed. This may also be called *feed speed.*

Example In milling operations, the feed of workpiece in each revolution of the cutter is given by V_f/n_t.

where, V_f = feed speed

n_t = rotational frequency of the tool.

The maximum cutting speed,

$$V = d_t.n_t$$

where, d_t = tool diameter.

For a constant feed, cutting speed can be doubled by doubling both the rotational frequency of the cutter, n_t as well as feed speed, V_f. The actual machining time will be reduced but the tool-wear rate will increase. However, production time may increase for carrying out frequent change of cutting tool. The production time will be high when low feeds and speeds are used.

There will be optimum condition for minimum production time. Similarly, there will be optimum condition for minimum production cost. The cost of use of machine tool and operator for longer duration will be high at low speeds and feeds. At high speeds and feeds, costs will be high due to high cost of frequent tool replacement. A compromise has to be worked out for achieving both minimum production time and production cost.

8.2 FEED SELECTION

In roughing operations, the selection of correct feed will depend upon the tool forces and power consumption of the machine tool. In finishing cut, the selection of correct feed will depend upon the acceptable surface finish. The production rate in rough machining can be increased by increasing feed rather than increasing speed as tool wear rate and tool life are highly affected by speed rather than feed. However, increase in feed rate is accompanied by increase in tool forces. The latter are not affected by cutting speed. Therefore, the maximum limits on feed increase will be decided by the maximum capacity of the machine tool to withstand tool forces. The guiding principle in selection of optimum condition in rough machining is that the feed should always be set at the maximum possible for the machine tool. The feed for finishing cut will be specified by the surface finish requirement.

8.3 SELECTION OF CUTTING SPEED

The selection of optimum cutting speed should ensure minimum production cost and minimum production time. The components of cost of machining a workpiece are:

1. *Labour cost* includes salaries and wages of workers directly involved in machining.
2. *Cost of machine tool operation* includes interest and depreciation of capital cost of the machine, cost of power consumption and maintenance.
3. *Overheads* include costs of building, land, office equipment and salaries of supervisors and managers.
4. *Material handling costs* include cost of loading and unloading of job, idle time of machine, operating cost of special loading devices.
5. *Cost of tooling* for machining per workpiece.
6. *Cost of resetting of tool* includes time for regrinding and resetting on the machine.

The effect of cutting speed on various costs is shown in Fig. 8.1. The cutting speed can be selected for most economical operations from this figure.

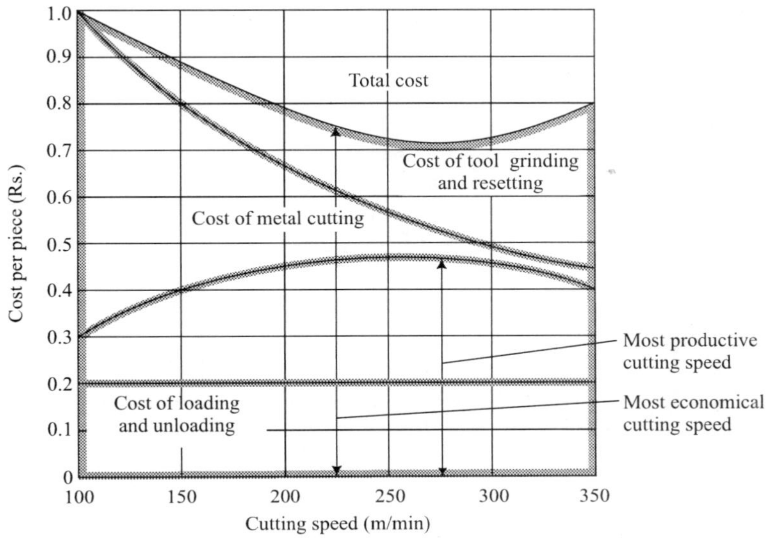

Fig. 8.1 Effect of cutting speed on total cost of product.

8.4 COSTS FOR SINGLE PASS TURNING OPERATIONS

The above costs for a simple turning operation can be estimated as follows.

1. **Machining time:** The actual time during which cutting action is taking place on a lathe,

$$T_m = \frac{L}{fN} \text{ (min)}$$

where, L = Length of bar to be turned (mm)

f = feed per revolution (mm)

N = spindle speed (rpm)

The cutting speed (surface speed of workpiece)

$$V = \frac{\pi DN}{1000} \text{ (m/min)}$$

where, D = diameter of workpiece (mm)

$$\therefore \quad T_m = \frac{\pi DL}{1000 fV} \text{ (min)}$$

Cost of cutting action = $C_u T_m$

where, C_u = Cost of machine operators including labour cost per unit time

2. **Job handling cost:** Job handling cost = $C_u T_h$

   where, T_h = Time of loading and unloading

3. **Tooling cost:**

   (i) *Throw-away carbide bits*

   Cost per cutting edge of tool,

$$C_e = \frac{\text{Cost of bit}}{\text{Number of cutting edges per bit}}.$$

   (ii) *Regrindable tools (HSS tools)*

$$C_e = \frac{\text{Cost of the tool}}{\text{Number of times tool can be reground}} + \text{cost of regrinding}$$

   (iii) Cost of tool per job = $Ce.\frac{T_m}{T}$

   where T = Tool life (minutes).

   (iv) *Tool setting cost*

$$\text{Cost of tool setting} = C_u \frac{T_m}{T}.T_s$$

   where T_s = Time required for setting the tool.

   (v) *Total machining cost*

$$C_p = C_u.T_m + C_u T_h + C_u \frac{T_m}{T} T_s + C_e \frac{T_m}{T}. \qquad \text{...(1)}$$

   In equation (1) T_h and C_u are constant. The job handling cost does not depend on cutting speed and feed. The cost factors containing the term T_m will be function of cutting speed and feed.

4. **Machining time**: Total machining time,

$$T_0 = T_m + T_h + \frac{T_m.T_s}{T}$$

5. **Rate of production**: The rate of production is the reciprocal of T_0.

$$\therefore \qquad R_p = \frac{1}{T_0} = \frac{1}{T_m + T_h + \frac{T_m.T_s}{T}}$$

8.5 OPTIMUM CUTTING SPEED IN TURNING FOR MINIMUM COST

The cost of turning operation from equation (1)

$$C_p = C_u\left[T_m + T_h + \frac{T_m T_s}{T}\right] + C_e\frac{T_m}{T}.$$

where, $T_m = \dfrac{\pi DL}{1000 fV}$

If feed, f is kept constant, T_m can be expressed as

$$T_m \propto \frac{1}{V} = \frac{B}{V} \quad ...(2)$$

where, $B = \dfrac{\pi DL}{1000 f}$.

From Taylor's equation for tool life,

$$VT^n = C$$

where, C = constant

$$\therefore \quad T = \left(\frac{C}{V}\right)^{1/n} \quad ...(3)$$

where n depends upon tool material

$\approx$ 0.125 for HSS

$\approx$ 0.25 to 0.3 for carbides

$\approx$ 0.5 to 0.7 for ceramics.

Eliminating T_m and T in equation (2) and (3),

$$C_p = C_u\left[\frac{B}{V} + T_h + \frac{BT_s}{c^{1/n}}V^{1/n-1}\right] + C_e\frac{BV^{1/n-1}}{c^{1/n}} \quad ...(4)$$

For optimum cutting speed V_{0pt},

$$\frac{\partial C_p}{\partial V} = 0$$

$$\therefore \quad C_u\left[-\frac{B}{V^2} + \left(\frac{1}{n} - 1\right)\frac{BT_s}{C^{1/n}}V^{(1/n-2)}\right] + \left(\frac{1}{n-1}\right)C_e\frac{BV^{(1/n-2)}}{C^{1/n}} = 0$$

$$\therefore \quad V = V_{opt} = \frac{C}{\left[\left(\frac{1}{n} - 1\right)\left(T_s + \frac{C_e}{C_u}\right)\right]^n}$$

The optimum tool life,

$$T_{opt} = \left(\frac{1}{n} - 1\right)\left(T_s + \frac{C_e}{C_u}\right)$$

Example 8.1 A batch of 1000 steel components are to be rough machined using a feed of 0.2 mm/rev. The size of component is 100 mm diameter and 250 mm length. The cost data is as follows:

Cost per cutting edge of throw-away carbide insert = Rs. 50

Resetting time per cutting edge = 1 minute

Total machining rate (including operator cost) = Rs. 300 / hr.

The Taylor tool life equation is

$$VT^{0.25} = 650$$

where V is in m/min and T is in minute

Calculate: (i) optimum cutting speed for minimum cost

(ii) optimum tool life.

(iii) total production cost. Assume loading and unloading time is 2 min per component and inital set-up time is 2 hours.

(iv) total production time is 2 hours.

(v) total production time for the total batch.

Solution:

Data $n = 0.25$ $\qquad C_e = \text{Rs. } 50$

$C = 650$ $\qquad C_u = \text{Rs. } \dfrac{300}{60} = \text{Rs. } 5/\text{min}$

$f = 0.2$ mm/rev.

$D = 100$ mm $\qquad T_s = 1$ min

$L = 250$ mm $\qquad T_h = 2$ min.

Batch size = 1000 $\qquad$ Initial set up time = 120 min.

(i) Optimum cutting speed for minimum cost,

$$V_{opt} = \frac{C}{\left[\left(\frac{1}{n} - 1\right)\left(T_s + \frac{C_e}{C_u}\right)\right]^n}$$

$$= \frac{650}{\left[\left(\frac{1}{0.25} - 1\right)\left(1 + \frac{50}{5}\right)\right]^{0.25}} = 271 \text{ m/min.}$$

(ii) From Taylor's equation,

$$VT^n = C$$

$$T^{0.25} = \frac{650}{271} = 2.398$$

$\therefore \qquad T = 33$ minutes.

or, $$T_{opt} = \left(\frac{1}{n} - 1\right)\left(T_s + \frac{C_e}{C_u}\right)$$

$$= \left(\frac{1}{0.25} - 1\right)\left(1 + \frac{50}{5}\right) = 33 \text{ min.}$$

(iii) Total production cost,

$$C_{\text{total}} = C_u(\text{time for initial setup}) + 1000\left[C_u T_m + C_u T_h + C_e \frac{T_m}{T} + C_u \frac{T_m}{T} T_s\right]$$

$$T_m = \frac{\pi DL}{1000 f V_{opt}} = \frac{\pi \times 100 \times 250}{1000 \times 0.2 \times 271} = 1.45 \text{ min.}$$

$$\therefore \quad C_{\text{total}} = 5 \times 120 + 1000\left[5 \times 1.45 + 5 \times 2 + \frac{50 \times 1.45}{33} + \frac{5 \times 1.45}{33} \times 1\right]$$

$$= \text{Rs. } 20267.$$

(iv) Total production time,

$$T_{\text{Total}} = \text{Total initial setup time} + 1000\left(T_m + T_h + \frac{T_m}{T} \times T_s\right)$$

$$= 120 + 1000\left[1.45 + 2 + \frac{1.45}{33} \times 1\right]$$

$$= 3614 \text{ min.}$$

$$= 60 \text{ hrs and } 14 \text{ min.}$$

8.6 OPTIMUM CUTTING SPEED IN TURNING FOR MAXIMUM PRODUCTION RATE

The production rate is the reciprocal of production time.

$$R_p = \frac{1}{T_p} = \frac{1}{T_m + T_h + \frac{T_m T_s}{T}}$$

From Taylor's equation for tool life,

$$VT^n = C$$

$$\therefore \quad T = \left(\frac{C}{V}\right)^{1/n} \qquad ...(5)$$

Substituting the value of T from equation (5)

$$R_p = \frac{1}{\left(\frac{B}{V}\right) + T_h + \frac{B}{V} T_s \Big/ (C/V)^{1/n}}$$

For maximum production rate,

$$\frac{\partial R_p}{\partial V} = 0.$$

$$V = V_{opt} = \frac{C}{\left[\left(\frac{1}{n} - 1\right)T_s\right]^n}$$

and, $$T_{opt} = \left(\frac{1}{n} - 1\right)T_s.$$

Example 8.2 Repeat example 8.1 by using maximum production rate as the optimizing criterion .

Solution: The given data.

$n = 0.25$ $C_e =$ Rs. 50

$C = 650$ $C_u =$ Rs. 5/min

$f = 0.2$ mm/rev. $T_s = 1$ min

$D = 100$ mm $T_h = 2$ min.

$L = 250$ mm

Batch size = 1000. Initial setup time = 120 min

(i) Optimum cutting speed for maximum production rate,

$$V_{opt} = \frac{C}{\left[\left(\frac{1}{n} - 1\right)T_s\right]^n}$$

$$= \frac{650}{\left[\left(\frac{1}{0.25} - 1\right) \times 1\right]^{0.25}} = 494 \text{ m/min.}$$

(ii) Optimum tool life,

$$T_{opt} = \left(\frac{1}{n} - 1\right)T_s$$

$$= \left(\frac{1}{0.25} - 1\right) \times 1 = 3 \text{ min}$$

Or, using Taylor's equation,

$$V_{opt} T_{opt}^{0.25} = 650.$$

$$T_{opt}^{0.25} = \frac{650}{494} = 1.315$$

$$\therefore \qquad T_{opt} = (1.315)^{1/0.25} = 3 \text{ min.}$$

(iii) $$T_m = \frac{\pi DL}{1000 f V_{opt}} = \frac{\pi \times 100 \times 250}{1000 \times 0.2 \times 494} = 0.795 \text{ min}$$

$$T_{opt} = T = 3 \text{ min}$$

$\therefore$ Total production cost

$$C_{\text{Total}} = C_u \times \text{Initial setup lime} + 1000\left[C_u T_m + C_u T_h + C_e \frac{T_m}{T} + C_u \frac{T_m}{T}.T_s\right]$$

$$= 5 \times 120 + 1000\left[5 \times 0.795 + 5 \times 2 + \frac{50 \times 0.795}{3} + \frac{5 \times 0.795}{35} \times 1\right]$$

$$= 600 + 1000\,(3.975 + 10 + 13.25 + 1.325)$$

$$= \text{Rs. } 29{,}150$$

(iv) Total production time,

$$T_{\text{Total}} = \text{time of initial set up} + 1000\left[T_m + T_h + \frac{T_m}{T} T_s\right]$$

$$= 120 + 1000\left[0.795 + 2 + \frac{0.795}{3} \times 1\right]$$

$$= 3180 \text{ min.}$$

$$= 53 \text{ hours.}$$

8.7 MACHINING AT MAXIMUM EFFICIENCY

The cutting speed for maximum efficiency (or maximum profit rate) may be different from that for maximum production rate or minimum cost.

The profit per unit,

$$P_u = S_p - C_p$$

where, S_p = Sale price or revenue per piece

C_p = Production cost per piece

The rate of profit,

$$P_r = \frac{S_p - C_p}{T_p}$$

where, T_p = Production time per piece.

The selling price depends upon market conditions i.e., supply and demand conditions. However, it may tentatively be assumed as constant.

$$\therefore \qquad P_r = P_u . R_p$$

where, $\qquad R_p$ = Rate of production

The cost of production can be divided into two components.

1. Fixed cost of establishment of infrastructure including machines and buildings.
2. Variable cost which is a function of production rate and includes consumption of materials, electricity, lubricating oil, coolant, etc.

   Therefore, rate of production cost,

   $$C_{pr} = A + E.R_p$$

   where, A = Fixed cost

   E = Variable cost

   The profit rate,

   $$P_r = -A + (S_p - E)R_p.$$

   Now, $$E = C_m + C_u\left[\frac{B}{V} + T_h + \frac{BT_s}{C^{1/n}}V^{1/n-1}\right] + C_e\frac{BV^{1/n-1}}{C^{1/n}}$$

   where, C_m = Cost of material used per piece.

$$\therefore \qquad P_r = -A + \frac{S_p - C_m}{T_m + T_h + \frac{T_m}{T}.T_s}$$

$$= -\frac{C_u\left[\frac{B}{V} + T_h + BT_s\frac{V^{1/n-1}}{C^{1/n}}\right] + C_e\frac{BV^{1/n-1}}{C^{1/n}}}{\left(T_m + T_h + \frac{T_m}{T}.T_s\right)}$$

where, $T_m = \frac{B}{V}$

and, $\qquad T = \frac{C^{1/n}}{V^{1./n}}$

$$\therefore \; P_r = \frac{S_p - C_m - C_u\left[\left(\frac{B}{V}\right) + T_h + BT_s\frac{V^{1/n-1}}{C^{1/n}}\right]}{(B/V) + T_h + BT_s\left(\frac{V^{1/n-1}}{C^{1/n}}\right)} - \frac{C_e\left(\frac{BV^{1/n-1}}{C^{1/n}}\right)}{\left(\frac{B}{V}\right) + T_h + BT_s\frac{V^{1/n-1}}{C^{1/n}}} - A$$

For maximum efficieary,

$$\frac{\partial P_r}{\partial V} = 0$$

The cutting speed for maximum efficiency,

$$V = V_{mp} = \frac{BC_e}{n(S_p - C_m)[T_{mp} - (1-n)(C_e T_h + (S_p - C_m)T_s]}$$

where, T_{mp} = Tool life for maximum efficiency.

The Taylor's equations is

$$V_{mp} T_{mp}^n = C$$

$$\therefore \quad T_{mp} = \left(\frac{C}{V_{mp}}\right)^{1/n}$$

$$\therefore \quad V_{mp} = \frac{BC_e}{n(S_p - C_n)\left[\left(\frac{C}{V_{mp}}\right)^{1/n} - (1-n)(C_e TV_n) + (S_p - C_m)TV_s\right]}.$$

This equation can be solved numerically using method of iteration.

QUESTION BANK NO. 8

1. How is feed selected for rough machining and finishing cut?
2. With the help of a suitable graph, discuss the effect of cutting speed on cost of production and rate of production.
3. Derive expression for optimum cutting speed in turning for
   (a) minimum production cost
   (b) maximum production rate.
4. What is machining at maximum efficiency? Derive an expression for cutting speed in turning for maximum profit.

9

COST ESTIMATION OF MACHINED COMPONENTS

Introduction; Material cost; Loading and unloading of machines; Tool and feed setting; Intermachine component movement; Tool wear and replacement cost; Allowances for tool engagement and disengagement; Machining costs, Question Bank.

9.1 INTRODUCTION

Cost estimation of machined components is required at two stages.

(i) A detailed and accurate cost estimation is carried out by the manufacturing and industrial engineers after the component has been designed and its production planned. Every step in production is known and cost can be estimated with high degree of accuracy.

(ii) A preliminary and quick cost estimation is carried out on the basis of minimum information available from the designer. It is required to ensure that ultimate design will avoid unnecessary manufacturing expense and the component will be manufactured under reasonable economic condition. An attempt can be made to minimize the manufacturing cost.

At this preliminary stage, the detailed design is not frozen. The workholding devices and tools used have not been worked out. Final decision has not been taken on the specific work material.

The simplest approach would be to specify the shape and size of the original workpiece and the quantity of material to be removed by machining. The cost can be estimated on the basis of material cost per unit weight and average cost of removal of material per unit volume by machinery.

However, this simple method does not take into account the unproductive costs involved in a series of machining operations.

Various costs contributing to the total cost of machined components are discussed.

9.2 MATERIAL COST

The cost of original workpiece is one of the most important factors in the total cost of a machined component. This cost will often form more than 50% of the total cost. This cost should be estimated with sufficient care. The material prices fluctuate considerably. These prices should be taken from material supplier. For rough estimate, the comparative cost data is given in Table 9.1. The volume of material required can be estimated by the designer. The cost of material can be estimated using the data given in table 9.1.

Table 9.1 Comparative cost of various raw materials.

Sl. No.	*Raw Material*	*Sp. Weight*	*Bar*	*Rod*	*Sheet <6 mm*	*Plate >6 mm*	*Tube*
	Ferrous						
1.	Carbon steel	7.83	1.00	1.00	0.70	0.80	1.80
2.	Alloy steel	8.58	1.50	1.50	2.35	–	–
3.	Stainless steel	7.83	3.00	3.00	4.90	4.90	–
4.	Tool steel	7.83	12.60	12.60	–	12.60	–
	Non-ferrous						
5.	Aluminium alloys	2.77	3.80	3.80	3.80	4.90	9.00
6.	Brass	8.58	1.75	2.35	3.75	3.75	3.75
7.	Nickel alloys	8.30	11.00	11.00	11.00	11.00	–
8.	Magnesium alloys	1.83	6.50	6.50	12.00	12.00	6.50
9.	Zinc alloys	6.37	3.00	3.00	3.00	3.00	–
10.	Titanium alloys	4.51	30.00	30.00	50.00	50.00	–

9.3 LOADING AND UNLOADING OF MACHINES

Non-productive costs are incurred due to time spent on loading and unloading of a workpiece into a machine tool. These times are function of weight of workpiece, type of machine tool and workholding device. The loading and unloading times for different weights and different workholding devices are given in Table 9.2. The time required for turning coolant on and off, cleaning of workholding or clamping device, etc. may be added to the figures given in the Table 9.2.

9.4 TOOL AND FEED SETTING

There are non-productive costs incurred on machine tool for setting of tool, feed, speed, change of speed, withdrawal of tool, etc. If different tools are used, there are costs for tool engagement and indexing. There are production costs for basic setup time and additional setup time for cutting tool. The total setup time should be divided by the batch size to estimate setup time per component. Such unproductive times for different machine tools are given in Table 9.3.

Table 9.2 Time in Seconds of Loading and Unloading of Different Weights of Workpiece.

Sl. No.	CLAMP/HOLDING DEVICE	WORKPIECE WEIGHT(KG)				CRANE
		0-0.2	0.2-4.5	4.5-14	14-27	
1.	U- Clamps and Angle plate	27.5	35	44	71	277
2.	Centres without dog	13.5	19	24	35	73
3.	Centres with dog	25.5	40	57	98	250
4.	Universal chuck	16	23	32	53	–
5.	Four jaw chuck	34	41	50	71	–
6.	Three-clamp on table	29	34	40	59	265
7.	Collet	10	15	21	–	–
8.	Three-clamp face plate	32	43	58	82	196
9.	Three-screw horizontal fixture	26	33	42	70	275
10.	Hand-held	1.5	6.5	12	–	–
11.	Jig	26	33	42	–	–
12.	Magnetic table	3	5	8	–	–
13.	Parallels	14	20	25	67	255
14.	Rotary table or Index plate	29	36	45	72	278
15.	V-Block	25	30	36	78	365
16.	Vice	16	19	24	40	174

Table 9.3 Tool Setting Times for Different Machine Tools

Sl. No.	Machine tool	Time		
		* Tool engagement (seconds)	Basic Tool setup (hours)	Additional Tool setup (hours)
1.	Horizontal band saw	–	0.2	–
2.	Manual turrent lathe	9	1.2	0.2
3.	CNC turret lathe	1.5	0.5	0.15
4.	Milling machine	30	1.5	–
5.	Drilling machine	10	1.0	–
6.	Horizontal boring machine	30	1.3	–
7.	Broaching machine	15	0.6	–
8.	Gear hobbing machine	40	0.9	–
9.	Grinding machine	20	0.6	–
10.	Internal grinding machine	25	0.6	–
11.	Machining center	10	0.7	0.55

* Average time to engage tool, engage and disengage feed, change speed, feed or tool.

9.5 INTER MACHINE COMPONENT MOVEMENT

Batches of partially machined components have to be moved between machine tools for further processing. A model was prepared to move stacks of pallets of workpieces using forklift trucks around the factory. The following expressions were developed.

The transportation time for a round trip by a forklift truck,

$$t_f = 25 + 3\,(lp + l_{rd})\,[\text{sec}]$$

where, l_p = length of pathway between machines (m)

l_{rd} = distance moved by truck to return [m]

In case $(l_p + l_{rd})$ in 139 m on average and there is a return trip with empty pallets for every trip with a load of full pallets, the total time,

$$t_f = 442 \text{ sec.}$$

The allowances for transportation time when added to loading and unloading time as given in Table 9.1, there will be significant time for large workpieces.

9.6 TOOL WEAR AND REPLACEMENT COSTS

A tool has to be replaced whenever it wears. Two type of costs are incurred.

(1) The cost of machine idle time while the worn tool is replaced by the operator.

(2) The cost of providing new cutting edge to worn tool or a new tool.

The tool wear and replacement cost is affected by changes in cutting speed. Similarly, the machining cost also depends upon cutting speed.

The cost of machining a feature in one component on one machine tool can be expressed as:

$$C_m = Mt_m + \frac{Q(Mt_a + C_t)t_m}{t} \qquad ...(1)$$

where, M = machine and operator rate

t_m = machining time (time the machine tool is operating).

Q = part of t_m for which a tool point is contacting the workpiece.

t = tool life while the cutting edge is contacting the workpiece.

t_{cr} = tool changing time.

C_t = cost of providing a new cutting edge.

The machining time,

$$t_m = \frac{k}{V} \qquad ...(2)$$

where, k = constant for a machining operation

V = Cutting speed.

The tool life t can be obtained from Taylor's tool-life equation

$$Vt^n = V_r t_r^n \qquad ...(3)$$

where, V_r and t_r are reference cutting speed and tool life respectively.

n = Taylor tool-life index which is a function of tool material
= 0.125 for high-speed steel tool.
= 0.25 for carbide tool.

Substituting the value of t_m and t from equations (2) and (3) into equation (1)

$$\therefore \quad C_{\min} = \frac{Mt_{mc}}{1-n} \qquad ...(4)$$

where, t_{mc} = machining time for optimum cutting speed.

If cutting speed for minimum cost is used, the tool replacement cost can be found out from the factor $(1/1 - n)$.

For high-speed steel tooling.

$(1/1 - n) = 1.14$

For carbide tools

$(1/1 - n) = 1.33$.

If optimum cutting speed connot be used due to power limitation, at least cutting speed should be reduced. It will give more savings in tool costs as compared to when feed is reduced. The correction factor $(1/1 - n)$ calculated for above condition will over-estimate the tool life.

Machining cost for use of maximum power

The machining cost,

$$C_p = Mt_{mp} + \frac{Q(Mt_{c_1} + C_t)t_{mp}}{t_{p_0}} \qquad ...(5)$$

where, V_{p_0} = cutting speed for use of maximum power.

t_{p_0} = tool life under maximum power condition

From Taylor's tool-life equation,

$$t_{p_0} = t_c\left(\frac{V_{p_0}}{V_c}\right)^{1/n} \qquad ...(6)$$

where, t_c = tool life under minimum cost condition.

$$= \left[\frac{1-n}{n}\right]Q\left(t_{c_1} + \frac{C_t}{M}\right) \qquad ...(7)$$

Substituting equations, (6) and (7) into equation (5), then equation (2) gives,

$$C_p = Mt_{mp}\left[1 + \left(\frac{n}{1-n}\right)\left(\frac{t_{mc}}{t_{mp}}\right)^{1/n}\right] \qquad ...(8)$$

9.7 ALLOWANCES FOR TOOL ENGAGEMENT AND DISENGAGMENT

During machining, tool is not cutting for whole of time during each cut, pass or operation. Allowances must be made for time spent during tool engagement and disengagement. There time allowances depend upon type of machining operation. Typical values of allowances as correction factors have been presented in Table 9.4. These allowances should be applied to actual machining time.

Table 9.4 Allowances for tool engagement (approach) and disengagement.

Sl. No.	*Operation*	*Allowance*	*Remarks*
1.	Lathe operations (Turning, facing, cutting off, boring, grooving, threading)	$t'_m = t_m + 5.4$ $t'_m = t_m + 2200d_a^2$	$d_a < 0.05$ $d_a \geq 0.05$
2.	Drilling (Twist)		
	– Approach	$t'_m = t_m\left(1 + 0.5\dfrac{d_a}{l_w}\right)$	
	– Start	$t'_m = t_m + \left(\dfrac{25.9}{V_f}\right)d_a^{1.67}$	
3.	Face and end milling	$l'_w = l_w + d_t + 0.066 + 0.011\ d_i$	
4.	Surface grinding	$l'_w = l_w + \dfrac{d_t}{2}$	
5.	Hole broaching	$l_t = 6 + 6d_a + 6l_w$	

where, t_m = machining time (s)

d_a = diameter of work surface, drilled hole, machined surface (m)

l_w = length of machined surface (m)

V_f = rate of surface generation (m^2/min)

= (speed × feed)

t'_m or l'_m = corrected value after allowance is applied.

9.8 MACHINING COSTS

The cutting conditions i.e., cutting speed, feed and depth of cut in single-point tool operation work be known for accurate estimation of actual machining time. For multi-point tool operations, cutting speed, feed speed, depth of cut and width of cut must be known. The data is available in machining data handbooks. For rough machining operations, a better estimate of machining time can be obtained from specific cutting energy, volume of material to be removed and power available for machining.

Cutting speeds, feeds and depth of cut generally used in different machining operations for common work materials are given in annexure I.

QUESTION BANK NO. 9

1. Briefly discuss the importance and stages of cost estimation of machined components.
2. Discuss the various elements of machining cost.
3. Write short notes on the following:
   1. Material cost.
   2. Loading and unloading time of machines.
   3. Tool and feed setting time.
   4. Time for inter machine movement of components.
   5. Tool wear and replacement costs.
4. What are actual cutting time and idle time in machining operations?

10

NONCONVENTIONAL MACHINING PROCESSES

Introduction; Advantages and applications; Classification; Comparative performance of cutting process; Ultrasonic machining; Water-jet machining; Abrasive jet machining; Chemical machining; Electrochemical machining; Electrolytic grinding; Electrical discharge machining; Wire electrical discharge machining; Laser-beam machining; Electron-beam machining; Plasma-arc cutting; Question Bank.

10.1 INTRODUCTION

In recent years, hard, high-strength and temperature-resistant materials have been developed for aerospace, nuclear engineering and other industries. These superalloys have high strength-to-weight ratio, hardness, and heat-resisting qualities. Some examples of these materials are hastalloy, waspalloy, nitralloy, nimonics, carbides, stainless steels, heat-resisting steels and many other high-strength-temperature-resistant (HSTR) alloys. The conventional edged tool machining, in spite of recent technological developments, is highly uneconomical. The drilling and milling processes have poor degree of accuracy and surface finish. Machining of these materials into complex shapes is difficult, time-consuming and sometimes impossible.

Nonconventional machining processes have emerged to overcome the above limitations of conventional machining processes. These processes do not employ a traditional tool for metal removal but utilize some form of energy for metal machining such as mechanical, electrical, chemical, electrochemical, thermo-electric, etc. Most of the new machining processes have been developed specifically for materials that are difficult to machine, but some of them have found use in the production of complex shapes and cavities in more readily machined materials.

10.2 ADVANTAGES AND APPLICATIONS

Nonconventional machinery processes have certain advantages over traditional processes and are used for the following reasons.

1. Machinability

Many materials having high hardness, thermal resistance and abrasive wear are difficult to machine by traditional methods. Many unconventional machining processes can be applied for such applications.

2. Complex Shape

Many shapes such as a sharp-cornered square hole cannot he produced by conventional machining methods. Nonconventional machining processes can be used for machining complex shapes with ease. Complex shapes can be produced even in materials different to machine.

3. Surface Defects

Conventional machining processes can cause surface cracks and residual stresses. Many unconventional machinery processes can remove material without above surface defects.

4. Precision

Many nonconventional processes can give high precision not available with conventional machining.

5. Miniaturization

Conventional machining may not be suitable for very fine details such as drilling of deep small sized holes. Many nonconventional processes can produce small and fine jobs.

6. Computer Integration

Many nonconventional processes can be computer-controlled and linked directly to CAD/CAM in integrated manufacturing system.

10.3 CLASSIFICATION OF NONCONVENTIONAL MACHINING PROCESSES

The need of special processes has been responsible for the invention of a number of machining processes such as ultrasonic machining (USM), spark erosion machining or electro-discharge machining (EDM), electrochemical machining (ECM), electrochemical grinding (ECG), ion-beam machining (IBM), laser-beam machining (LBM), etc.

The different nonconventional machining processes can be broadly classified into following groups according to energy source used.

1. Mechanical Processes

The three main processes that utilize a mechanical action for removing material are:

- Ultrasonic machining (USM).
- Abrasive-jet machining (AJM).
- Water-jet machining (WJM).
  This includes abrasive water-jet machining.

The schematic diagram of these processes is shown below:

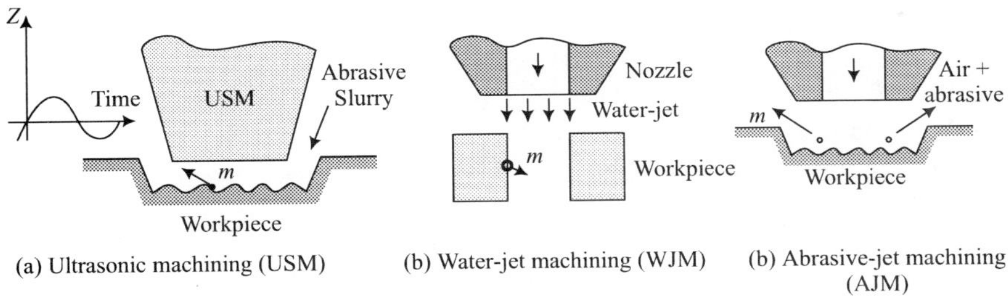

Fig. 10.1 Mechanical processes.

2. Chemical Process

The chemical processes are chemical cutting, milling, blanking, etc. The schematic diagram of the chemical process is shown below.

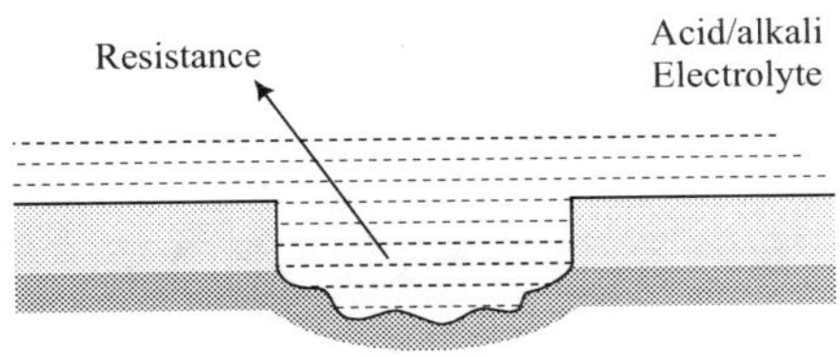

Fig. 10.2 Chemical machine (CHM).

3. Electrochemical Processes

These processes include electrochemical machining (ECM) and electrochemical grinding (ECG), deburring, honing, etc. The schematic diagram is shown.

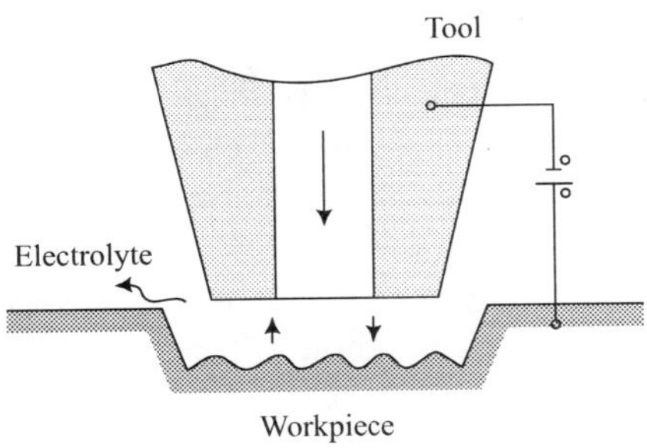

Fig. 10.3 Electrochemical machine (ECM).

4. Electrothermal Processes

These processes include:

- Electro-discharge machining (EDM).
- Laser-beam machining (LBM).
- Electron-beam machining (EDM).
- Plasma-beam machining (PAM).
- Ion-beam machining (IBM).

The schematic diagram of important electrothermal processes is shown.

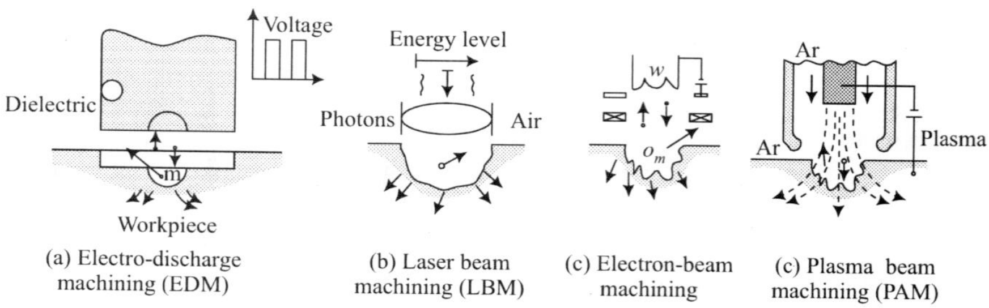

Fig. 10.4 Electro-thermal processes.

10.4 COMPARATIVE PERFORMANCE OF CUTTING PROCESSES

The selection of nonconventional machining process depends upon the following parameters.

- Physical parameters.
- Properties of work material and the shape to be machined.
- Process capability or machining characteristics.
- Economic consideration.

10.4.1 Material Application

The conventional machinery processes vary in applications depending upon the strength and hardness of the material to be machined. The nonconventional machining processes cover most of the metals and alloys. The material applications of various nonconventional methods are given in Table 10.1.

The applications of nonconventional machining processes are also influenced by shape and size of the product, i.e., holes, through holes, surfacing, through cutting, special applications.

10.4.2 Process Capability

The process capability or machining characteristics can be analyzed with respect to:

Table 10.1 Material application.

Process	*Alumi-nium*	*Steel*	*Super alloy*	*Tita-nium*	*Refract-ories*	*Ceramics*	*Plastic*	*Glass*
Ultrasonic machining (USM)	P	F	P	F	G	G	F	G
Abrasive-jet machining (AJM)	F	F	G	F	G	G	F	G
Electrochemical machining (ECM)	F	G	G	F	F	N	N	N
Chemical machining (CHM)	G	G	F	F	P	P	P	F
Electro-discharge machining (EDM)	F	G	G	G	G	N	N	N
Ion-beam machining (IBM)	F	F	F	F	G	G	F	F
Laser-beam machining (LBM)	F	F	F	F	P	G	F	F
Plasma-arc machining (PAM)	G	G	G	F	P	N	P	N

Index: *G* = Good *P* = Poor *F* = Fair *N* = Not applicable

- Metal removal rate.
- Tolerances maintained.
- Surface finish.
- Depth of surface damage.
- Power required for machining.

Process capabilities of different processes are given in Table 10.2.

10.4.3 Economics

The economics of the various processes are analyzed by considering:

- Capital cost.
- Tooling cost.
- Consumed power cost.
- Metal removal rate efficiency.
- Wear of tooling.

The cost of removing material is an important factor for the selection of processes. The economic parameters for various processes are compared in Table 10.3.

The following *conclusions* can be made:

- Conventional machining processes remove material faster.
- Power consumption is less for conventional processes.

Table 10.2 Process capabilities.

Process	*MRR* (mm^3/min)	*Tolerance* (*micron*)	*Surface finish* (*micron CLA*)	*Depth of surface damage* (*micron*)	*Power* (*watts*)
Conventional machining	50,000	50	0.5-5	25	3,000
Ultrasonic machining (USM)	300	7.5	0.2-05	25	2,400
Abrasive-jet machining (AJM)	0.8	50	0.5-1.2	25	250
Electrochemical machining (ECM)	15,000	50	0.1-2.5	5	100,000
Chemical machining (CHM)	15	50	0.5-2.5	5	—
Electro-discharge machining (EDM)	800	15	0.2-1.2	125	2,700
Ion-beam machining(IBM)	1.6	25	0.5-2.5	250	150 ((Averg.) 200 (Peak)
Laser-beam machining (LBM)	0.1	25	0.5-1.2	125	2 (Averg.) 2000 (Peak)
Plasma-arc machining (PAM)	75	125	Rough	500	50,000

Table 10.3 Economic parameters.

Process	*Metal removal rate* (cm^3/s)	*Specific cutting energy* (GJ/m^3)	*Equipment cost per unit (MRR)*
Conventional machining	14	3	1
Abrasive-jet machining (AJM)	0.0001	10,000	–
Plasma-arc machining (PAM)	1.5	10	–
Electrochemical machining (ECH)	1.0	500	60
Electro-discharge machining (EDM)	0.1	150	40
Ultrasonic machining (USM)	0.0005	150	600
Ion-beam machining (IBM)	0.001	1500	–40,000
Laser-beam machining (LBM)	0.0001	150,000	160,000

- The cost material removing is much higher for nonconventional processes as compared to conventional processes.
- Nonconventional processes are the only possible technologies for machining hard materials.

10.5 ULTRASONIC MACHINING (USM)

Ultrasonic machining is used for precision machining of hard brittle materials. A grit-loaded slurry is circulated between the workpiece and the tool. The tool vibrates at ultrasonic frequency of 20 to 40 kHz and gradually fed into the workpiece. Material is removed by the abrasive action of the slurry and impact force due to vibration of the tool. The tool has the same shape as the groove to be machined. Analysis of the mechanism of material removal indicates that the process can be called *ultrasonic grinding* (USG).

10.5.1 Ultrasonic Machining Equipment

The ultrasonic machining equipment is shown in the schematic diagram. The main elements are:

- A high frequency generator.
- A transducer which utilizes the piezoelectric or magneto-strictive effect.
- A transformer, which is shaped to amplify the vibration output of the transducer.

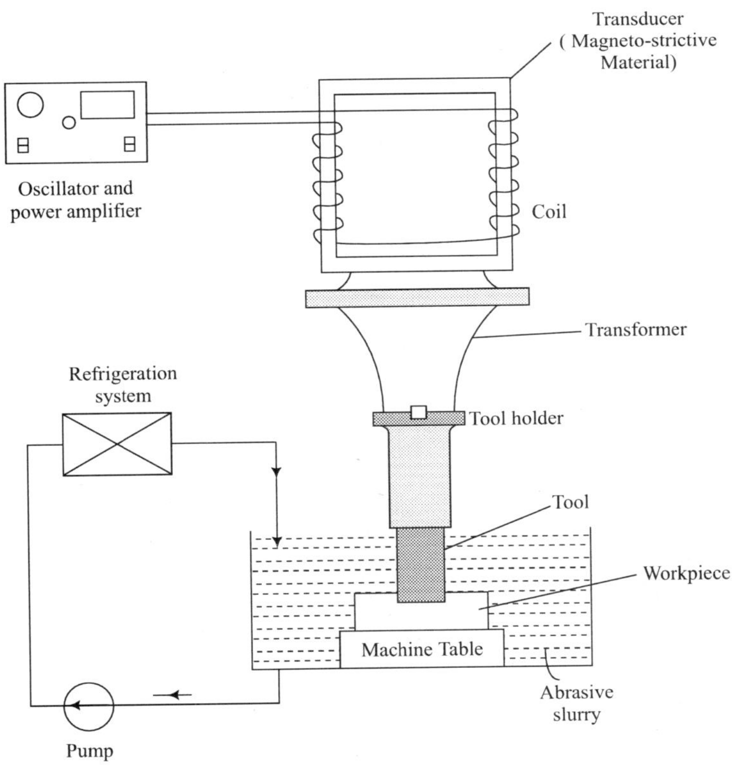

Fig. 10.5 Ultrasonic machining equipment.

- A tool and tool holder.
- Abrasive slurry which is cooled and circulated into the small gap between the tool and workpiece.

1. Electronic oscillator and amplifier: It is also called generator. It converts the available electrical power of low frequency (50 Hz) to high frequency power of the order of 20 kHz. This is supplied to the transducer.

2. Transducer: It converts electrical energy into mechanical motion. There are two types of transducers.

- **Piezoelectric transducer:** Crystals of quartz or polycrystalline ceramics vibrate when applied to alternating voltage. The amplitude of vibration is maximum at the resonant frequency of the crystal. Such transducers convert electrical energy into vibration energy at 96% efficiency without cooling.
- **Magnetostrictive transducer:** Ferromagnetic materials such as nickel and nickel alloys are magnetized and application of alternating current to the coils surrounding the material causes vibration at the applied frequency. The energy conversion efficiencies are only 25 to 30% and water cooling is required to remove waste heat from the transducer.

3. Transformer or concentrator: The amplitude of vibration of transducer is inadequate for cutting purposes. The vibration is transmitted to the cutting tool through a mechanical concentrator or velocity transformer which provides intense vibration of desired amplitude at the tool end. The transformer is of taper shape and reduction in cross-sectional area causes amplification of the vibration. Amplitude may be increased by 600%.

4. Tool holder: The tool holder also helps to increase the amplitude at the tool face in the range of 0.013 to 0.1 mm.

5. Tools: The shape of the tool is the inverse of the cavity to be eroded in the workpiece. Stainless steel, brass, mild steel and other ductile materials are used to minimize tool wear. The tools are fastened to tool holder by brazing to reduce possible fatigue problems.

10.5.2 Abrasive Slurry

The slurry consists of small abrasive particles mixed with water or oil in the concentration of 30 to 60% by weight. The commonly used abrasives are: boron carbide, silicon carbide, diamond dust and aluminium oxide (alumina).

Boron is the most expensive material and is used for cutting of tungsten carbide, tool steel and gems. Alumina is best for cutting glass and ceramics. Silicon is most commonly used.

The abrasive slurry is circulated to the tool-workpiece interface by pumping. A refrigerated cooling system is used to cool the slurry to a temperature of 5 to 6°C. The slurry is kept in a tank so that cutting zone is dipped in the slurry.

The liquid (water or oil) should have the following characteristics.

- Good wetting characteristic.
- Low viscosity.
- High thermal conductivity.
- Anticorrosive property.
- Density equal to that of abrasive.
- Low cost.

Grit Size

The grit size is usually about the same as the amplitude of vibration being used. It varies between 200 and 2000 grits. Coarse grades are good for roughing, whereas finer grades are used for finishing. Fresh abrasives cut better. Therefore, slurry must be replaced periodically.

10.5.3 Process Variables and their Effects

The machining rate or cutting rate and accuracy of machining depend upon certain process variables. The cutting rate depends upon:

- Grain size of abrasive.
- Abrasive material.
- Concentration of slurry.
- Amplitude of vibration.
- Frequency.

A. Machining Rate

The maximum speed of penetration in soft and brittle materials such as soft ceramics is of the order of 20 mm/min, but for hard and tough materials, the penetration rate is lower.

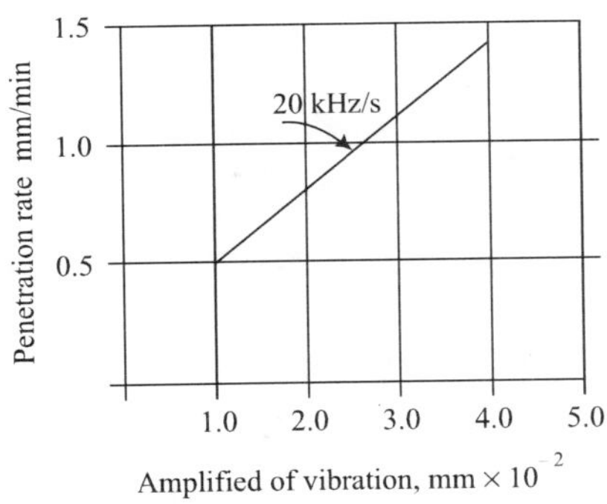

Fig. 10.6 Effect of amplitude of tool oscillation on machining rate.

1. Effect of vibration amplitude and frequency: The cutting rate is proportional to the square of the amplitude of vibration for many materials. However, it is also influenced by slurry concentration.

Increase in frequency increases the number of blows on grain particles which in turn impinge on the metal surface. The machining rate is directly proportional to the frequency of vibration.

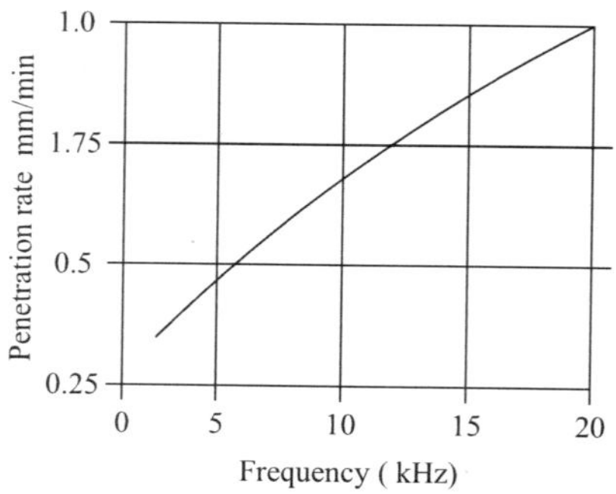

Fig. 10.7 Effect of frequency of vibration on machining rate.

2. Effect of abrasive material and grit size: Characteristics of abrasive particles such as size, shape, material and hardness of grains affect the metal removal rate. Increase in grit size increases machining rate till the grit size becomes nearly equal to the amplite of vibration. Beyond this, the increase in grit size decreases the machining rate. This is illustrated in the figure.

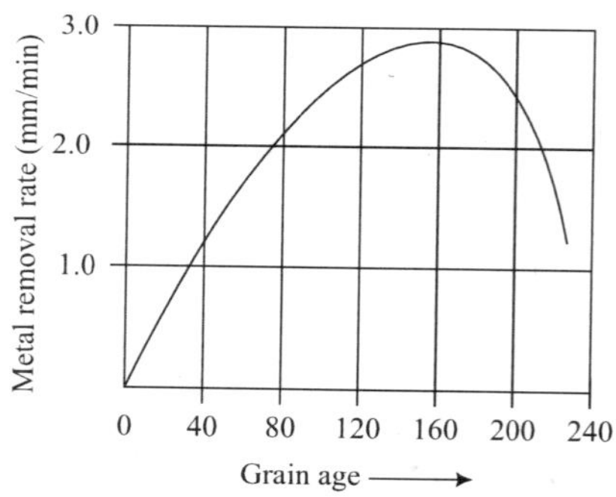

Fig. 10.8 Effect of grit size on machining rate.

3. Effect of slurry composition: Slurry is a mixture of abrasive grains and a carrier fluid, generally water. The concentration of grain particles in the slurry and the type of abrasive has effect on machining rate. The slurry should flow easily. The grits get worn out and machining rate falls rapidly requiring periodic replacement of worn out grits.

The viscosity of the carrier fluid helps dampen oscillations of the grits, thereby reducing the energy with which the grits hit the surface of workpiece. The viscosity of carrier fluid adversely affects the machining rate.

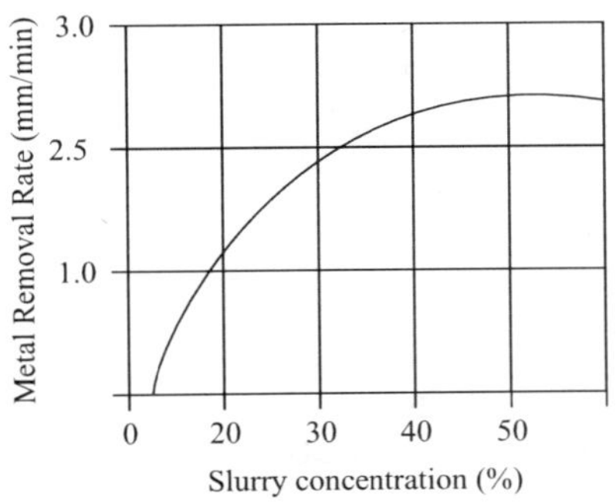

Fig. 10.9 Effect of slurry concentration.

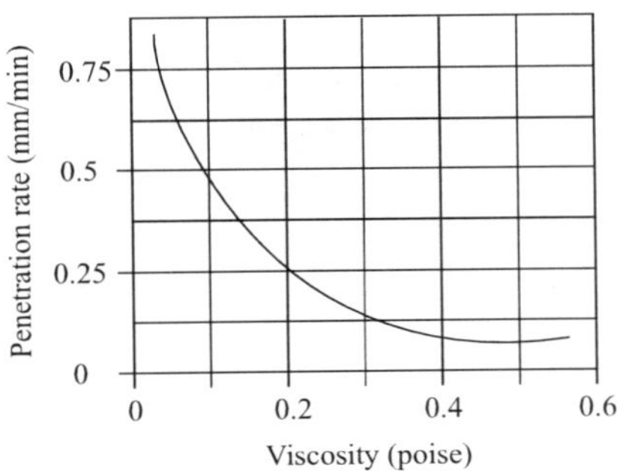

Fig. 10.10 Effect of viscosity of carrier fluid.

4. Effect of static load: The static load on the tool affects the machining rate. The machining rate increases with the increase of static load till it reaches a maximum value after which the machining rate decreases with the further increase of static load.

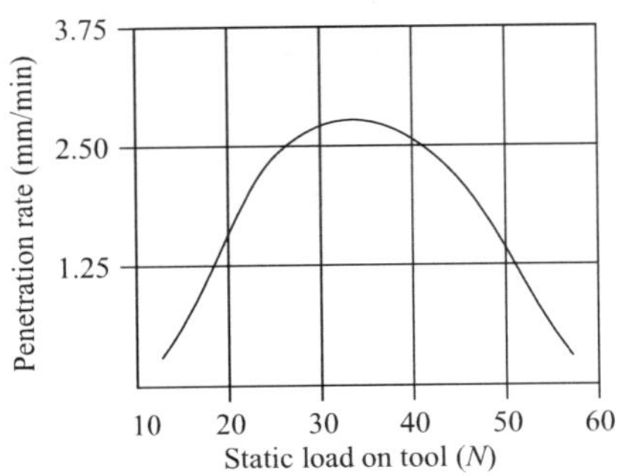

Fig. 10.11 Effect of static load on tool.

5. Workpiece material: Machining rate depends upon material properties such as hardness, brittleness and toughness.

6. Machining area: Machining rate depends upon machining area such as:

- Shape of groove.
- Area of groove.
- Depth of hole.

B. Accuracy of Machining

Dimensional accuracy upto $\pm$ 0.005 mm is possible with USM, and surface finish down to Ra value of 0.1-0.125 μm can be obtained. A minimum corner radius of 0.10 mm is possible in finish machining. Holes as small as 0.076 mm can be drilled, and the upper limit on cavity size is approximately 75 mm. Machining depths of 50 mm or more are possible. In practice depths are usually less than 5 mm.

Ultrasonic machining is a non-thermal process. The machining surface is not altered or damaged. This process can be successfully used for machining of components which are highly stressed in service. Examples are turbine blades and ceramic or ceramic coated components.

10.5.4 Applications

USM machines are available from a light portable type of 20 W rating to heavy machines of 2 kW input. The simplicity of the process makes it economical for a wide range of applications. USM has been successfully employed for the following applications.

- Machining of hard and brittle materials. These materials cannot be machined by conventional processes. Examples: glass, ceramic, tungsten carbide, diamond, zirconia, graphite, etc.
- Dies of tungsten carbide and diamond for wire drawing, forging and extrusion process.
- Drilling of non-circular holes in glass and similar materials.
- Drilling of holes of any shape in teeth without creating pain. This is a useful practice by dentists.
- Threading in components of hard metals and alloys. This is achieved by rotating and translating the workpiece or the tool.
- Machining of holes of various shapes by using appropriate tools. Other types of shapes can be manipulated by giving suitable motion to workpiece during cutting.
- Drilling, grinding, profiling and milling operations on conducting and non-conducting materials.
- Drilling holes of various shapes with straight or curved axes.

10.5.5 Limitations

The major limitations of USM processes are as follows:

- Low machining rates as compared to conventional machining process. The maximum metal removal rate is 3 mm^3/s.
- Power consumption is very high as compared to conventional machining processes.
- It is difficult to machine very deep holes, as the slurry movement is restricted. The depth of cylindrical holes is presently limited to 2.5 times the diameter of the tool.
- It is difficult to design the tool to get precise dimensions on the workpiece as tool wear is comparable to metal removal rate. Wear of the tool increases the angle of hole and sharp corners become rounded.
- The process is limited to machining of small workpieces.
- There is high tool wear in the process. Tool replacement is essential in the production of accurate dimensions especially blind holes.

10.5.6 Recent Developments

In USM processes, slurry has to be continuously fed which limits the machining rate and dimensional tolerance while drilling deep holes. Recently, a new development in ultrasonic machining has taken place where no slurry is used. The tool is impregnated with diamond dust or other hard abrasive grits. The tool is oscillated at ultrasonic frequencies as well as rotated. If it is not possible to rotate the tool, the workpiece may be rotated.

The advantages of USM without slurry are:

- Faster cutting rate of materials such as glass and ceramics.
- Machining of alumina and other difficult-to-machine materials at economic speeds.
- Higher accuracy in machining compared to conventional USM with slurry. The comparison is shown in the Fig. 10.12.
- The process is more adaptable for automation.
- The tool can be used as milling cutter for machining different shapes.
- Holes upto 75 mm deep have been drilled in ceramics without any fall in machining rate. This was not possible with conventional USM.
- The hole dimensions can be kept within $\pm$ 0.125 mm.

10.6 WATER-JET MACHINING

The material is removed due to erosion effects of a high-velocity, small-diameter jet of water. The process is used mainly for drilling and slitting of sheets and slabs. Intricate shapes can be easily cut with a very narrow kerf width.

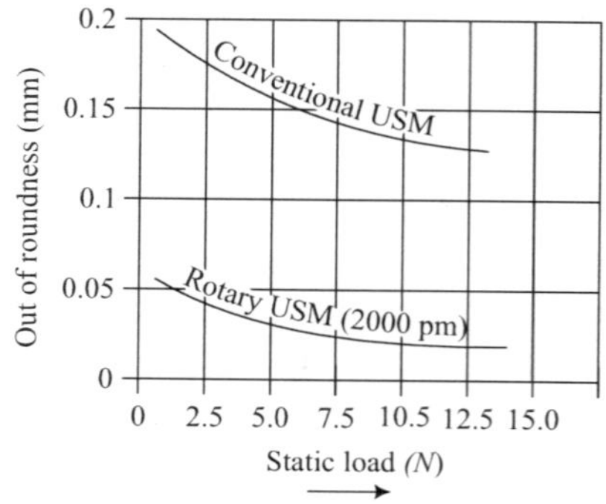

Fig. 10.12 Comparison of conventional and rotary USM.

10.6.1 Water-jet Machining Equipment

The schematic diagram of water-jet machining equipment is shown. The equipment consists of the following:

- **Water supply system:** It consists of a pumping unit, hydraulic amplifier with outlet pressure of 350-400 MPa, on-off valve and delivery piping.
- **Nozzle:** The nozzle contains a saphine insert to form a small-diameter jet of 0.07 to 0.5 mm. The positioning unit moves the nozzle or workpiece to bring about the cutting action. The distance between nozzle and workpiece is in the range of 3 to 25 mm. The positioning unit can be operated manually or with numerically controlled worktable motion. The nozzles can also be moved with an industrial robot.

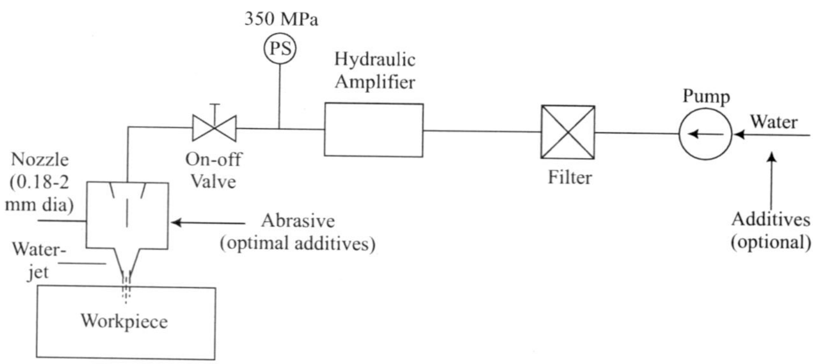

Fig. 10.13 Water-jet and abrasive water-jet machining.

10.6.2 Abrasive Water-jet Machining

Water-jet machining is used for cutting soft non-metallic materials such as paper, wood, leather and foam. The machining rates are high. There is no thermal damage to the workpiece.

The water-jet machine can be used as **abrasive water-jet machine** for machining hard materials. Abrasive particles are introduced into the stream after the nozzle. The common abrasive materials are garnet, silicon and silicon carbide grains.

10.6.3 Performance of Water-jet Machine

It is used to cut soft, non-metallic material sheets in multilayer stacks. The cutting rate depends upon:

- Material of workpiece.
- Thickness of workpiece.
- Jet pressure.
- Nozzle diameter.
- Cutting speed.

Typical water-jet cutting rates for different materials are given in Table 10.4.

Table 10.4 Water-jet cutting rates.

Material (mm)	*Thickness (MPa)*	*Pressure (mm)*	*Nozzle diameter (m/min)*	*Speed*
ABS plastic	2.2	379	0.15	1.0
Cardboard	1.4	379	0.10	6.1
Cheese	100	379	0.20	22.8
Foam rubber	76	207	0.10	300.0
Leather	1.5	275	0.13	76.2
Plexiglass	3	379	0.13	0.9
Wood	3.2	379	0.13	1.0

10.6.4 Advantages of Water-jet Machining

- The force level on the workpiece is low. Therefore, there is no damage to crushable materials during cutting.
- The kerf width is low, approximately 0.28 mm more than nozzle diameter. It is, therefore, possible to cut close nesting shapes from sheet materials.
- Predrilling of starting holes is not needed.
- The jet can pierce the material at any location.
- Cutting can be carried out in any direction.
- Surface finish of cut edges is good with no burrs and no thermal damage.
- The tolerances are within ±0.1 mm to 0.2 mm. These depend upon type and thickness of workpiece material.

10.6.5 Performance of Abrasive Water-jet Machining

Abrasive water-jet machining can cut a wide range of materials of various hardness, reflectivity and conductivity. Most of the materials upto a thickness of 100 mm can be cut. The cutting speeds for various materials are given in Table 10.4. The main advantages of this method of machining are:

- Kerf widths are only 1.5 mm to 2.3 mm.
- There is no thermal or mechanical damage to the surface of workpiece.
- The brittle materials can be cut without damage and cut edges free of cracks. Examples: acrylic, graphite, silica glass.
- It can cut porous materials such as concrete upto a thickness of 300 mm.
- There is no delamination during cutting and no ragged fibres protruding from the cut edges. Therefore, abrasive water-jet machining has special application for trimming of cured fibre composite and laminated materials.

10.7 ABRASIVE-JET MACHINING (AJM)

A high-speed stream of abrasive particles carried by a high pressure gas or air is impinged on the surface of workpiece. The metal is removed due to erosion caused by the impact of high-speed abrasive particles as a jet. The gas stream flows through a nozzle at pressures upto 0.7 MPa and a velocity of 300 m/s. The chipping action produced is very effective on hard and brittle materials. Examples: glass, silicon, tungsten, ceramics, etc.

10.7.1 Abrasive-jet Machine

The abrasive-jet machine is shown in Fig. 10.14. The air or gas is compressed and mixed with abrasive powder in the mixing chamber of nozzle. The jet diameter is 0.12 mm to 1.25 mm. The delivery nozzle is made of saphire or tungsten carbide. The useful life of nozzle is about 30 hours.

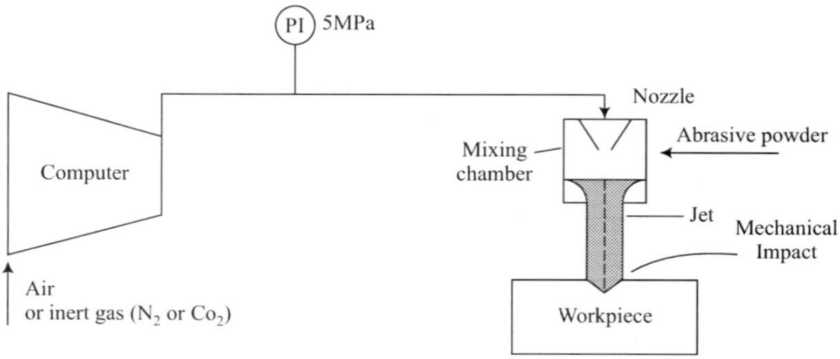

Fig. 10.14 Abrasive-jet machine.

The abrasive powder is made of aluminium oxide, silicon carbide and sodium bicarbonate mixture. The particle sizes vary from 10 to 50 μm.

10.7.2 Performance

- Cutting rate depends upon the size of abrasive particles, their shape and velocity. The particle size is between 10 to 50 μm. Sharp edged particles give high cutting rate. Pneumatic pressure of 2.5 bar to 10 bar is used.
- The distance of nozzle from the work surface affects the cutting rate. Maximum cutting rates are obtained when the nozzle tip distance is 7 to 13 mm from workpiece.
- A typical material removal rate is 16 mm/min for cutting glass.
- The effect of abrasive-jet pressure and grain size on the material removal rate is shown.
- *Accuracy:* An accuracy of ±0.1 mm can be easily obtained. Close control of various parameters can give a dimensional tolerance of ±0.05 mm.

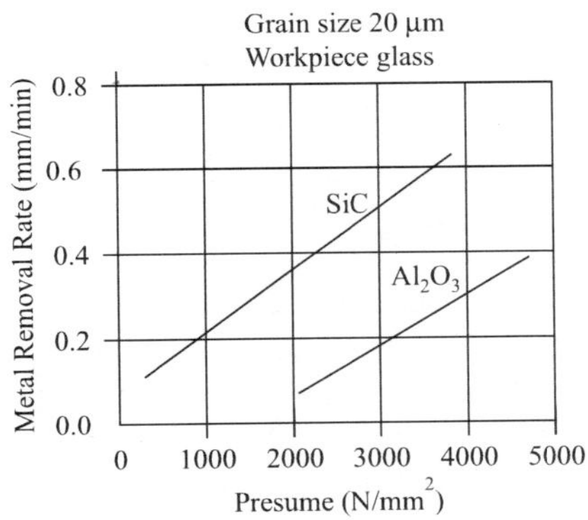

Fig. 10.15 Effect of jet-pressure on metal removal rate.

10.7.3 Applications

Abrasive-jet machining finds applications in the following areas:

- Cleaning and polishing of plastics (nylon and teflon) components.
- Etching of markings on glass cylinders.
- Frosting of interior surface of glass tubes.
- Production of shallow crevices, deburring and intricate shapes in hard brittle materials.
- Cutting slots, contouring and drilling in metal foils and thin sections of ceramics and glass.
- Micro-grit blasting for cleaning and polishing.

- Intricate holes in electronic components such as resistor paths in insulators and pattern in semiconductors.
- Engraving of characters on toughened-glass automobile windows.

10.7.4 Advantages

The advantages of AJM are:

- Intricate shapes and holes can be made in materials of high hardness and toughness.
- Low capital cost.
- Ability to cut fragile and heat sensitive materials. There is no damage as no heat is generated and overall forces generated are low.

10.7.5 Disadvantages

The main disadvantages of this process are:

- It has limited applications as material removal rate is low.
- Poor machining accuracy.
- High wear rate of nozzle.
- Abrasive grains can stick into soft materials. Additional cleaning is required.

10.8 CHEMICAL MACHINING

A strong acid or alkali solution is used to dissolve material selectively, resulting in controlled removal of material. A *mask* of chemically resistant coating is selectively applied to the workpiece to protect the surfaces that are not be machined.

There are four types of chemical machining processes.

1. Chemical milling: The erosion of material is used to produce channels and pockets in the surfaces of workpiece or to remove material from large areas of the surfaces of workpiece to reduce weight.

2. Chemical blanking: Through holes and slots are produced normally in thin sheet material.

3. Chemical engraving: Chemical engraving is used for production of printed circuit cards and engraving of intricate designs.

4. Gaseous chemical machining or hot chlorine machining: A jet of reactive gas is applied on the machining zone for special purpose chemical machining such as deburring of parts. The gas used is chlorine.

10.8.1 Chemical Milling

Chemical milling is used to produce complex profiles. There are five steps in chemical milling:

1. Cleaning or part preparation.
2. Masking or application of chemically resistant material.
3. Etching: Exposing workpiece to the etchant by dip or spray.
4. De-masking: Removal of mask material by stripping and cleaning the part.
5. Finishing, including inspection and other processes.

1. Cleaning or part preparation: The components are thoroughly cleaned and degreased by immersion in trichloroethylene vapour or some alternative chemical cleaner followed by washing in clean water. The alternative cleaning chemical may be a mild alkaline solution at 80°-90°C. One of the roughest methods of cleaning is to coat the component all over by spraying or dipping in the chemical solution. This removes dust and oil. The cleaning ensures a good adhesion of the coating or marking agent.

2. Masking: After cleaning, the component is coated with a cut and peel maskant by brushing, dipping or spraying (upto 0.2 mm). The maskant can be a neoprene based fluid or some plastic solution impervious to the action of the etching agent. It is mildly heated for drying. The desired shape is cut on the maskant with a scribe knife. The unmachined portions of the maskant are peeled off. A template is used to portray the desired machining shape within tolerances. Masking materials for different metals are given in Table 10.5.

Table 10.5 Etchants and maskant materials for different metals.

Material	*Etchant*	*Temperature, °C*	*Etching rate (mm/min)*	*Maskant*
Aluminium	NaOH, $FeCl_3$	90	0.03 and 0.02	Acrylonitrite butyl and neoprone rubber
Nickle	$FeCl_3$	45-50	0.025	Neoprone
Magnesium	HNO_3	<38	0.035	Polymers
Copper	$FeCl_3$, $CuCl_3$	45-50	0.05 and 0.03	Neoprone
Gold	HCl: HNO_3 = 3:1	<38	0.035	Polymers

3. Etching: The component is dipped completely into a tank of etchant which will dissolve the exposed metal. With control of time, temperature and solution strength, accuracies in the range of ± 0.01 mm can be achieved on shallow depths of cut. The surface finish is around 5 μm.

A variety of materials are used for dissolving the workpiece material. The selection of etchant depends upon the following factors.

- The workpiece material.
- The mask material.
- The depth of etch required.
- The surface finish required.

- The material removal rate required.
- The heat-treated condition of the material.
- The metallurgical properties required.
- Economic considerations.

Recommended etchants along with etching rate are given in the above table.

4. De-masking: After etching to the desired depth, the etchant is removed by washing thoroughly. The entire masking is stripped from the component. The component is cleaned completely.

5. Finishing: The part is inspected as per requirement. The surface of aluminium component may be anodised or treated temporarily with a temporary protective agent as required.

10.8.1.1 Types of masks

Three basic types of masks are used.

Cut-and-Peel Masks: A film of chemically resistant material is applied by dipping the workpiece or by spraying or flow coating. A dry film thickness upto 0.13 mm is produced. The rubbers used are vinyl, styrene and butadine. The rubber film is cut and peeled selectively. The cutting may be done manually using a scriber and template or by laser cutting. Manual scrubbing of mask achieves an accuracy of ± 0.13 to ± 0.75 mm. Depth of 13 mm cut can be made on thick mask material achieved by dipping or spraying.

Screen-Resist Masks: The mask material is applied on the workpiece surface by printing by using stencils and a fine polyester or stainless steel mask screen. Screen printing is a fast and economical method of applying mask material. The accuracy achieved is ± 0.05 to ± 0.18 mm and etching depth is restricted to 1.5 mm.

Photo-Resist Masks: Photochemical machining uses photo-resist masks. This widely used process produces intricate and finely detailed shapes. The workpiece is coated with photo-resist material. A master transparency is held against the workpiece and exposed to ultraviolet light. The light activates photo-resist material in those areas which are transparent portions in the master negative. The areas corresponding to opaque parts of the negative are dissolved during etching process. The accuracies achieved are ± 0.025 to ± 0.005 mm.

10.8.2 Chemical Blanking

In chemical milling, if the metal is totally removed from certain areas, the process is called chemical blanking. The process is also called chem-blanking, photo forming, photo-fabrication or photo-etching. Thin sheets and foils of any metal can be processed but it is not recommended to use material less than 2 mm thick. The steps of process are:

- Cleaning, degreasing and pickling by acids or alkalis.
- Drying of component.

- Dipping, whirl coating or spraying of photo-resist materials.
- Drying and curing.
- Placing transparency or photographic plate.
- Exposure to light of particular frequency.
- The image is developed as in photography.
- The unexposed portions are dissolved during developing.
- Spraying with chemical etchant in a machine or dipping in etchant solution.
- The finished part after 1 to 15 minutes is removed and etchant is removed by rinsing.

Advantages

- Printed circuit cards, intricate designs can be produced.
- Thin materials of the order of 0.005 mm can be etched.
- High accuracy of ± 0.015 mm can be maintained.
- Automatic photographic technique can be used for high production rate.

10.8.3 Advantages of Chemical Machining

- Burr-free components are produced.
- The machined components are not distorted.
- High surface finish.
- Most difficult-to-machine materials can be processed.
- Several components can be produced simultaneously.

10.8.4 Disadvantages of Chemical Machining

- Metal removal rate is slow.
- Metal thickness less than 2 mm cannot be machined easily.
- Highly skilled technology.
- The equipment can be damaged by corrosive etchant.
- Handling of chemicals is inherently dangerous.
- The fumes produced during machining are a major *health* hazard.
- Elaborate safety arrangements are required.

10.8.5 Applications of Chemical Machining

- Lightening of aircraft wing panels and structural members. Material is removed from selected areas by chemical machining using cut-and-peel masking.
- Printed circuit boards. The unwanted copper deposits on board is removed photo-chemically.

- Blanking of hard and brittle materials where conventional methods are not suitable.
- Delicate shapes in thin materials. Conventional blanking dies can be costly.

10.9 ELECTROCHEMICAL MACHINING

Electrochemical machining is based on Faraday's laws of electrolysis. The metal is removed by the controlled dissolution of the anode of an electrolytic cell. If two suitable metal poles are placed in a conducting electrolyte and direct current is passed through them, the metal on the positive pole gets depleted and its material is deposited on the negative pole.

The cathode is shaped as mirror image of the finished workpiece. The workpiece is connected to the positive supply. The cathode (tool) is connected to the negative terminal and is advanced towards the anode (workpiece) through the electrolyte. The electrical circuit is completed. Metal is removed from the workpiece by electrical action and cathode shape is reproduced on the workpiece. The electrolyte is pumped at high pressure through the gap between the workpiece and tool to carry away the heat generated and metal debris produced during cutting.

10.9.1 Electrolysis Process

The electrons flow from workpiece to the tool through power supply. The following chemical reactions take place:

1. At the cathode: The electrolysis process liberates hydroxyl ions (negatively charged) and free hydrogen.

$$M^+ + \bar{e} \rightarrow M \text{ (Metal)}$$

$$2H^+ + 2\,\bar{e} \rightarrow H_2 \text{ (Hydrogen evolution)}$$

2. At the anode: The hydroxyl ions combine with the metal ions of the anode to form insoluble metal hydroxides. Thereby metal is removed from the anode.

$$M \rightarrow M^+ + \bar{e} \text{ (Metal dissolution)}$$

$$2H_2O \rightarrow O_2 + 4H^+ + 4\,\bar{e} \text{ (Oxygen evolution)}$$

$$2Cl^- \rightarrow Cl_2 + 2\,\bar{e} \text{ (Halogen gas evolution)}$$

The metal dissolution reaction is the main reaction that occurs. This process continues and the cathode (tool) reproduces its shape in the workpiece (anode). The tool does not contact the workpiece.

10.9.2 Electrochemical Machining Plant

The ECM process is illustrated in Fig. 10.16. The workpiece (a conductor of electricity) is placed in a tank on the machine table and is connected to the +ve

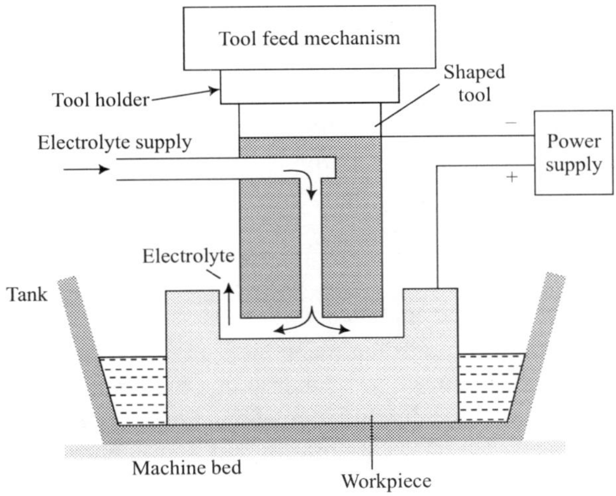

Fig. 10.16 Electrochemical Machining.

terminal of a high-current, low-voltage dc supply. The tool electrode, shaped to form the required cavity in the workpiece is connected to the –ve terminal of dc supply. The electrolyte flows through the gap between the tool and workpiece. It is pumped back to the working zone. The tool is held in a holder which is moved by a servomotor to maintain a constant gap between the tool and the workpiece during machining.

The metal is dissolved by the action of electrical current from the workpiece very rapidly in the region where the tool and workpiece are closest as the electrical resistance is minimum. The form of tool is reproduced on the workpiece as the tool is fed gradually towards the workpiece.

The electrolyte serves two important functions in ECM.

- The electrolyte provides the medium for electrolysis.
- It removes heat generated in the working region due to high current flow through the electrodes and electrolyte.

The electrolyte is cleaned of debris formed during metal removal and recirculated.

10.9.3 Elements of ECM

Important elements of ECM are:

- Electrolyte.
- Cathode tool.
- Anode workpiece.
- DC power supply.

Electrolyte: The main *functions* of an electrolyte in ECM are:

- Carrying electric current between the tool and workpiece.
- Removal of machining and other insoluble debris from cutting region.
- Dissipation of heat produced in the operation.

The main *characteristics* of an electrolyte include:

- Good electrical conductivity.
- Non-toxity and chemical stability.
- Non-corrosive property.
- Low viscosity and high specific heat.

The electrolyte flows through the machining gap. Its velocity is low when moving adjacent to anode and a thin boundary layer is created. The metal ions pass through the boundary layer by diffusion and the metal removal rate depends upon the ion movement. Therefore, a uniform and thin boundary layer over the anode surface is essential.

In order to avoid crowding of hydrogen gas and debris at the electrode surface, the electrolyte must flow with a high velocity of 30 to 60 m/s. If the machining gap is filled with bubbles of hydrogen gas, the machining operation stops.

The characteristics and applications of various electrolytes are given in Table 10.6.

Table 10.6 Characteristics and applications of electrolyte solutions.

Group	*Electrolyte*	*Concentration (% age by weight in aqueous solution)*	*Properties*	*Applications*
Natural salts	Common salt (NaCl)	5-20	Inexpensive, non-poisonous, pilting corrosive, moderate m.r.r.	Alloy and unalloy steel
	Sodium chlorate ($NaClO_3$)	20-45	Inflammable, high m.r.r.	Steel
	Sodium nitrate ($NaNO_2$)	Upto 12	Poisonous, moderate m.r.r.	Copper alloys
Acid	Sulphuric acid (H_2SO_4)	Upto 10	Highly corrosive, poisonous	Nickel, chromium, cobalt alloys
Alkali	Sodium, hydrate (NaOH)	Upto 10	Corrosive	Tungsten and molybdenum

The most common electrolyte solution is brine (NaCl) in the concentration of 0.1 to 0.25 kg/litre of water. It is inexpensive and non-poisonous. Its electrical conductivity is fairly constant. However, it is corrosive and large amount of sludge is formed. Higher power consumption is needed to push the high viscosity solution through narrow gap.

Flow Characteristics of Electrolyte

1. In case of rectangular electrode, the *flow velocity* of electrolyte can be calculated from the following equation.

$$U = \frac{V^2 l}{r h^2 \rho_e C_e (t_s - t_\infty)} \text{ (m/s)}$$

where, V = voltage applied (V)

l = length of electrode (m)

r = specific resistance (ohm –m)

h = gap length (m)

ρ_e = density of electrolyte (kg/m^3)

C_e = specific heat of electrolyte (J/kg–K)

t_∞ = ambient temperature (K)

t_s = boiling temperature of electrolyte (K)

2. The *flow rate* of electrolyte per unit time through the gap is given by:

$$\dot{Q} = U\,b\,h \text{ (m}^3\text{/s)}$$

where, U = flow velocity (m/s)

b = width of workpiece (m)

h = gap length (m)

2. Tool Material

The main *requirements* of a tool material are:

- Good conductor of electricity.
- Rigid to withstand fluid pressure.
- Corrosion resistance against electrolyte.
- Machinability to any shape.

Properties of some of the tool materials are given in Table 10.7.

Table 10.7 Properties of tool materials (comparative Index).

Property	Material			
	Copper	*Brass*	*Stainless steel*	*Copper-tungsten*
Electrical resistivity	1.00	4.00	53.00	8.00
Stiffness	1.60	1.00	1.90	2.20
Machinability	6.00	8.00	2.50	1.00
Thermal conductivity	25.00	7.50	1.00	10.00

The other materials are aluminium, graphite, bronze, platinum and tungesten carbide. The tools are made by electroformining and cold forging. The most important factor in ECM tool design is the provision of a suitable passage through the tool for efficient electrolyte flow through the cutting gap and to prevent stragnation areas.

3. DC Power Supply

The electric current has to overcome the resistance of the electrolyte.

$$I_c = \frac{V_s}{R_e} = \frac{AV_s}{r_e a_g}(A)$$

where, I_c = Electrical current (A)

V_s = Supply voltage (V)

R_e = Resistance due to electrolyte (Ohm)

r_e = specific resistivity of electrolyte (Ohm—m)

$$= \frac{R_e a_g}{A}$$

a_g = gap between tool and work surface (m)

A = area of work surface exposed to electrolyte (m^2)

Therefore, the gap width,

$$a_g = \frac{\eta_e V_s}{r_e \rho V_f}$$

where, η_e = current efficiency

$$= \frac{\text{Actual metal removed}}{\text{Theoretical metal removal}}$$

= fraction of the current actually effective in electrolysis.

ρ = density of work material (kg/m^3)

V_f = Tool feed speed (m/s)

Conclusions

- Gap width decreases with increase of feed speed.
- Higher feed speed or smaller gap give higher accuracy of production.
- Gap width increases with increase of applied voltage Vs and accuracy of production increases.
- Increase in gap width causes increase in resistance to current flow.
- Higher temperature leads to increase in resistivity r_e.

The common supply voltage in ECM ranges from 5V to 20V. Lower voltage is used for finish machining and higher voltage for rough machining.

The effect of flow rate on current density for different supply voltage is shown in Fig. 10.17.

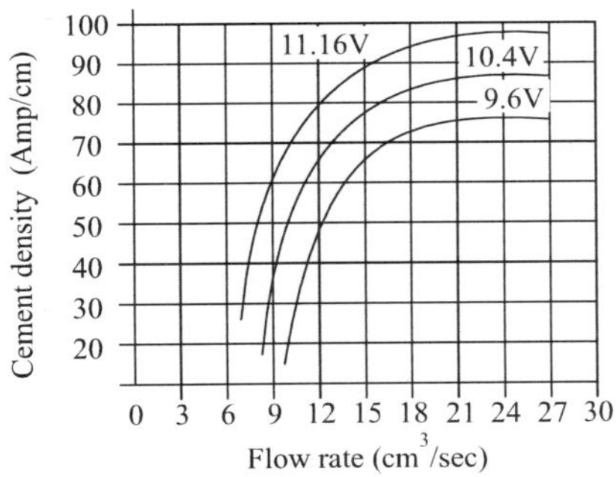

Fig. 10.17 Effect of flow rate on current density.

10.9.4 Metal Removal Rate

The metal removal rate is a function of electrochemical equivalent of workpiece material. The quantitative unit of electricity is called Faraday. It is the amount of electricity that reduces one gram-equivalant weight of a substance at the cathode and oxidises one gram-equivalent weight of a substance at the anode.

$$1 \text{ Faraday} = 96500 \text{ Coulombs}$$
$$= 6.02 \times 10^{23}\,\bar{e}$$

From Faraday's law of electrolysis, the mass of ions dissolved in kg ion.

$$W = \frac{EIt}{F}$$

where, F = Faraday constant
= 96,500 Coulombs
= 26.8 Amp-hr.
t = time in seconds
I = Electrical current in Amp.
E = Electrochemical equivalent of work material

$$= \frac{N}{n}$$

N = Atomic weight of metal
n = Valency.

Assuming 100% current efficiency and steady flow condition, the metal removal rate in terms of height of removed layer.

$$\text{MRR} = \frac{W}{A\rho t} \text{ (m/s)}$$

where, A = Machined area (m^2)

$$\rho = \text{Density of workpiece (kg/m}^3\text{)}$$

$$\therefore \quad \text{MRR} = \frac{E\,I\,t}{FA\rho t} = \frac{EI}{FA\rho}$$

$$\therefore \quad \text{MRR} = \frac{EJ_c}{F\rho} \text{ (m/s)}$$

where, J_c = Current density

$$= \frac{I}{A}\left[\frac{A}{m^2}\right]$$

Taking $A = 1 \text{ m}^2$

$$\therefore \quad \text{MRR} = \frac{EI}{F\rho} \text{ (m}^3\text{/s)}$$

The tool feed rate, V_f is given by

$$V_f = \frac{V_s}{r_e a_g} \times \frac{E}{F\rho} \text{ (m/s)}$$

where, V_s = supply voltage (V)

r_e = specific resistivity of electrolyte (ohm – m)

a_g = gap between tool and work surface (m)

The current should not be excessive to produce heat and break down the composition of electrolyte. The normal current density is used in the range of 1.5 MA/m^2 to 4.5 MA/m^2.

The material removal rates for a range of materials are given in Table 10.8.

Table 10.8 Metal removal rates.

Material	*Density (g/cm³)*	*Removal rate* (cm³/min)
Aluminium	2.7	1.9
Cobalt	8.9	1.9
Copper	8.9	3.9
Iron	7.8	2.1
Lead	11.3	5.0
Molybdenum	10.1	1.8
Nickel	8.9	1.9
Titanium	4.5	1.9
Tungsten	19.3	0.8
Uranium	19.0	1.8
Zirconium	6.5	1.9

10.9.5 Accuracy of Machining

The accuracy of ECM depends upon:

- Supply voltage, V_s.
- Tool feed rate, V_f.
- Temperature of electrolyte.
- Concentration of electrolyte.

An accuracy of ±0.02 to 0.04 mm can be maintained in ECM. The internal radii greater than 0.2 mm and external radii of 0.05 mm can be achieved. Taper of 0.010 mm for 10 mm depth and side over-cut is about 0.1 to 0.2 mm, surface finish is 0.2 to 0.8 μm (CLA) and no burrs or sharp edges are left on the workpiece.

10.9.6 Applications

The main applications of ECM are:

- Machining of hard-heat-resistant alloys.
- Cutting cavities in forging dies.
- Drilling of holes.
- Machining of complex profiles, e.g., turbine blades and aerospace components.
- Machining of tungsten carbide.
- Production of nozzles in alloy steels.
- All conducting materials can be machined.

10.9.7 Advantages of ECM

The main advantages of ECM are:

(i) The material removal rate is higher for high-strength-temperature-resistant (HSTR) materials as compared to that of conventional machining processes.
(ii) Residual stress and work-hardened layer thickness are very low.
(iii) Complex profiles can be machined which are beyond the capability of conventional machining processes.
(iv) High surface finish.
(v) Absence of tool wear.
(vi) Extremely thin metal sheets can be processed without distortion.

10.9.8 Disadvantages of ECM

The main limitations of ECM are:

(i) The specific power consumption is very high. It is nearly 100 times more than turning and milling of steel.

(ii) Non-conducting materials cannot be machined.

(iii) Special preventive measures are required to protect the machine against corrosion and rusting.

10.9.9 Numerical Examples

Example 10.1 In an ECM process, for machining of iron, a metal removal rate of 1cm^3/min is desired. Determine the current necessary. Assume atomic weight of iron to be 56, its valency 2 and density 7.8 gm/cm^3. Faraday constant may be taken as 96,540 Coulombs.

Solution:

$$\text{Atomic weight of iron, } N = 56$$

$$\text{Valency, } n = 2$$

$$\text{Density of iron, } \rho = 7.8 \text{ g/cm}^3$$

$$\text{MRR} = \frac{EI}{F\rho} \text{ [cm}^3\text{/s]}$$

$$\therefore \quad I = \frac{(\text{MRR})F\rho}{E}$$

$$\text{MRR} = 3 \text{ cm}^3\text{/min}$$

$$F = 26.8 \text{ ampere-hr} = 1608 \text{ amp-min.}$$

$$E = \frac{N}{n} = \frac{56}{2} = 28$$

$$\therefore \quad I = \frac{3 \times 1608 \times 7.8}{28} = 1343.8 \text{ A.}$$

Example 10.2 Estimate the metal removal rate (in cc/hr) of an alloy containing 18% cobalt, 62% nickel and 20% chromium during Electrochemical Machining (ECM) with a current of 500 amperes. The desnity of the alloy is 8.28 gm/cc. The following data is available:

Metal	*Gram Atomic Weight*	*Valency*
Cobalt	58.93	2
Nickel	58.71	2
Chromium	51.99	6
Faraday's Constant		96,500 Coulombs/mole

Solution: $F = 26.8$ amp-hr.

$I = 500$ A.

$\rho = 8.28$ gm/cc.

$$E = \frac{N}{n} = \frac{58.93}{2} \times 0.18 + \frac{58.71}{2} \times 0.62 + \frac{51.99}{6} \times 0.2$$

$$= 25.2367$$

$$\text{MRR} = \frac{EI}{F\rho} = \frac{25.2367 \times 500}{96500 \times 8.28} = 0.01579\ [\text{m}^3/\text{s}]$$

$$= 56.85\ \text{m}^3/\text{hr}$$

$$= 56.85 \times 10^6\ \text{cc/hr}$$

Example 10.3 A dc power supply of 18 V and 5000 A was used in ECM using copper electrode and sodium chloride solution. A tool-work gap of 0.5 cm constant was maintained. The data for iron anode may be taken as:

Atomic weight = 56

Valency = 2

Density = 7.87 g/cm^3.

Calculate the machining rate and electrode feed rate for the following data for electrolyte.

Specific resistance = 5.0 ohm-cm

Current efficiency = 100%

Faraday constant = 96,500 Coulombs.

Solution:

$$\text{MRR} = \frac{EI}{F\rho}$$

$$= \frac{(N/n)I}{F\rho} = \frac{(56/2)}{96500} \times \frac{5000}{7.87}$$

$$= 0.1843\ \text{cc/s}$$

$$= 1.843 \times 10^{-7}\ \text{m}^3/\text{s}$$

Electrode feed rate,

$$V_f = \frac{V_s}{r_e a_g} \times \frac{E}{F\rho}$$

$$= \frac{18}{5 \times \frac{0.5}{10}} \times \frac{28}{96500 \times 7.87} = 2.6545 \times 10^{-3}\ \text{cm/s}$$

$$= 1.59\ \text{mm/min.}$$

Example 10.4 An ECM with copper electrode and NaCl as electrolyte is used for machining of iron. The ECM has following dimensions:

$$\text{Electrode area} = 2.54 \times 2.54 \text{ cm}$$

$$\text{Electrode-work gap} = 0.0254 \text{ cm.}$$

The electrolyte properties are:

$$\text{Specific heat} = 4.17 \text{ J/g-K}$$

$$\text{Density} = 1 \text{ g/cm}^3$$

$$\text{Specific resistance} = 3 \text{ ohm-cm}$$

(a) Estimate the permissible fluid flow velocity, if the maximum permissible temperature of the electrolyte is 95°C and ambient temperature is 35°C. Applied voltage is 12 V.

(b) Calculate the maximum removal rate if the permissible current density was 160 am/cm^2.

Solution:

(a) Fluid flow velocity,

$$U = \frac{V_s^2 l}{r_e a_g^2 \rho_e C_e (t_s - t_\infty)}$$

$$U = \frac{(12)^2 \times 2.54}{3 \times (0.0254)^2 \times 1 \times 4.17(95 - 35)} = 755.3 \text{ cm/s}$$

$$= 453 \text{ m/min.}$$

(b) Metal removal rate,

$$\text{MRR} = \frac{EJ_c}{F\rho} = \frac{28 \times 160}{96{,}500 \times 7.86} = 0.0059 \text{ cm/s}$$

$$= 3.54 \text{ mm/min.}$$

10.10 ELECTROLYTIC GRINDING

Electrolytic grinding is a combination of electrochemical machining and conventional grinding. A rotating abrasive wheel which is a metal-bonded diamond wheel is used as tool electrode. The wheel is a good conductor of electricity. The electrolyte is fed between the wheel and the work surface in the direction of wheel rotation. The electrolyte is fed at the wheel periphery and is carried past the work surface by the action of work rotation. A constant gap is maintained between the wheel and the work surface.

The schematic diagram of electrolytic grinding is shown. Total metal removal rate is the sum of metal removal rate of electrolytic grinding and metal removal rate by mechanical abrasion. The metal removal rate of electrolytic grinding depends upon I, E and η_e.

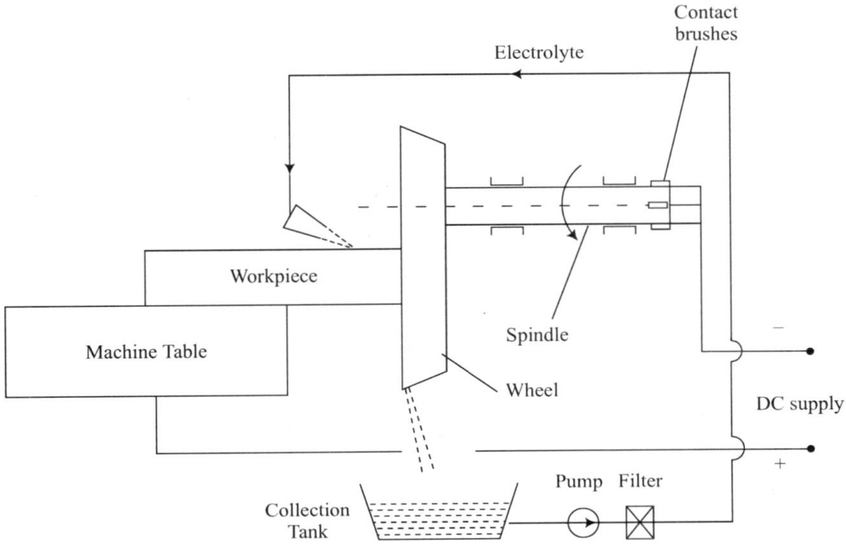

Fig. 10.18 Electrolytic grinding.

For, $J_c = 1\ \text{MA/m}^2$

$\eta_e = 100\%$

In practice metal removal rate by electrolytical process is 90% of the total. Only 10% of the total metal is removed by mechanical abrasion.

10.10.1 Accuracy of ECG

The following accuracy can be maintained by ECG:

1. Tolerance: ±0.02 mm
2. Surface finish : 0.2 - 0.4 μm on carbide
   0.4 - 0.8 μm on steel.
3. Minimum radius : 0.2 mm

10.10.2 Applications of ECG

1. Grinding of hardened steel, cemented carbides.
2. Resharpening and reconditioning of carbide tools, and difficult-to-grind materials.
3. Cutting of thin sections and thin-wall tubes of difficult materials without distortion or burr.

10.10.3 Advantage of ECG

The main advantages of ECG are:

1. Work is completely free of burrs and scratches
2. No heat cracks or distortions as no heat is developed.
3. Little pressure is exerted on the workpiece.
4. No wheel wear. This saving in wheel wear constitutes a major economy.
5. Wheel can be used for long periods without dressing.
6. High metal removal rates upon hard materials.

10.10.4 Disadvantages of ECG

1. High cost of ECG system.
2. High power consumption
3. Metal removal rate is limited to 15 mm^3/s

10.11 ELECTRICAL DISCHARGE MACHINING

The principle of erosion of metals by an interrupted electric spark discharge between the tool (cathode) and workpiece (anode) is used in electric discharge machining. The process is also called spark erosion, electro-erosion or spark machining. The sudden pulse of high intensity discharge can give rise to very high temperatures and possibly shock waves in the vicinity of spark. The surface layer in tine with the spark generally melts and is removed from the site by the pressure wave and cavitation. The repetitive spark discharges are produced from a pulsating dc power supply, with a dielectric fluid flowing between the workpiece and the tool. Each spark generates a localized high temperature of the order of 12000°C in its immediate vicinity. This heat causes part of the surrounding dielectric fluid to evaporate; it also melts and vaporizes the metal to form a small crator on the work surface. Since, the spark always occurs between the points of the tool and workpiece that are closest together, the high spots of the work are gradually eroded, and the form of the tool is reproduced on the workpiece. The dielectric fluid carries away the condensed metal globules formed during the process. As the metal is eroded, the tool is fed toward the workpiece by a servo-controlled feed mechanism.

Each pulse in the EDM cycle lasts only a few micro-seconds. Repeated pulses, at the rate of 100,000 per seecond, result in uniform erosion of material from the workpiece and from the electrode. As the process progresses, the electrode is advanced by the servo drive toward the workpiece to maintain a constant gap distance until the final cavity is produced.

10.11.1 Electro-Discharge Machining Plant

The main elements of electro-discharge machining plant are shown. The main elements consist of:

1. DC power supply unit.
2. Servo-control mechanism.
3. Dielectric fluid system.
4. Tool electrode.

The tool is mounted on the chuck attached to the machine spindle whose motion is controlled by a servo-controlled feed drive. The workpiece is placed in a tank filled with a dielectric fluid to a depth of 50 mm above the work surface to eliminate the risk of fire. The tool and workpiece are connected to a pulsating dc power supply. Dielectric fluid is circulated under pressure by a pump through holes in the tool electrode. A spark gap of 0.025 to 0.05 mm is maintained by the servomoter.

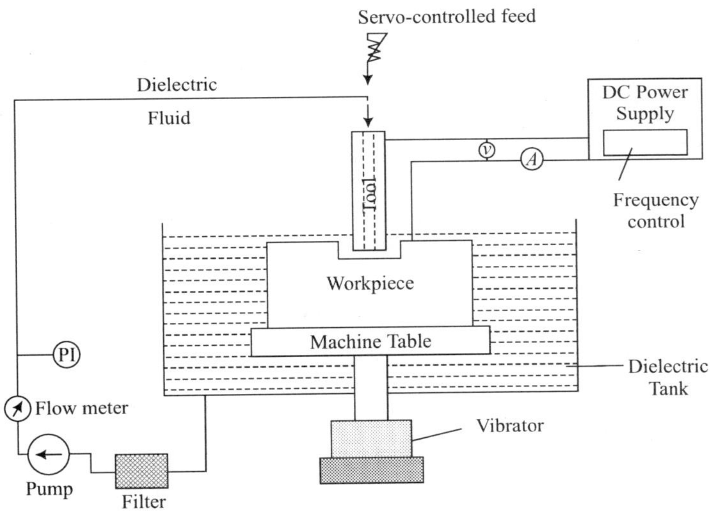

Fig. 10.19 Electro-discharge machining plant.

1. DC Power Supply Unit

Solid-state rectifiers convert the input power into continuous dc power. A bank of power transistors are switched by a digital multivibrator oscillator circuit and controls the flow of dc power. The high-power pulse output is applied to the tool and workpiece to produce the sparks responsible for material removal. The modern pulse generators possess the means of accurate control over discharge duration, pulse time and current. These factors determine the overcut and hence accuracy and surface finish.

2. Servo-control Mechanism

The servo-mechanism controls the infeed of the tool into the workpiece. EDM tool feed control is shown. It consists of a reversible servomotor connected between one terminal of the condenser (D) and the slider of a variable resistance (B).

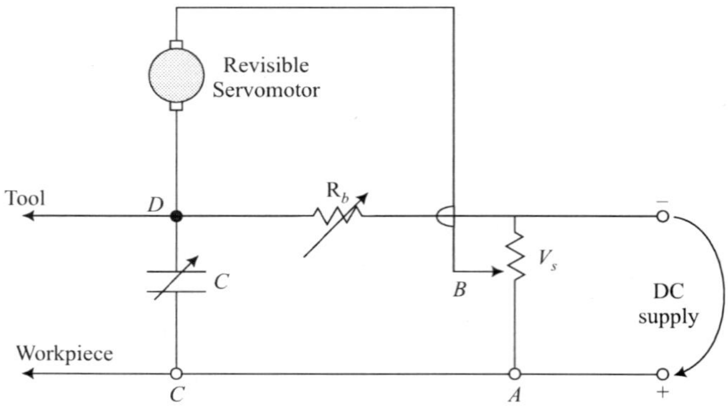

Fig. 10.20 EDM Power Supply.

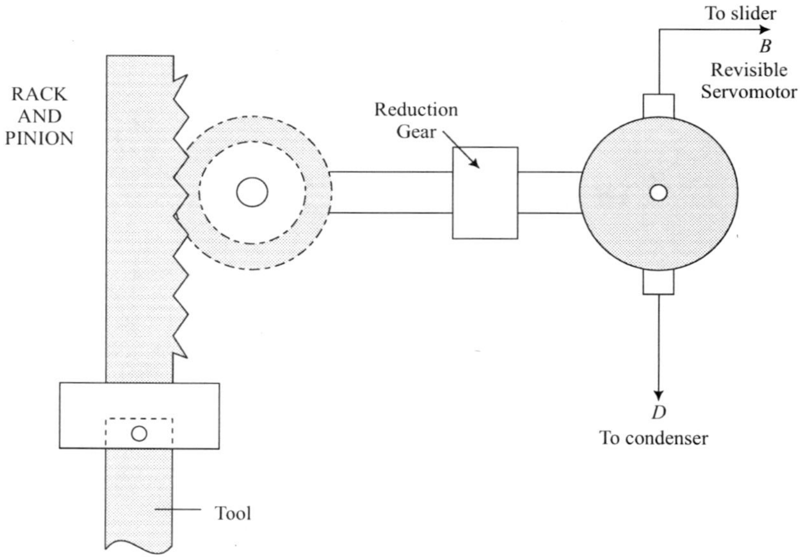

Fig. 10.21 EDM Tool Feed Control.

Electro-hydraulic-servo control gets its input signal as the voltage difference between a selected reference voltage and actual voltage across the gap. The signal is amplified and the tool, as it wears out a little, is advanced by hydraulic control. A short circuit across the gap causes the servo to reverse the motion of the tool until the current gap is established: A constant gap of about 0.025 to 0.05 mm is maintained. The exact gap can be varied by changing the setting of the variable resistance.

3. Dielectric Fluid System

The essential requirements of a dielectric fluid are as follows:

1. Should remain electrically non-conducting until the required breakdown voltage has been reached.
2. Once the breakdown voltage has been reached, it should break down electrically in the shortest possible time.
3. After the occurrence of discharge, it should rapidly quench the spark or deionize the spark gap.
4. Effective coolant.
5. Should wash away the swarf particles from working gap.
6. Good fluidity.
7. Cheaply and easily available.

The common dielectric fluids are hydrocarbon oils which satisfy the above requirements. The main fluids used are paraffin, kerosene, transformer oil, try-ethyle glycol.

The dielectric fluid should be fed through the tool. It should be filtered before reuse, so that chip contamination of the fluid should not affect the machining accuracy. There are three methods of flushing of dielectric fluid to ensure free circulation.

1. Pressure flushing.
2. Section flushing.
3. Side flushing.

4. Tool Material

The EDM electrode shape determines the shape of the hole or cavity that is sunk into the workpiece. The main requirements of tool material are as follows:

1. Electrically conductive
2. Low wear rates.
3. Machinable easily.
4. Provides good surface on the workpiece.

The properties of some EDM tool materials are given in the Table 10.9.

Each spark discharge removes material from both the workpiece and the electrode and consequently the electrode wears during the machining pocess. The material of electrode is selected on the basis of *wear ratio*. This is the ratio of volume of metal lost from the tool and the volume of metal removed from the work. Wear ratio depends upon the tool and work materials.

For example:

Tool	Workpiece	Wear ratio
Brass	Brass	0.5
Brass	Hardened carbon steel	1.0
Brass	Tungsten carbide	3.0

Table 10.9 Physical Properties of Electrode Materials.

Sl. No.	*Properties*	*Copper*	*Graphite*	*Tungsten*	*Iron*
1.	Melting point, °C	1083	Sublimes	3395	1535
2.	Boiling point, °C	2580	>4000	>5930	>2800
3.	Thermal conductivity with Ag = 100%	94.3	00.3	296	16.2
4.	Electrical conductivity with Ag = 100%	96.5	0.1	48.1	16.2
5.	Strength, N/mm^2	241	34	3134	275
6.	Modulus of elasticity, $N/mm^2 \times 10^3$	124	5.9	351	186
7.	Heat of Vaporization, Cal/cm^3	12740	~20,000	~22680	~16900

The wear ratio should be 0.1. This can be achieved by the following methods:

1. Ratio of melting point of tool material and melting point of the workpiece. This ratio should be high for high wear ratio.
2. The following materials have wear ratio in the increasing order: graphite, copper-graphite, copper, brass.

10.11.2 Process Parameters

Metal removal rates, surface finish and accuracy depend upon the following process parameters:

1. Electrical current

The metal removal rate increases with current flow but surface finish decreases.

2. Spark voltage

The spark voltage has similar effect as current flow.

3. Spark frequency

The frequency range of modern EDM machining is from 180 Hz for rough machining to several hundred kilohertz for fine machining. The volumetric removal rates vary from 0.001 to 0.1 cm^3/hr.

4. Spark-gap width

The accuracy of process highly depends upon spark-gap width; the smaller the gap, the higher the accuracy. The small spark gap results in lower working voltage and lower metal removal rate. Typical spark gaps range from 0.012 to 0.05 mm. As the gap decreases, efficient flushing of gap becomes difficult.

5. Undercut

The shape of the area of the cavity produced in the workpiece should theoretically be the same as that of the tool. But this is not true because of overcut. Overcut is the distance, the spark will penetrate the workpiece from the tool and remove metal from the workpiece. The overcut is generally 0.025 to 0.2 mm on all surfaces. It causes the radius of cavity in the workpiece larger than radius of the tool nose. The radius of projections on the workpiece becomes lesser than radius of the cavity of tool.

The overcut is a function of spark voltage and decreases with frequency as shown.

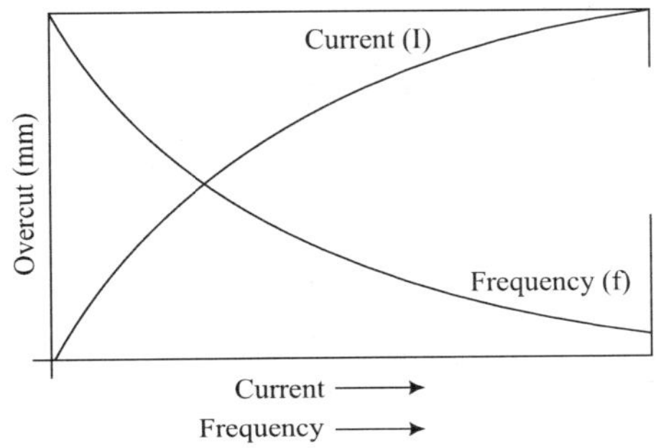

Fig. 10.22 Change of overcut with current and frequency.

10.11.3 Metal Removal Rate (MRR)

The volume of metal removed per second is proportional to the number of sparks per second, N and energy content of the sparks. The number of sparks per second is given by

$$N = \frac{1}{RC} \ln\left(\frac{E}{E - Vd}\right)$$

where, R = charging resistance

C = capacitance

E = supply voltage

V_d = discharge voltage

$$\text{MRR} \propto \left(N \times \frac{1}{2} C V_d^2\right) = \lambda \frac{V_d^2}{R} \ln \left| \frac{1}{1 - \frac{V_d}{E}} \right|$$

where, λ = constant of proportionality

$$\text{MRR} = \lambda \frac{E^2}{R} \left(\frac{Vd}{E}\right)^2 \ln\left[\frac{1}{1 - \frac{V_d}{E}}\right]$$

For maximum value of MRR, the ratio V_d/E lies between 0.7 to 0.9. For a given value of E, the optimum value of V_d can be found out. The control circuit of the servo motor is made compatible with V_d.

The number of sparks per second may vary between 3000 and 10,000.

The material being cut will also affect MRR and is inversely proportional to melting point of metal.

$$\text{MRR} = \frac{2.4}{(\text{Melting point}°\text{C})^{1.25}}$$

Therefore, EDM will cut aluminium much faster than steel.

10.11.4 Accuracy

The following accuracy can be achieved by EDM.

(1) Tolerance in normal production = ±0.05 mm

(2) Minimum tolerance = ±0.003 mm

(3) Taper value = 0.005 to 0.05 mm/100 mm depth

(4) Overcut = 5 to 100 μm

(5) Surface finish on iron = 0.4 μm.

10.11.5 Applications of EDM

The process may be used economically for the following applications on workpieces, which are good conductor of electricity.

(1) Machining of dies for forging, blanking, extrusion, etc.

(2) Drilling of micro-holes in fuel injector nozzles.

(3) Machining of hydraulic valve spools.

(4) Slitting of hard alloys.

(5) Manufacture of fragile components.

(6) Thread cutting, helical profile milling, rotary forming and curved hole drilling.

10.11.6 Advantages

EDM is very popular due to following advantages.

(1) Applicable to machining of all electrically conducting metals and alloys and does not depend upon melting point, hardness, toughness and brittleness.

(2) Complicated shapes.

(3) Machining time is less than conventional machining.

(4) Hardened workpiece can be machined without any heat treatment.

(5) Fragile and slender workpiece can be machined without distortion.

(6) Hard and corrosion resistant surfaces for die making can be developed.

(7) The surface finish with cratering for lubrication increases die life.

10.11.7 Disadvantages

The following are the limitations of the process:

(1) Profile machining of complex contours is not possible.

(2) Machining time is very long.

(3) Surface and metallurgical properties are affected by heating of workpiece.

(4) Tool wear rate very high.

(5) High specific power consumption.

10.12 WIRE ELECTRICAL–DISCHARGE MACHINING

Wire electrical-discharge machining is an adoption of the electrical-discharge machining and works on the same principle as EDM. In place of shaped tool, an endless wire travelling through the workpiece is used as electrode. The discharge takes place between the wire and workpiece. As the wire travels forward, cutting proceeds. A slot is made through the entire thickness of the workpiece.

The wire electrode is drawn from a supply reel and collected on the take-up reel. Fresh wire is delivered to the work area continuously. The wire is guided by a diamond or saphire guides and kept straight and under tension. This avoids tapering of workpiece.

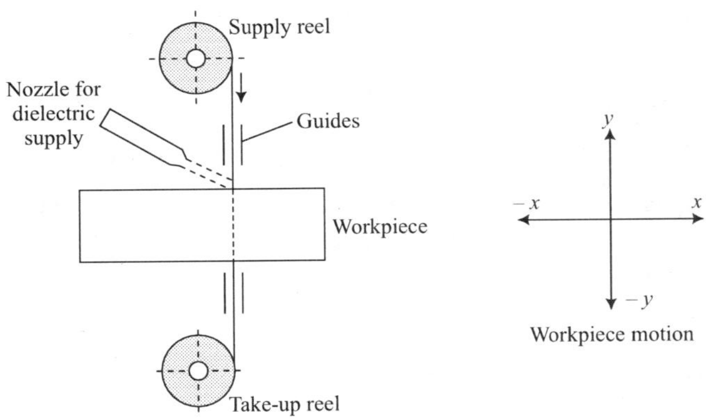

Fig. 10.23 Wire-EDM.

This is a new process finding wide applications especially in tool making field. The main elements of wire-EDM are as follows:

(1) Power supply

(2) Dielectric fluid
(3) Electrode wire
(4) CNC control

10.12.1 Power Supply

The power supplies for wire-EDM are essentially the same as for basic EDM. The current carrying capacity of the wire limits currents to 10A but less than 20A. The spark frequencies are 1 MHz to give a fine surface finish on the workpiece. High-frequency dc pulses are delivered to the wire and workpiece, causing spark discharges in the narrow gap between the two.

10.12.2 Dielectric Fluid

A stream of dielectric fluid is directed coaxially with the wire to flood the gap between wire and the workpiece. Deionized water is used as dielectric fluid because of the following advantages:

(i) Low viscosity to ensure adequate flow.
(ii) No fire hazards.
(iii) High cooling rates.
(iv) High material removal rates.

10.12.3 Electrode Wire

Copper and brass wires are commonly used for the electrodes with diameter 0.15 to 0.3 mm. For high precision cutting a fine wire of molybdenum steel of diameter 0.03 to 0.15 mm is used.

The spark gap ranges from 0.025 to 0.05 mm.

10.12.4 CNC Control

The workpiece is moved under CNC relative to wire. Complex-shaped profiles can be cut through sheet and plate materials. Adaptive control is necessary to avoid short circuits. Gap-voltage is senced to ensure machining as per programmed path.

10.12.5 Machining Accuracies

(1) Linear cutting rates 38 to 115 mm/hr in 25 mm thick steel.
(2) Accuracies of cut-out profile ±0.007 mm.
(3) External corners of 0.038 mm radius.

10.12.6 Applications

Wire-EDM is used for the following:

(1) Fabrication of process stamping dies, extrusion dies, powder compaction dies.

(2) Profile gauges and templates.

(3) Complicated cut-outs in difficult-to-machine metals without the need of grinding.

(4) Manufacture of electrodes for EDM.

(5) Manufacture of punches, dies and stripper plates.

10.13 LASER-BEAM MACHINING

Light Amplification by Stimulated Emission of Radiation or commonly known as Laser has many scientific and industrial applications. Laser-beam machining can be used for cutting, drilling, slotting and engraving of most materials. The material is removed by means of a high-energy coherent beam of light of a single wavelength. The laser beam is focussed by lens on a very small spot by a nozzle. The highly concentrated light at the focus produces energy densities above 100 W/mm^2. Lasers of this power capacity can vaporize or melt most materials. Therefore, cutting and drilling can be achieved without applying any mechanical force on the workpiece. The heat-affected zone is also very small.

The working principle of a laser-cutting head is shown in the figure 10.24. The lense focuses the laser-beam on to the workpiece through a gas cutting nozzle. A short duration pulse from the laser causes a small volume of material of workpiece to melt and partially vaporize. The explosive escape of the vapour removes most of the melted material in the form of a spray of molten droplets. Multiple pulses can be used to penetrate thick materials that cannot be drilled with one pulse.

10.13.1 Types of Lasers

Solids, liquids and gases can be used as lasing materials. In metal working, the solid state and gas lasers are generally used. The crystal rods are made from the following materials:

(i) Man-made ruby made of aluminium oxide with 0.05 per cent chromium.

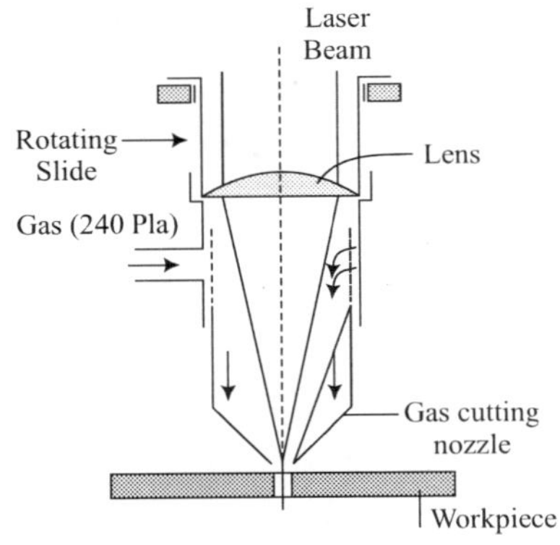

Fig. 10.24 Laser-cutting head.

(ii) Calcium fluoride crystals doped with neodymium (Ca + Fe_2 + Nd)

(iii) Nd-Yag laser consists of a single crystal rod of yttrium aluminium garnet with ions of neodymium. Flash tubes are mounted in a reflecting cavity parallel to the lasing rod. The principle of operation is explained in the following figure.

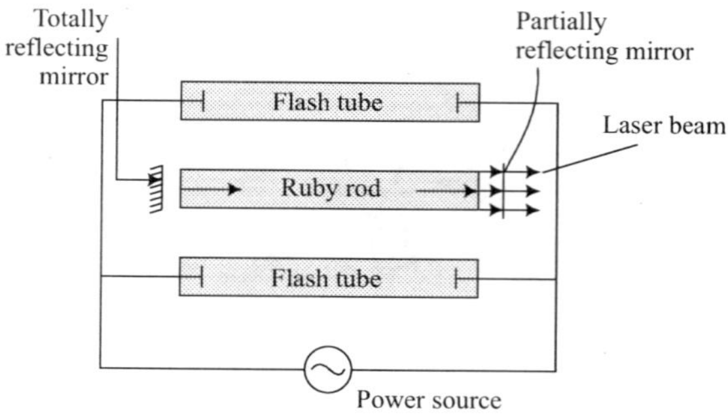

Fig. 10.25 Ruby Laser.

In gas lasers, He-Ne, Argon, CO_2 gases are used. A single gas or a mixture of gases is filled in an optically transparent tube. The power source is flash tube or electric discharge between rods. The power output depends upon the length of tube. A CO_2 laser can develop 50W/m length of tube.

A typical high-power CO_2 laser is shown in the figure. CO_2 gas generates photon. N_2 gas is introduced to sustain the action. He is added for intra-cavity cooling which is needed for stabilization of discharge.

The main properties of industrial lasers used for drilling and cutting are given in Table 10.10. More than 70% of lasers used for industrial application are Nd:YAG or CO_2 lasers.

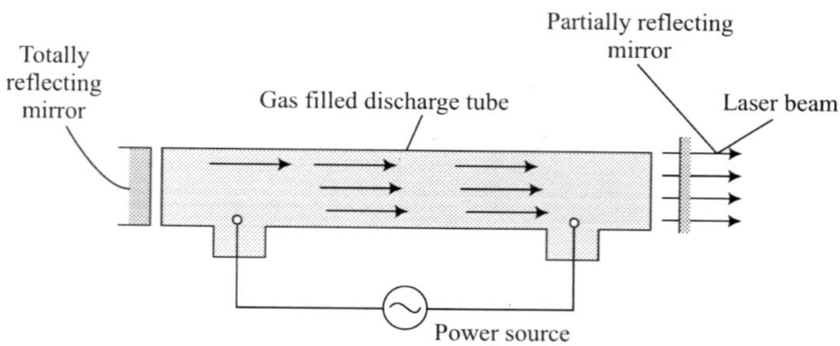

Fig. 10.26 Gas Laser.

Table 10.10 Characteristics of industrial laser.

Laser Type	*Mode of operation*	*Power (W)*	*Applications*
Ruby	Pulsed	50	Holography, drilling, balancing
Nd : Glass	Pulsed	50	Fabrication, drilling diamond
Nd : YAG	CW/Pulsed	600	
Alexandrite	CW/Pulsed	100	
CO_2	CW/Pulsed	25,000	Welding and Cutting of metals, scribing ceramics, drilling vaporization.

10.13.2 Material Processing with Lasers

High-rate pulsing or continuous operation is used for cutting operations.

(1) Percussion drilling of small holes

The laser-beam is focussed on the workpiece surface. A short direction pulse from the laser melts and evaporates a small volume of material. Multiple pulses can be used to penetrate thick materials that cannot be drilled with one pulse.

Holes produces by laser are:

(i) Tapered.
(ii) Approximately round.
(iii) Hole size highly repeatable.
(iv) Depth of recast layer is very small.
(v) The drilling time 1ms to 2 s depending upon workpiece thickness.
(vi) Max. size of hole which can be drilled 1.3 mm diameter.

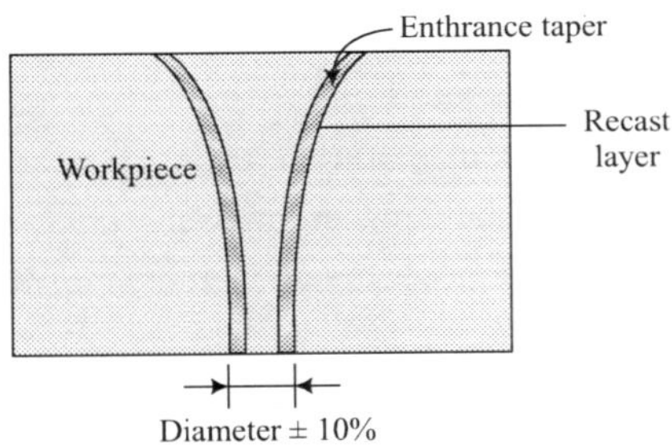

Fig. 10.27 Hole Drilled by Laser.

(2) Cutting or trepanning operations

Rapid pulses or continuous operations of laser is used in addition a jet of gas is introduced to assist the melting and cleaning of molten material. The following gases are used.

(i) Oxygen is used for carbon steel for oxidation, quicker melting operation and high cutting speeds. The cutting edges left are oxidised and heat-affected zone is enlarged.

(ii) Air is cheaper to use but cutting rate is reduced due to less amount of oxygen present in the air.

(iii) Argon is used for shielding the cut surfaces against oxidation. The cutting speed is lower by 25% as compared to air.

10.13.3 Machining Rate

The maximum machining rate in LBM depends upon the material thickness and laser power. Lower cutting rates give better quality of the cut surfaces. The material removal rate is of the order of 4000 mm^3/hr.

The cutting rate,

$$C = k\frac{P}{EAt_1} \quad \text{(mm/min)}$$

where,
P = Laser power incident on workpiece surface, [W]
E = Vaporization energy of material, [W/mm^2]
A = Area of laser-beam at focal point, [mm^2]
t_1 = Thickness of material, [mm]
k = Material constant, [mm/min]

The amount of energy required to raise a volume of metal to its vaporization point,

$$E = \rho V\frac{C_P(t_m - t_\infty) + C_v(t_b - t_m) + h_f + h_{fg}}{\eta} \quad \text{[J]}$$

where,
ρ = Density of material, [kg/m^3]
V = Volume to be evaporated, [m^3]
C_p = Specific heat at constant pressure, [J/kg-K]
C_v = Specific heat at constant volume, [J/kg-K]
t_m = Melting point, [°C]
t_∞ = Ambient temperature [°C]
t_b = Boiling point [°C]
h_f = Latent heat of fusion [J/kg]
h_{fg} = Latent heat of vaporization [J/kg]
η = Efficiency of the process

Example 10.5 A hole of 1.3 mm diameter is to be drilled by LBM in a steel sheet of 1mm thick. Estimate the energy and power density required for the following data.

Density of steel, $p = 8 \text{ g/cm}^2$.

Specific heat, $C_p = C_v = 0.462$ J/g-K.

Melting point, $T_m = 1808$K.

Boiling point, $T_b = 3000$K.

Ambient temperature, $T_\infty = 20°C + 273 = 293$K.

Latent heat of fusion, $h_f = 280$ J/g.

Latent heat of vaporization, $h_{fg} = 6846$ J/g.

Process efficiency, $\eta = 70\%$

Solution:

The volume to be evaporated,

$$V = \frac{\pi}{4} d^2 t = \frac{\pi}{4} (1.3)^2 \times 1 \times 10^{-3} \text{ cm}^3 = 0.00132732 \text{ cm}^3$$

The amount of energy required,

$$E = \rho V \frac{C_P(T_m - T_\infty) + C_v(T_b - T_m) + h_f + h_{fg}}{\eta}$$

$$= 8 \times 0.00132732 \frac{[0.462(1808 - 293) + 0.462(3000 - 1808) + 280 + 6846]}{0.7}$$

$= \mathbf{127\ J}$ **Ans.**

Assume a short pulse of 10^{-5} second duration.

The power required,

$$P = \frac{E}{10^{-5}} = \frac{127}{10^{-5}} \text{ W.}$$

The power density,

$$\frac{P}{A} = \frac{127 \times 10^5}{\frac{\pi}{4}(0.13)^2} = \mathbf{9568 \times 10^5\ W / cm^2}. \quad \textbf{Ans}$$

10.13.4 Applications of LBM

(1) Laser is best used for a variety of cutting as well as drilling operations.

(2) The accuracy of profile cutting is ±0.1 mm with numerical control or photo-electric tracer.

(3) Used for drilling small holes and cutting complex profiles of ceramics or other hard and thin materials.

(4) Used for engraving and partial cutting.

(5) Steel metal trimming, blanking and resistor trimming.

(6) Mass micro-machining production.

10.13.5 Advantages of LBM

(1) Any solid material which can be melted without decomposition can be cut with laser-beam.

(2) There is no direct material contact between tool and workpiece and no contamination.

(3) Non-metals as well as metals can be machined.

(4) It is possible to drill and cut areas not easily accessible.

(5) The heat affected zone is small.

(6) Very small holes can be machined.

(7) There is no tool wear.

(8) Rubber, plastics and other soft materials can also be machined.

(9) High cutting speeds, narrow kerf and straight edge can be achieved by adjusting the focussing of beam intensity.

(10) Highly complex shape may be cut and precise locations can be achieved.

(11) High productivity, worker safety, remote processing.

10.13.6 Limitation of LBM

(1) It cannot be used to machine Al, Cu and other alloys with high heat conductivity or high reflectivity.

(2) Low overall efficiency of the order of 10 to 15%.

(3) Limited to thin sheet.

(4) Low material removal rate.

(5) Holes machined are not round and straight.

(6) Life of flash lamp is short.

(7) Cost is high.

(8) Difficult to control the output energy of laser precisely.

10.14 ELECTRON–BEAM MACHINING (EBM)

A focussed beam of high-velocity electrons strikes the workpiece. The kinetic energy of electrons is converted into heat. There is rapid melting and vaporization of the workpiece material. The electrons can be accelerated to velocities as high as 75% of the velocity of light.

10.14.1 Electron-Beam Gun

The schematic diagram of an electron-beam gun is shown. It consists of the following elements:

(1) A tungsten filament is heated to emit electrons.

(2) The electrons are accelerated with the help of high potential difference.

(3) The electron beam is focussed with the help of negative grid cup.

(4) The electrons pass through anode tube and finally focussed by a magnetic lens or coil.

(5) A magnetic deflection coil is used to control the beam. The beam can be focussed to a fine point or spread over an area depending upon its applicaiton. This focussing varies the energy density.

(i) For machining, power density required is 10^7 W/mm^2.

(ii) For welding, power density required is 10^6 W/mm^2.

(iii) For metalizing, power density required is 10^4 W/mm^2.

(iv) For melting, power density required is 10^3 W/mm^2.

(v) For heat treatment, power density required is 10 W/mm^2.

(6) The work is done in a high vacuum chamber (10^{-5} mm of Hg) to eliminate scattering of beam of electrons.

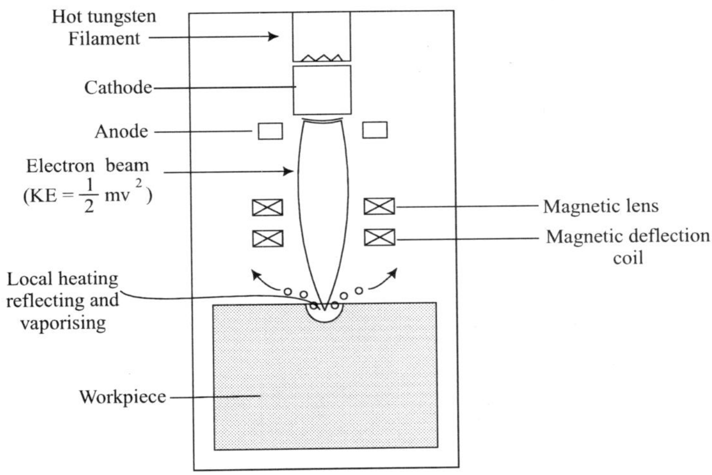

Fig. 10.28 Electron-beam Gun.

10.14.2 Electron-Beam Drilling

The complete action of electron-beam drilling consists of the following four steps.

(1) The sharply focussed electron beam strikes the workpiece surface. There is local heating and fusion of the surface.

(2) The high power density of the electron beam produces a deepening metal vapour capillary encircled by molten material.

(3) There is a through hole and electron beam may also penetrate into auxiliary base material.

(4) The high vapour pressure of auxiliary material, the molten part of workpiece is expelled.

A single pulse is used to produce a single hole. Multiple pulses may be required for thick materials.

10.14.3 Applications and Advantages

(1) Suitable for drilling a large number of small holes.

(2) Drilling of holes can be done in hard materials where conventional machines cannot be used.

(3) EBM can process a wide range of materials.

(4) Thermal effects are limited due to high energy density and short duration of beam-workpiece interaction.

(5) The heat-affected zone is only upto 0.025 mm in depth.

(6) Hole diameters of 0.1 to 1.4 mm can be drilled in materials of 10 mm thickness.

(7) Brittle and fragile materials can be processed without fracture as no force is applied to the workpiece.

(8) Holes can be drilled at shallow angles of the order of 20°.

(9) Holes upto 1000 holes/cm^2 can be drilled.

10.14.4 Limitations of EBM

(1) The equipment is very expensive.

(2) Floor-to-floor machining time is high as it is necessary to work in vacuum.

10.15 PLASMA-ARC CUTTING

A high-temperature plasma stream strikes the workpiece and there is rapid melting. It is used mainly for cutting thick sections of electrically conductive materials.

Any gas is heated to very high temperature. The gas dissociates into free electrons, ions and neutral atoms. This state of gas is called plasma and is electrically conductive.

10.15.1 Plasma-Arc Torch

A plasma-arc torch is shown in the figure. An electric arc is maintained between the electrode (–ve) and workpiece (+ve). The primary plasma gas is heated to 33,000°C and delivered through a nozzle at a pressure of 1.4 MPa at a very high velocity. The secondary gas or water is admitted to constrain the plasma jet and clean the cut of molten material. When water is used it vaporises to produce a thin layer of steam to constrict the plasma jet. Water ensures high quality cuts, high nozzle life and small heat-affected zone.

10.15.2 Plasma-Arc Cutting Speeds

The maximum recommended cutting speed for plasma-arc machining depends upon the following factors:

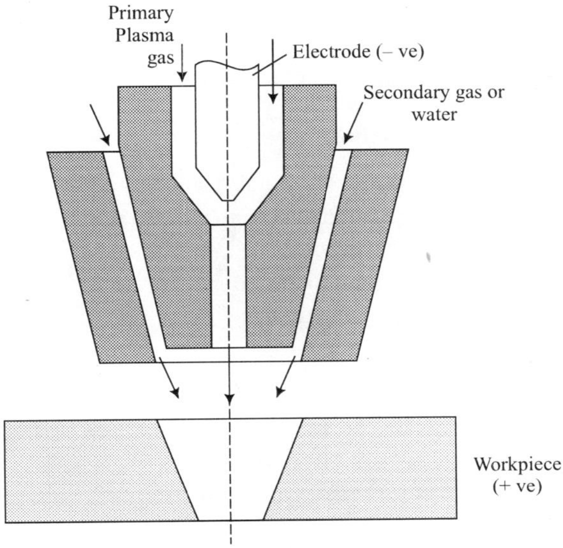

Fig. 10.29 Plasma-arc torch.

(1) Material thickness.

(2) Plasma current.

(3) Quality of cut surfaces desired.

(4) Material of workpiece.

Plasma-arc cutting speed for different sheet materials is given in Table 10.11.

Table 10.11 Plasma-Arc Cutting Speeds.

Material	*Thickness (mm)*	*Arc Current (A)*	*Cutting Speed (mm/min.)*
Stainless Steel	75	800	380
	130	1000	150
Aluminium	75	900	760
	180	1000	180
Brass	13	400	1780
Titanium	13	400	2285

The empirical relation can be built as follows:

$$V = p_1 T_s^{p_2}\left[P_3 \ln\left(\frac{1}{p_4}\right) + p_5\right]$$

where, V = Cutting speed [m/min]

T_s = Sheet thickness [mm]

I = Plasma-arc [A]

$P_1, \ldots p_5$ = constants dependent on material and surface quality required.

10.15.3 Applications and Advantages

(1) Plasma-arc cutting is used in metal fabrication and plate industries.

(2) Plasma-arc nozzle can be fitted to cutting machines with numerical as well as optical-line following controls.

(3) Complex profiles can be readily cut.

(4) High cutting speeds can be achieved.

(5) Plate thickness upto 200 mm can be cut.

(6) Tolerances of ± 0.8 mm can be achieved.

(7) The heat-affected zone varies between 0.7 and 5 mm in thickness.

QUESTION BANK NO. 10

(1) What is unconventional machining process? What are the limitations of conventional machining process? What is the need of unconventional machining process?

(2) Give classification of nonconventional machining processes.

(3) List advantages and applications of nonconventional machining processes.

(4) Giving neat sketch, explain the working of ultrasonic machining process. Also, discuss its scope of applications.

(5) Explain the effect of the following factors on cutting rate in ultrasonic machining.

(i) grit size

(ii) frequency of tool oscillation

(iii) amplitude of tool oscillation

(6) What is Abrasive-Jet Machining (AJM) process? Explain the effects of stand-off distance and abrasive grit size on material removal rate in AJM.

(7) What is Electro-discharge Machining (EDM) process? Obtain the expression of material removal rate in EDM process in terms of process parameters.

(8) Explain the role of the following in Electro-discharge Machining process:

(i) Dielectic media.

(ii) Capacitance of the circuit, and

(iii) Resistance of the circuit.

(9) Write a brief note about applicaiton of laser beams for machining.

(10) Explain the mechanics of material removal in Electrochemical Machining (ECM) process.

(11) What is Laser-beam Machining (LBM) process? With the help of a suitable sketch, explain its working. Also give scope of Laser-beam Machining application.

(12) How is the machining rate in ECM affected by the following?

(i) Rate of flow of electrolyte.

(ii) Temperature of electrolyte.

(iii) Applied voltage between tool and workpiece.

(13) Explain the cutting action in electron-beam machining. Describe at least three applications of the process.

(14) Write a brief note on Plasma-Arc Cutting process.

11

DESIGN FOR MACHINING

Introduction; Standardization; Selection of Work Material, Shape of Work Material, Shape of Component; Assembly of Components; Accuracy and Surface Finish; Machined Features; Design Guidelines; Question Bank.

11.1 INTRODUCTION

The manufacturing costs can be reduced and productivity improved by the following techniques:

1. Application of improved materials, tools and processes.
2. Effective industrial organization, factory layout, material handling and assembly techniques.
3. Automation for greater efficiency.
4. Product design.

An engineer designs a product for:

1. Performance
2. Appearance
3. Reliability
4. Technologicability (Machinability)

Technologicability is based on design principles to simplify the machining of workpiece and reduce manufacturing costs of the components. Nearly 80 to 90% of machines are required to machine metal, therefore, it is very important to reduce this process by using "near net shape" processes. This approach will help to reduce material consumption especially in large volume production.

The design for machining will involve the following steps.

1. Standardization of components.
2. Selection of raw material for workpiece.
3. Shape of components.
4. Assembly of components.

5. Accuracy and surface finish.
6. Cost estimation for machined components.

11.2 STANDARDIZATION

The following rules should be followed for designing of components to reduce machining and manufacturing costs.

(i) Rule 1: Use of standard components

The first rule in designing for machining (reduction) is to use standard components as much as possible. There are many small components which are manufactured in large quantities and standard sizes. These should be used wherever possible. Information of standard components is available in catalogues available from suppliers. The cost of these bought-out components is much lower than that of similar non-standard components when manufactured. Examples are nuts, bolts, washers, screws, seals, bearings, gears, sprockets, etc.

(ii) Rule 2: Minimization of machining by preshaping the workpiece

The manufacturing process such as casting, welding, extrusion, deep drawing, blanking, forging can be used for preshaping the workpiece so that machining is minimized. The justification for preshaping is more important for large batch sizes. For small batch sizes, the desired shape may be produced more economically by machining.

The designer should use preformed workpiece which are designed for similar previous jobs. The necessary patterns for casting or tools and dies for metal-forming processes may already be available.

(iii) Rule 3: Standardization of machined features

Standardized machined features such as screw threads, keyways, sealings for bearings, splines, etc. should be incorporated in the product design. Appropriate tools, jigs and fixtures required for standardized machined features would be available which can reduce manufacturing costs considerably. Information on standard features can be taken from machinery hand books and other reference books.

11.3 SELECTION OF WORK MATERIAL

The applicability of a material will depend upon the final function of the component and is decided by such factors as strength, resistance to wear, appearance, corrosion resistance, etc. After initial selection of the work material on the basis of above factors, the designer should also consider factors that help to reduce final cost of the component. These factors may include applicability, cost, availability, machinability and amount of machining required.

These are conflicting requirements and the final optimum choice is generally a compromise. A least-expensive work material may always result in minimum cost for the components. An expensive material may be more machinable and less expensive to machine. Clearly, the designer should select work materials that will result in minimum component cost.

11.4 SHAPE OF WORK MATERIAL

The work material can be classified as:

(i) **Partially formed before machining:** Forgings, castings and welded structures belong to this group.

(ii) **Rolled Products:** Metals are generally available as plate, sheet, bar or tube in wide range of standard sizes. The designer should check on the standard shapes and sizes from the national standards and suppliers. The selected material should require minimum machining. Standard material shapes and range of sizes are listed below:

Material	*Size*
1. Plate	6-75 mm thickness
2. Sheet	0.1-5 mm thickness
3. Round bar or rod	3-200 mm diameter
4. Hexagonal bar	6-75 mm (face to face)
5. Square bar	9-100 mm side
6. Rectangular bar	3 × 12 to 100 × 150 mm
7. Tubing	5 mm dia. to 1 mm wall to 100 mm dia. 3 mm wall

There are two types of components:

1. **Rotational Components:** These are machined from a circular or hexagonal bar or tube on a machine tool by applying rotary primary motion to the workpiece. Partially formed workpiece can also fall into this category.
2. **Non-rotational Components:** The remaining components are manufactured from square or rectangular bar, plate or sheet.

11.5 SHAPE OF COMPONENT

Classification System for Machined Components

The classification system developed by Optiz is based on the geometric code numbers and has the following features.

1. **Basic component shapes:** The basic component shapes have been classified as rotational and non-rotational.

   (i) *Rotational components* are those whose basic shapes can be machined on lathes, boring machines, cylindrical grinders and other machine tools by applying a rotary primary motion to the workpiece. The workpiece is *rotated*.

(ii) *Non-rotational components* are all other shapes than rotational, the machining of basic shape is achieved by giving *linear* motion to the workpiece. The machine tools required may be milling machines, planers, shapers, surface grinders.

2. **Dimensions of basic shape of rotational components:** The basic shape of a rotational component is a plain cylinder with dimensions sufficiently large to enclose the final shape. These components can be further sub-divided into the following:

(i) $\left(\frac{L}{D}\right) \leq 0.5$ are called disks.

(ii) $0.5 < \left(\frac{L}{D}\right) \leq 3$ are called short cylindrical components.

(iii) $\left(\frac{L}{D}\right) > 3$ are called long cylindrical components.

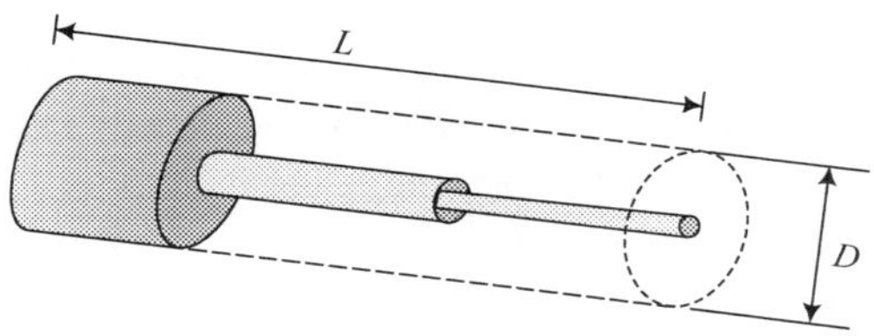

Fig. 11.1 Basic Dimensions of Rotational Components.

The method of changing the basic shape into final dimensions by various machining processes will depend upon the above sub-classification.

3. **Dimensions of basic shape of non-rotational components:** The basic shape of a non-rotational component is a right-rectangular prism with dimensions sufficiently large to enclose the final shape. In labelling the basic rectangular prism, the largest side is given the length A and the smallest the length C. The non-rotational components can be further sub-divided into the following categories.

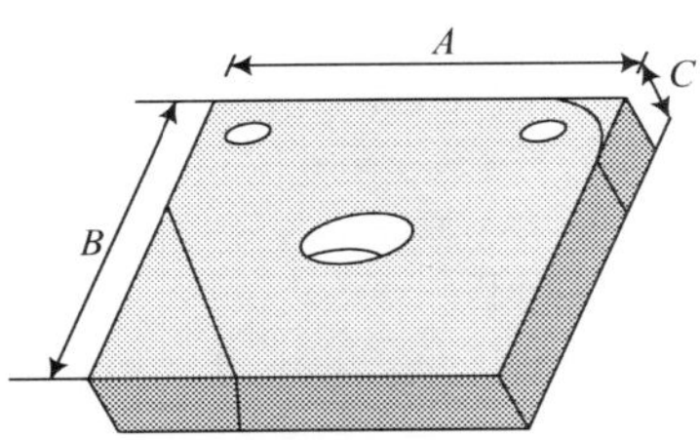

Fig. 11.2 Basic Dimensions of Non-rotational Components.

(i) $\left(\frac{A}{B}\right) \leq 3,\ \left(\frac{A}{C}\right) \geq 4$ are called flat components.

(ii) $\left(\frac{A}{B}\right) > 3$ are called long components.

(iii) $\left(\frac{A}{B}\right) < 3,\ \left(\frac{A}{C}\right) < 4$ are called cubic components.

The method of changing the basic shape into final dimensions of the component by various machining processes will depend upon the above sub-classification.

4. **Other advantages of classification systems:** The various additional advantages of classification system of components are:
   1. Identification of machining processes required to produce the component.
   2. Optimization of machining process.
   3. Categorization and grouping of components having similar features and requiring similar sequences of machining operations.
   4. Standardization of components and avoiding specifying machined features the company is not equipped to handle.
   5. Efficient planning the layout of machines in the factory to reduce the handling and transfer of components as much as possible.
5. **Supplementary information:** The classification system can have codes for the following additional information.
   1. Basic dimensions of the component.
   2. Material to be used.
   3. Form of raw material to be used.
   4. Accuracy of the machined surfaces required.

11.6 ASSEMBLY OF COMPONENTS

Most machined components must be assembled and the design for ease of assembly should take care of the following aspects.

1. It should be physically possible to assemble the components.

   Example: The screw thread on a bolt, stud or screw should be same as the mating thread on a nut or screwed hole on a component into which the bolt or screw is to be inserted.
2. Each operating machined surface on a component should have a corresponding machined surface on the mating component.

Examples: (i) For bolting of flanges on castings together, the area around the bolt holes should be machined perpendicular to the hole (spot-faced) to provide proper seating for the bolt heads, nuts or washers. (ii) The internal corners should not interfere with the external corner on the mating component.

3. The tolerances on mating surfaces should be correctly specified. Incorrect specification of tolerances can make assembly difficult or even impossible.

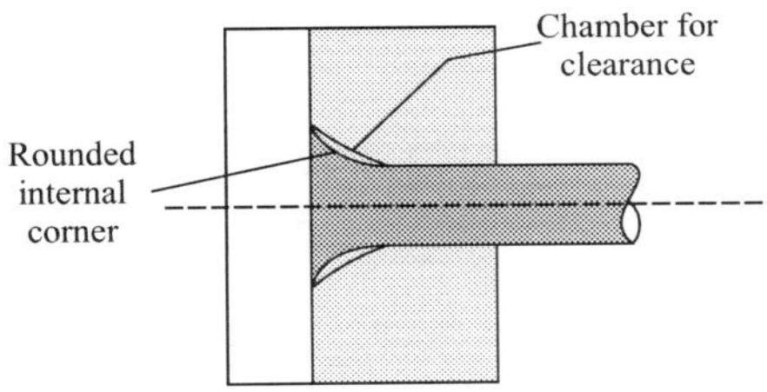

Fig. 11.3 Rounded corners and chambers.

11.7 ACCURACY AND SURFACE FINISH

The designer should specify the widest tolerances and roughest surface that would give acceptable performance for operating surfaces. When determining the accuracy and finish of machined surfaces, it is necessary to take into account the function intended for the machined surface. The specification of too-close tolerances or too-smooth surfaces can add to unnecessary manufacturing costs. If adequate accuracy and finish can also be achieved on a lathe used for rough machining, the use of cylindrical grinding could be saved for finishing process.

1. **Tolerances:** The following guide on difficulty of machining to obtain tolerances should be used.

Tolerance Range	*Difficulty in machining*
(i) 0.127 to 0.25 mm	Easily obtained.
(ii) 0.025 to 0.05 mm	Difficult, production cost increases
(iii) 0.0127 mm or greater	Very difficult, good condition of equipment, skilled operators are required. Significant increase in production cost.

2. **Surface finish:** General range of surface finish that can be obtained in different machining operations is given below. A surface finish of 1 μm or better requires separate finishing operations which substantially increases costs.

Machining operations	*Range of surface roughness (μm)*
1. Sawing	6.3 to 25
2. Turning, shaping, milling, boring	2 to 6.3
3. Drilling	2 to 5
4. Reaming	1 to 4
5. Grinding	0.5 to 2
6. Honing	0.125 to 0.5
7. Laping, Polishing	0.025 to 0.3

There is a relationship between machining time and surface finish.

The machining time,

$$t_m = \frac{0.181_w}{n_w (Ra\, R\varepsilon)^{0.5}}$$

where, R_a = arithmetical-mean surface finish (μm) $= \dfrac{0.1031 f^2}{R\varepsilon}$

where, f = feed

R_e = corner of the cutting tool

l_w = length of workpiece

n_w = rotational speed of workpiece

The machining time and hence the machining cost is inversely proportional to the square root of surface finish. The machining costs rise rapidly when low values of surface finish are required.

Smooth and accurate surfaces are obtained by finish grinding. The surface to be ground should be accessible, raised and should not interact to form internal corners.

11.8 MACHINED FEATURES

Different machine tools have capability to provide different machined features. These machined features can be categorized as follows for different working tools.

1. Features readily machined with standard tools.
2. Features expensive to machine even though standard tools can be used.
3. Features extremely difficult to machine that require the use of special tools or fixtures.
4. Features impossible to machine.

The relative difficulty of providing some features depends on the component's *L*/*D* ratio. The difficulty of providing internal features increases as this ratio increases for rotational components. For non-rotational components also the relative difficulty of providing features depends on the proportions of the component.

In the design of components, the following should be ensured.

1. Surfaces to be machined must be accessible when the workpiece is gripped in the work-holding device.
2. When the surface of the workpiece is being machined, the tool and tool-holding device must not interfere with the remaining surfaces on the workpiece.

The common machined features on rotational components which can be readily machined on different machine tools are listed in Table 11.1. Similarly, common machined features on non-rotational components which can be readily machined on different machining tools are listed in Table 11.2.

Table 11.1 Examples of Machined Features readily provided on Rotational components.

Machine Tool	*Name of Feature*	*Example*
Lathe or vertical boring machine	Stepped bars or shafts	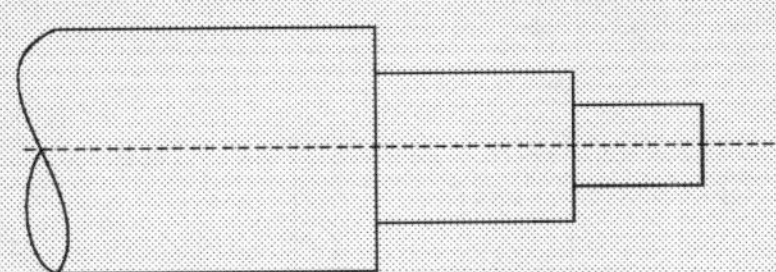
	Internal holes	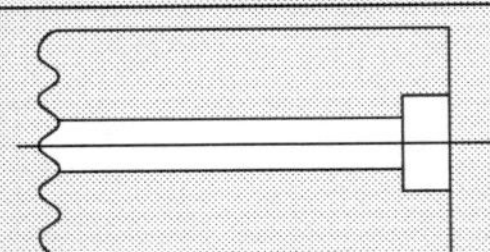
Drilling machine	Radial and auxiliary holes	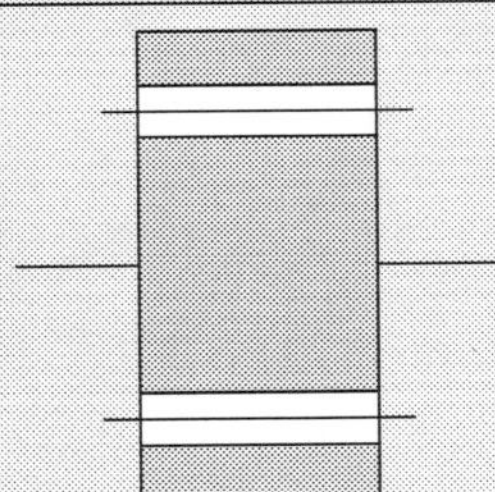 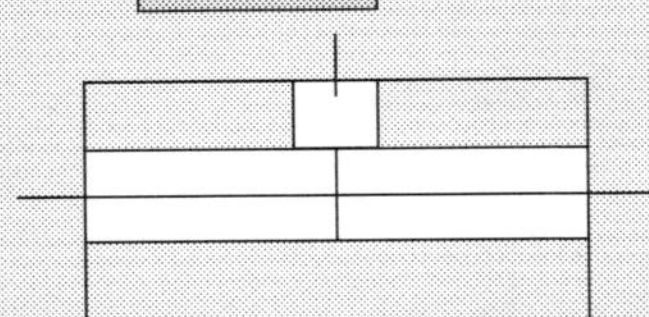
Milling and Shaping machines	External and internal keyways	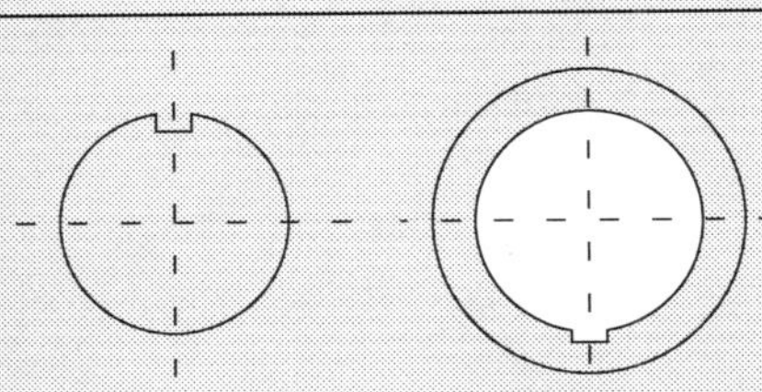

Table 11.2 Examples of Machined features readily provided on non-rotational components.

Machine Tool	*Name of Feature*	*Example*
Milling, shaping or planning machines	Plane parallel surface	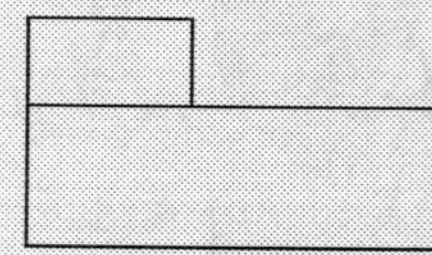
	Slots or grooves	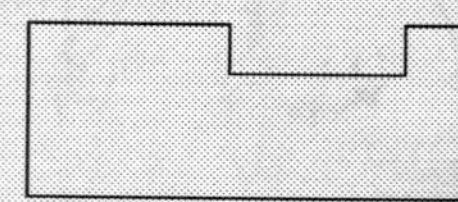
Drilling Machines	Through and blind Cylindrical holes	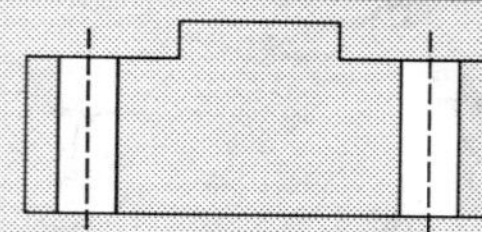 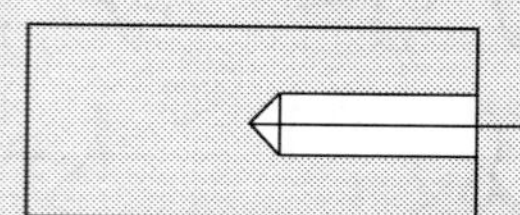
Lathe, boring, jig boring machines	Cylindrical bores	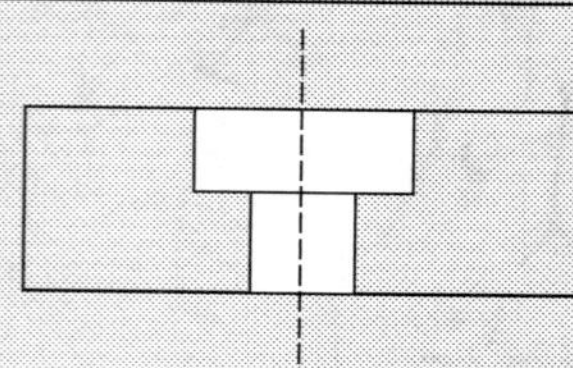

11.9 DESIGN GUIDELINES

The designer should keep in mind the following guideline for designing a machined component.

1. **Standardization**
   1. Use standard components as much as possible.
   2. Casting, forging, welding, etc. should be used for preshaping the workpiece.
   3. Standard preshaped workpieces should be used where possible.
   4. Standard machined features of the company should be used where possible.
2. **Raw material**
   1. The cost of component (cost of raw material plus cost of production) should be minimized by proper selection of raw material.

2. The rolled material should be used in standard form as available.

3. **Component design**

(a) *General*

1. The component should be designed, so that it can be machined completely on one machine tool.
2. The component should be designed, so that no machining is needed on unexposed surfaces of the workpiece when it is gripped in the holding device.
3. The machined features which cannot be handled by the company should be avoided.
4. The component should be designed, so that the workpiece when gripped in the holding device is sufficiently rigid to withstanding machining forces.
5. The tool, tool holder, workpiece and holding device should not interfere with one another during machining.
6. Design main and auxiliary holes as cylindrical with *L*/*D* ratio which can be machined with standard drills or boring tools.
7. The axillary holes should be parallel or normal to the axis of the workpiece or reference surface and related by a drilling pattern.
8. The ends of blind holes should be conical.
9. In tapped blind holes, threading should not run to the bottom of the hole.
10. Bent holes or dog-leg holes should be avoided.

(b) *Rotational components*

1. The cylindrical surfaces should be concentric and plane surfaces should be normal to the component axis.
2. The diameters of external step should increase onwards from the exposed face of workpiece.
3. The diameters of internal grooves should decrease onwards from the exposed face of workpiece.
4. Do not use internal machining for long components.
5. Do not use components with very large or very small *L*/*D* ratios.
6. The radii of internal corners on the component should be specified equal to radius of the corner of rounded tool.

(c) *Non-rotational components*

1. A reference and work holding base must be specified.
2. It will be preferable if all exposed surfaces of the component form a series of mutually perpendicular plane surfaces parallel to and normal to the base.

3. The internal corners normal to the base should have a radius equal to the radius of the rounded tool.
4. In the machined pockets, the internal corners normal to base should have as large a radius as possible.
5. It will be preferable to accommodate all plane surface machining (slots, grooves, etc.) to one surface of the component.
6. Cylindrical bores in long components should be avoided.
7. Use work material performed to the required cross-section and avoid machined surfaces on long components.
8. Extremely long or extremely thin components should be avoided.
9. The main bores should be normal to the base in flat or cubic components. The cylindrical surfaces should be in the decreasing order of diameter from the exposed face of the workpiece.
10. Blind bores should be avoided in large cubic components.
11. Internal machined features should be avoided in the cubic box-like components.

4. **Assembly of Components**
   1. It should be possible to assemble the components.
   2. Every operating machined surface on a component should have a corresponding machined surface on the mating component.
   3. The internal corners on a component should not interfere with a corresponding external corner on the mating component.
5. **Accuracy and Surface Finish**
   1. The widest tolerances and roughest surface should be specified that will give acceptable performance.
   2. The surfaces to be finish-ground should be raised and should never intersect to form internal corners.

QUESTION BANK NO. 11

1. Discuss the important rules of standardization for reducing machining and manufacturing costs.
2. Discuss the role of the following for reduction of machining costs.
   (i) Shape of work material.
   (ii) Shape of final component.
3. Discuss the machined features which can be readily produced on:
   (i) Rotational components.
   (ii) Non-rotational components.
4. Prepare design guidelines for rotational and non-rotational components.

5. What recommendation should be given for the following aspects while designing a component?
   (i) Selection of raw material.
   (ii) Assembly of components
   (iii) Specification of accuracy and surface finish.
6. What are general guidelines for component design.

Part II

Cutting Tools

12

NOMENCLATURE AND GEOMETRY OF CUTTING TOOLS

Introduction; Cutting edge and chip flow angles, System of cutting-tool nomenclature, Mathematical relationship between various rake systems and normal rake system, Question Bank.

12.1 INTRODUCTION

Cutting-tool nomenclature means systematic naming of the various parts and angles of a cutting tool. The surfaces on the point of a tool bear definite relationship to each other that are defined by angles. The principles underlying cutting tool angles are the same for single-point tool, multi-point tool or a grinding wheel.

The most important features of a cutting tool are the two cutting edges and the faces and flanks that intersect to form the cutting edges. The nomenclature of important features of a single-point cutting tool is shown in Fig. 12.1. The major cutting edge(s) removes the bulk metal and minor cutting edge(s) controls the final surface finish of workpiece.

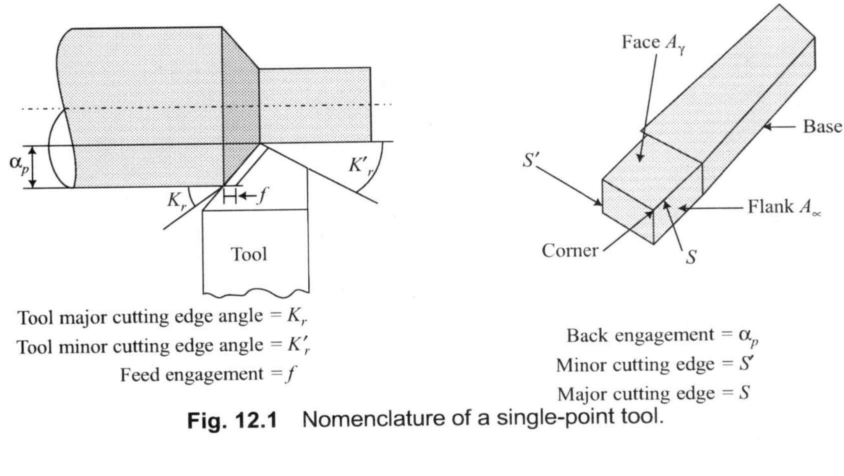

Fig. 12.1 Nomenclature of a single-point tool.

Following systems are used for geometric arrangements and nomenclature.

1. British Maximum Rake System
2. American Standards Associated System
3. German System
4. ISO-recommended system

12.2 CUTTING EDGE AND CHIP FLOW ANGLES

The oblique cutting operation of a tool is shown in Fig.12.2. Cutting-edge inclination, λ_s is the angle between the cutting edge and a normal to the direction of cutting. It is measured in the plane of new workpiece surface. In orthogonal cutting, $\lambda_s = 0$. The chip flows up the tool face. The chip-flow angle λ_γ is the angle of chip flow with a line normal to chip edge and is measured in the plane of tool face.

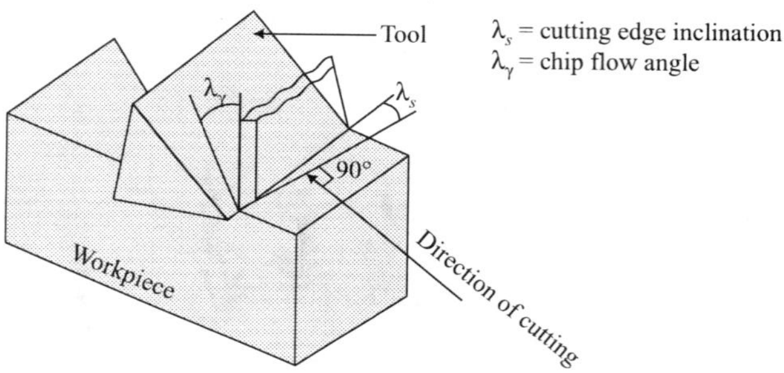

Fig. 12.2 Basic cutting operation.

12.2.1 Stabler's Chip Flow Law

If the chip does not change in width, the stabler's chip-flow law states that cutting-edge inclination is qual to chip flow angle.

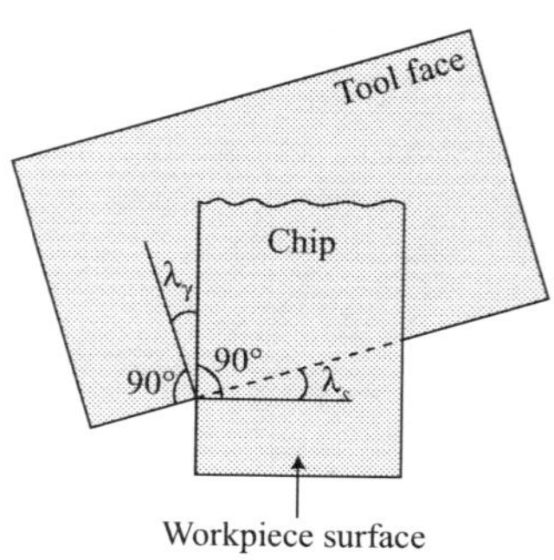

Fig. 12.3 Stabler's chip-flow law.

$$\lambda_s = \lambda_\gamma \qquad ...(1)$$

The following conclusions can be made from the law.

1. Equation (1) gives adequate approximation. However, in practice, there may be some variation to the law.
2. The cutting-edge inclination is a major parameter in the design of cutting tools as chip flow direction depends upon this inclination.
3. The system of tool nomenclature should include the definition of cutting-edge inclination to give physical meaning to the cutting process.
4. The system of nomenclature should help to set the specified tool angle onto the tool grinding vice or jig.

12.2.2 Tool Angles

Important tool angles as specified by American system are shown in Fig. 12.4.

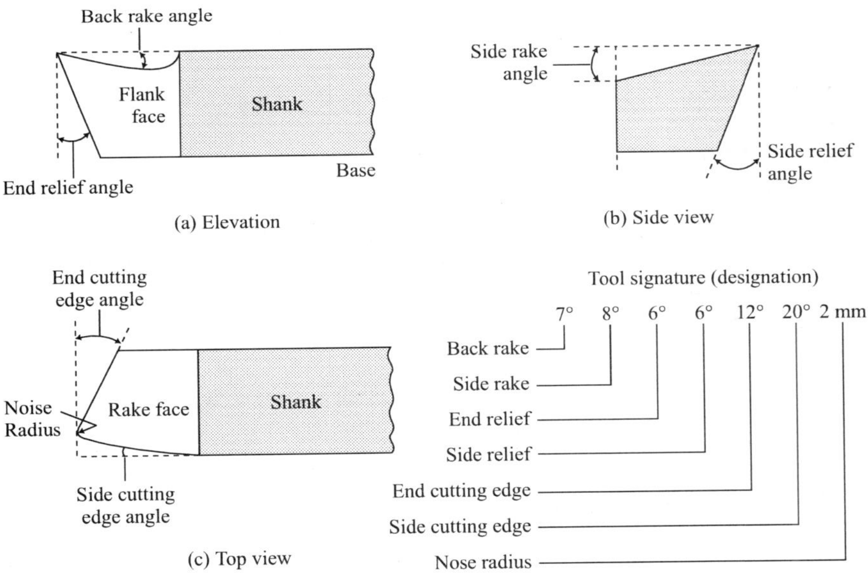

Fig. 12.4 Tool Angles as per American system.

When the tool is engaged, cutting takes place over the side cutting edge. The nose and some part of end cutting edge also help cutting. The chip flows over the rake face. The side and end flanks of the tool are relieved to avoid contact with machined surface. The two relief angles are called side relief angle and end relief angle respectively.

The rake face is inclined with respect to base. Back rake angle is the angle of inclination of the rake face with respect to the base. Side rake angle is the angle of inclination of rake face with respect to tool base.

12.2.3 Cutting Tool Signature

The signature is a sequence of numbers listing the various angles, in degrees, and the size of the nose radius in mm. This numerical method of designation has been standardized by the American Standard Association.

The seven elements that comprise the signature of a single point cutting tool is always stated in the order shown in Fig. 12.4.

12.3 SYSTEM OF CUTTING-TOOL NOMENCLATURE

12.3.1 British System

The rake angle is specified as the steepest slope of rake face. The maximum rake angle is the angle between the rake face and the tool base measured in a plane normal to both the rake face and the tool base. The main basis of the system was that chip flows in the direction of steepest slope which is not true.

The main advantage of the system that the specified rake angle can be easily set on the tool grinding vise or fixture for grinding the rake face. The chip flow angle and effective rake angle cannot be easily estimated in this system. The cutting edge inclination has no physical relationship with rake angle.

12.3.2 American System (ASA System)

The geometry of the rake face is expressed in terms of back rake angle and side rake angle. The back rake angle is the angle between the rake face and the base of the tool measured in a plane normal to the base of the tool and parallel to the longitudinal axis of the tool. The side rake angle is the angle between the rake face and the tool base measured in plane normal to the plane in which the back rake angle is measured (see Fig. 12.4).

The main advantage of the system is to calculate the setting angle of tool grinding vise easily from the following relationship

$$\tan \gamma_{set} = \cos \gamma_s \tan \gamma_b$$

where, γ_{set} = setting angle

γ_s = side rake angle

γ_b = back rake angle.

The main drawback of the system is that specified angles are not related to actual position of cutting edge.

12.3.3 German System (DIN System)

The rake face is specified by back rake angle and side rake angle as in American system. However, the planes for measurement of angles are different.

The back rake angle is the angle between rake face and tool base and measured in a plane normal to base plane but parallel to trace of the side cutting edge in the base plane. The side rake angle is the angle between the rake face and the base plane and measured in a plane normal to the trace of the side cutting edge in the base plane.

The advantage of the system is that two rake angles are related to respective cutting edges. The drawback of the system is that setting angle for tool grinding vise has to be calculated.

12.3.4 Normal Rake System (ISO System)

The tool normal rake angle γ_n is the angle between the tool face A_γ and the tool reference plane P_r and measured in the cutting edge normal plane perpendicular to the cutting edge. The normal rake angle and clearance depend upon resultant cutting speed angle as shown in Fig. 12.5.

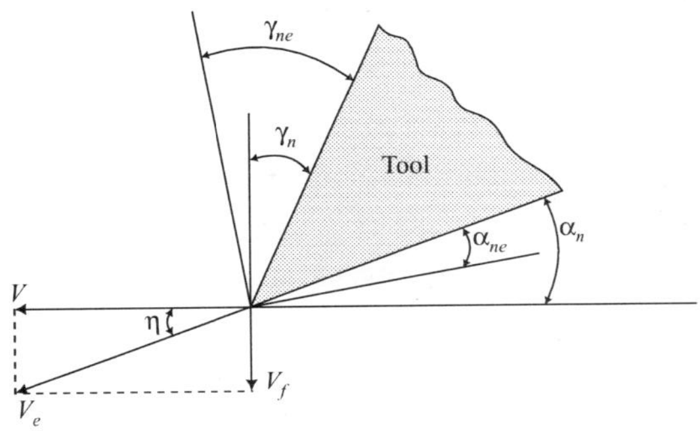

Fig. 12.5 Effective rake angles and clearance angles.

γ_n = tool normal rake angle

γ_{ne} = effective normal rake angle

α_n = tool normal clearance

α_{ne} = effective normal clearance

V = cutting speed

V_f = feed speed

V_e = resultant cutting speed.

η = resultant cutting speed angle.

Normal rake system was worked out by stabler on the basis that there is a good correlation between the cutting forces and normal rake angle in oblique cutting. The tool rake face can be ground by directly setting the angles specified by this system on tool grinding vise.

12.4 MATHEMATICAL RELATIONSHIPS OF VARIOUS RAKE SYSTEMS INTO NORMAL RAKE SYSTEM

The following angles affect the cutting process in the normal rake system.

1. The angle of inclination of side cutting edge with respect to cutting velocity direction.
2. The rake angle normal to cutting edge.

For conversion of the various systems, it is necessary to find

– angle of obliquity, I

– normal rake angle with side cutting edge, α_n.

(a) American system and normal rake system

$$\tan \alpha_n = \cos \gamma + \tan \alpha_{as} + \sin \gamma \tan \alpha_{as}$$

where, α_{as} = rake angle as per American system.

(b) British system and normal rake system

$$\tan \alpha_n = \sin I \tan (\gamma + \theta_{bm})$$

where θ_{bm} = Angle between maximum rake plane with tool axis in British system

(c) German system and normal rake system

$$\tan \alpha_n = \tan \alpha_{gs} \cos \alpha_{gb}$$

where, α_{gs} = side rake angle as per German system

α_{gb} = back rake angle as per German system

Example 12.1 The tool signature in American system is 8-10-6-6-12-2-30-0. Determined angle of obliquity and normal rake angle as per normal system.

Solution:

The back rake angle, $\alpha_{ab} = 8°$

side rake angle, $\alpha_{as} = 10°$

cutting edge angle, $\gamma = 30°$.

Nose radius = 0 mm.

The angle of obliquity,

$$\begin{aligned} I &= \tan^{-1} (\cos \gamma \tan \alpha_{ab} - \sin \gamma \tan \alpha_{as}) \\ &= \tan^{-1} (\cos 30° \tan 8° - \sin 30° \tan 10°) \\ &= \tan^{-1} (0.1919) \\ &= \mathbf{10.86°} \quad \textbf{Ans.} \end{aligned}$$

The normal rake angle,

$$\alpha_n = \tan^{-1}(\cos\gamma \tan\alpha_{as} + \sin\gamma \tan\alpha_{ab})$$
$$= \tan^{-1}(\cos 30° \tan 10° + \sin 30° \tan 8°)$$
$$= \tan^{-1} 0.229.$$
$$= \mathbf{12.9°} \quad \textbf{Ans.}$$

QUESTION BANK NO. 12

1. Explain nomenclature of a single-point tool with the help of suitable sketches.
2. Explain stabler's chip flow law.
3. What is tool signature system? Show different tool angles as per American system.
4. What are different systems of cutting tool nomenclature. Discus the advantages and disadvantages of various systems.
5. Explain normal rake system. Why this system has been adopted by ISO?
6. List various mathematical expressions of conversion for various rake systems into normal rake system.

13

TOOL MATERIALS

Introduction : Desirable properties of tool materials; Types of Tool Materials; Characteristics of Cutting Tool Materials; Question Bank.

13.1 INTRODUCTION

The cutting tools are required to operate under high loads at elevated temperatures well above 1000°C. In addition, severe frictional conditions occur between the tool and chip and between the tool and the newly machined workpiece surface.

The following factors influence the performance of cutting tools:

1. Relative hardness of the tool and work material.
2. Chemical compatibility of the tool and work material.
3. Cutting temperatures.
4. Type of machining operation, i.e., continuous or intermittent.
5. Rigidity and condition of machine tool.
6. Condition of the surface of workpiece, i.e., presence of scale, rust, abrasive particles.

The main requirements for cutting tool materials are:

1. Wear resistance.
2. Hot hardness.
3. Toughness.

All the above properties are not present simultaneously in a material. For example, a material with high wear resistance and high-temperature resistance has low toughness. The selection of tool material is generally a compromise between separate requirements and their relative importance depends on the specific applications.

13.2 DESIRABLE PROPERTIES OF TOOL MATERIALS

13.2.1 Wear Resistance

Wear resistance of a tool material is its ability to retain its shape and sharpness for a sufficiently long time. This properly should be retained during machining a given workpiece at a given cutting speed. Wear of tool is caused by the following mechanisms.

(i) ***Abrasion:*** When a chip flows over the rake face of the cutting tool under high pressure, abrasion of material takes place. Abrasion is also caused by the rubbing action of the newly machined surface with the tool flank. The rate of abrasion depends upon the hardness of work material and cutting temperature.

(ii) ***Adhesion:*** When the tool material adheres with the chip or newly machined surface, there is gradual loss of tool material. This is called wear of tool due to adhesion.

(iii) ***Diffusion:*** The atoms of hard alloy constituents of tool material may transfer into chip or work material. This diffusion of tool material results in its weakening. The weakened layer of tool material is removed by abrasive surface.

13.2.2 Hot Hardness

The tool material must remain harder than work material at elevated operating temperatures. Hot hardness is the ability of tool material to retain its hardness even at elevated temperatures. It is desirable to develop tool material with high hot hardness. These materials can operate at higher cutting speeds resulting in high productivity.

A tool should also fully recover its hardness at room temperature after machining at high temperature. This is called recovery hardness.

13.2.3 Toughness

The tool material should be tough with high strength and ductility. It should be able to withstand shocks and vibration due to impact loads and intermittent cutting operations. There should be no breakage of tool. However, it has been found out that materials having high hot hardness and wear resistance are brittle with less toughness.

13.2.4 Cost and Ease of Fabrication

The cost and ease of fabrication should be within reasonable limits. It should be possible to regrind, weld/braze/fix to tool holder easily.

13.2.5 Thermo-physical Properties

The tool material should possess high thermal conductivity to remove heat quickly from the cutting region. It should have low coefficient of thermal expansion. It should not distort during heat treatment.

13.3 TYPES OF TOOL MATERIALS

Last century has seen many developments in the field of tool materials with increasing wear resistance, hot hardness and toughness properties to meet the demands of high-speed machining and hard machining.

1. **Carbon Steels:** The earliest tool materials used were hardened carbon steels. These steels are restricted to low cutting speeds and temperatures. The hardness of carbon steel depends upon the martensite present which softens at 250°C and above. Carbon steel tools are presently used in wood working and machining of soft materials.
2. **High-Speed Steels:** The heat treatment process was discovered and used for producing high-speed steel-cutting tools. These steels had high wear resistance and higher metal removal rates were possible by the adoption of higher cutting speeds.
3. **New Tool Materials:** Developments in metallurgical science and technology have helped to produce new tool materials such as cast alloys, cemented carbide, sintered oxides or ceramics. At the present time more than 50% of cutting tools (mostly multipoint) used world over are based on cemented carbide materials, with 40% (mostly single-point) being made from high-speed steels and the remaining 10% from all of the other material classes. In terms of the per cent of material removed, cemented carbides represent 2½ to 3½ times the amount removed by high-speed steels.
4. **Tool Coatings:** Recent developments in tool materials have included the coating of high-speed steel and tungsten carbide tools with thin, high-wear resistant layers to improve performance.

Major classes of tool materials are given in Table 13.1. These have been arranged in order of increasing temperature resistance and hot hardness. Unfortunately, an increase in hardness is usually accompained by decrease in impact strength (toughness) and increase in tool manufacturing problems.

Table 13.1 Major Classes of Tool Materials.

Sl. No.	*Class of material*	*Wear resistance*	*Hot hardness*	*Toughness*
1.	Carbon Steels	Lowest ↓	Lowest ↓	Highest ↑
2.	High-Speed Steels			
3.	Cast alloys			
4.	Tungsten Carbides			
5.	Cermets			
6.	Titanium Carbides			
7.	Ceramics			
8.	Polycrystalline diamond and cubic boron nitride			
9.	Single-crystal diamond	Highest	Highest	Lowest

13.4 CHARACTERISTICS OF CUTTING TOOL MATERIALS

The main characteristics of tool materials are discussed below.

13.4.1 High Carbon Steel

The main characteristics of high carbon tool steels are:

(i) These are plain carbon steels containing 0.6 to 1.5% carbon.

(ii) The refinement of grain size and increase in hardness is achieved with the addition of small percentage of silicon, manganese, chromium and vanadium.

(iii) These are easy to manufacture and cutting edge can be easily sharpened.

(iv) These steels can be used with operating temperature very low. They lose hardness rapidly at temperatures greater than 200°C.

(v) They have low wear resistance and hot hardness.

(vi) They are used in making cutters, twist drills, turning and form tools for machining soft or free cutting materials like wood, magnesium, brass and aluminium.

(vii) They are also widely used for manufacture of hard tools like taps, files, reamers, chisels, hacksaw blades, etc.

(viii) Properties can be considerably enhanced by alloying with chromium.

13.4.2 High-Speed Steel (HSS)

(i) High-speed steels are alloy carbon steels with tungsten (~18%) and chromium (~4%). Other alloying elements are cobalt, vanadium or molybdenum.

(ii) There are more than 20 grades of HSS tool materials in common use with their own advantages and limitations.

(iii) These can be divided into three main groups: tungsten based, molybdenum based and molybdenum-cobalt based alloys. The composition of some of standard grades of HSS are given in Table 13.2.

(iv) Addition of vanadium increases abrasion resistance and cobalt increases hot hardness.

(v) HSS are manufactured by conventional alloy steel manufacturing process, powder metallurgy process and electro-slag refining process.

(vi) Properties of HSS tools are significantly influenced by heat treatment process which must be carried out very carefully. During the heat treatment process, carbides of alloying elements are formed and dispersed throughout the material as very hard particles. Cobalt, however, dissolves to substitute for iron atoms in the crystal matrix.

(vii) High-speed steel-cutting tools are relatively inexpensive and tough but have limited hot hardness. These can be used for cutting temperatures upto 550°C.

Table 13.2 Composition and Properties of Standard High-Speed Steels.

Grade	Composition (%)						Property rating 1 to 10 (highest)			
	C	W	Mo	Cr	V	Co	Wear resistance	toughness	Hardness	Cost
1. Conventional Steels										
T_i	0.75	18.00	–	4.00	1.00	–	4	8	5	5
M_1	0.80	1.75	8.50	3.75	1.15	–	4	10	5	3
M_2	0.85	6.00	5.00	4.00	2.00		5	10	5	3
2. Cobalt Steels										
M_{33}	0.88	1.75	9.50	3.75	1.15	8.25	5	5	8	5
T_5	0.80	18.00	–	4.25	2.00	8.00	5	4	8	6
T_6	0.80	20.00	–	4.50	1.75	12.00	5	2	9	8
3. High Vanadium Steels										
M_3	1.05	6.00	5.00	4.00	2.40	–	6	6	6	4
M_4	1.30	5.50	4.50	4.00	4.00	–	9	6	6	4
T_{15}	1.50	12.00	–	4.50	5.00	5.00	10	9	9	6
4. High Hardness Cobalt Steels										
M_{42}	1.10	1.50	9.50	3.75	1.15	8.25	6	9	9	5
M_{44}	1.50	5.25	6.50	4.25	2.00	12.00	6	3	10	6

(viii) High-speed steel can be hot-rolled, forged to rough shapes, then machined, heat treated and then finish ground.

(ix) The surface of HSS tool can be further hardened by ball rolling. The rake and flank faces of the tools are finely polished to remove all burrs and uneveness. The tool life is appreciably increased by polishing.

(x) It is a widely used material for manufacture of monolithic tools such as drills, taps, milling cutters, etc. for machining of steels of upto 350 BHN hardness. There is, however, an increasing tendency to use HSS inserts for turning. The inserts are clamped, brazed or welded to a carbon or low-alloy steel body. The carbon steel shanks are friction welded to HSS cutting ends of drills and narrow HSS strips are electron-beam welded to low-alloy steel bands for making band saws. Indexable HSS inserts are mechanically clamped in tool holders.

(xi) HSS cutting tools are available with a thin coating of titanium nitride to increase wear resistance. This is discussed in detail in # 13.4.3.

13.4.3 Coated High-Speed Steel Tools

This is a recent development. The high-speed steel provides a relatively ductile, shock resistant core of the tool while the coating is applied to the core which has very high wear resistance and low friction.

(i) Thin layer (2 to 6 μm) coating of a refractory metal carbide or nitride is applied. The main coating materials used are titanium carbide, titanium nitride, hafnium nitride and alumina.

(ii) A thin (1 to 2 μm) coating of titanium nitride (TiN) has an equivalent hardness of Rockwell C80-85. This coating has exceptional wear resistance and lower friction than the uncoated high-speed steel.

(iii) There are two methods of coating used:

- Physical Vapour Deposition (PVD). Particles of the coating are deposited physically onto the surface of workpiece kept in high vacuum condition at temperatures in the range of 200-500°C.
- Chemical Vapour Deposition (CVD). The tools are heated in a sealed reaction chamber to around 1000°C in an inert atmosphere. Gaseous hydrogen and volatile compounds supply the metallic and non-metallic constituents. It is a thermochemical process and is a more commonly used method.

(iv) The cost of coated tool is 2 to 5 time the cost of plain HSS tool.

(v) The life of a coated tool is increased by 5 to 10 times the life of plain HSS tool.

(vi) Coated HSS tools are used for cutting not only general materials but also hard-to-machine alloys of Cr–Mo steel. Hard machining can replace grinding for the manufacture of complex shapes on one machine tool and using single tool.

13.4.4 Cast Alloys

(i) Cast alloy tools are cast into their final shape.

(ii) These alloys do not contain iron and contain primarily cobalt (40-55%), chromium (25-35%), tungsten (1.5-3%) and carbon (0-5%).

(iii) The carbide phase is about 25 to 30% by volume of the matrix.

(iv) The tools are cast in graphite chilled molds to produce a fine grained hard surface made up of complex carbides.

(v) The core of the tool is a tough carbide enriched material.

(vi) These tools can be used at higher speeds than HSS tools.
These alloys retain hardness upto temperatures of 750°C.

(vii) These materials have low coefficient of friction and do not form built-up edge during machining.

(viii) These are used for machining of cast iron, malleable iron and hard bronzes.

(ix) These are useful for making form tools as they are castable.

13.4.5 Cemented Carbide Tools

These are the most commonly used cutting tools at present.

(i) A mixture of 94% tungsten powder and 6% carbon by weight is prepared. This mixture is combined with cobalt, compacted and sintered in a furnace at 1400°C. The cobalt acts as binder phase. The percentage of hard carbide particles varies from 60 to 95%. This process of mixing and sintering or cementing is called powder metallurgy and the product is called sintered or cemented.

(ii) A large variety of cemented carbide tools are manufactured as per ISO standard classification for their use for machining different materials. Standard classification of cemented carbide tool materials is given in Table 13.3.

Table 13.3 Standard Classification of Cemented Carbide Tool Materials.

ISO Designation	*Work Material*	*Composition %age*				*Machining Applications*
		WC	*Co*	*TiC*	*TaC*	
P 01	Steel	64	3	25	8	Finishing
P 10	Steel	76	6	12	6	Finishing
P 20	Steel					Roughing
P 30	Steel	82	8	8	2	Roughing
P 40	Steel	70	12	6	12	Interrupted Cuts
P 50	Steel					Interrupted Cuts
M 10	Steels, cast irons, non-ferrous metals					Finishing
M 20	Steels, cast irons, non-ferrous metals					Finishing
M 30	Steels, cast irons, non-ferrous metals					Finishing
M 40	Steels, cast irons non-ferrous metals					Finishing
K 01	Cast Irons	97	3	—	—	Finishing
K 10	Non-ferrous metals	96	4	—	—	Finishing
K 20	Non-metallic	94	6	—	—	Finishing
K 40	Non-metallic	94	6	—	—	Roughing

Applications: P series for long chipped metal and steels.
M series for stainless steels and heat resistant alloys.
K series for short chipping materials, cast irons, steels and non-ferrous alloys.
01 : Light finishing.
:
:
50: Heavy rough machining.

(iii) Different manufacturers use different proportions of tungsten carbide (WC), cobalt (Co) and other alloying elements and also different particle size, mixing procedures and sintering temperatures and pressures to produce tools of the same classification.

(iv) The properties of carbide tools are highly dependent on cobalt content. Hardness of tool material decreases and toughness increases with the increase of cobalt.

(v) The hardness of cemented carbide also depends on grain size. The finer the grain size, greater the hardness.

(vi) The cemented carbides can maintain high hardness values at temperatures as high as 1200°C. These can be used at much higher cutting speeds than HSS or cast-alloy tool materials.

(vii) Cemented carbides are not tough and cannot be shaped after sintering. Therefore, these are used in the form of small tips or inserts which are brazed or clamped to steel shanks. The clamped-on inserts are thrown away after all the cutting edges have been worn. These are called disposable or throwaway inserts.

(viii) Straight tungsten carbides are the strongest and most wear resistant but are subject to rapid crating when machining steels. Tantalum carbide and titanium carbide are added to improve resistance to crating.

(ix) New grades of cemented carbide with increased densities and smaller grain size powder have been developed.

(x) Research is taking place into "nano-phase" carbides with particle size of the order of 0.1-0.2 μm.

(xi) Coating of aluminium and zirconium oxides deposited on tool surface at high temperature retard diffusion wear of the tool. A very thin coating of titanium carbide or hafnium nitride can improve the resistance to crater wear of tungsten cabide tools.

13.4.6 Cermet Tools

(i) These are cemented materials that use hard particles other than tungsten carbide. Hard particles include titanium carbide, titanium nitride, titanium carbonitride and molybdenum carbide.

(ii) The metal binder phase is a mixture of cobalt and nickel.

(iii) These are produced by powder metallurgy as similar to that for tungsten carbide.

(iv) These are more wear resistant and are of same hardness as tungsten carbide.

(v) Strength, toughness and thermal shock resistance are lower.

(vi) These can handle high cutting speeds with moderate feeds.

(vii) These are used for semi-finish and finish milling.

13.4.7 Ceramic Tools

(i) Ceramic tools are very hard with high compressive strength and low thermal conductivity.

(ii) They are extremely brittle and cannot be used where shock and vibration occur.

(iii) These are made from sintered aluminium oxide (Al_2O_3), SiAlON (a combination of silica, aluminium, oxygen and nitrogen) and silicon nitride (Si_3N_4).

(iv) Various powders are mixed together and sintered at about 1700°C.

(v) Sintered oxides can be used at cutting speeds two to three times those employed with tungsten carbides.

(vi) Aluminium oxide is the most common ceramic tool and performs best for finishing and semi-finishing with non-interrupted cuts.

(vii) Ceramic cutting tools are used for turning and milling of cast iron, superalloys and finishing of hardneed steels.

(viii) Composite ceramic tools consisting of Al_2O_3 matrix reinforced by tiny silicon carbide whiskers are used for roughing and finishing of nickel-based alloys, cast iron and steels. These tools have very high fracture toughness.

(ix) They are mostly available in the form of throwaway inserts.

13.4.8 Diamond

(i) The single crystal natural or synthetic diamond is used as cutting tool material.

(ii) Diamond has both hard and soft axes. If it is used in the direction of a soft axis (or parallel to clearage place), it wears of quickly.

(iii) Diamond has hardness more than any other material. It is chemically inert and has high thermal conductivity. Oxidation of diamond starts at about 450°C and can crack. Therefore, diamond tool is kept flooded with coolant during cutting and light feeds are used.

(iv) The diamond is extremely brittle. There should be minimum vibrations during cutting operation. The machine tool, tool holder and clamping of diamond tool to the tool shanks must be rigid.

(v) These are used for machining of non-ferrous metals like aluminium, brass, copper and bronze. It is also used for non-metallic materials like plastics, epoxy resins, hard rubber, glass and precious metals like gold, silver and platinum. These should not be used for machining ferrous metals.

(vi) Diamonds can be used for high cutting speeds. The feed should be limited to 0.02 to 0.1 mm/sec and depths of cut to about 0.5 mm.

13.4.9 Polycrystalline Tools

(i) Polycrystalline tools are fabricated by compacting small particles of diamond (synthetic or natural) or cubic boron nitride (CBN) under high pressure temperature conditions.

(ii) These are more homogeneous with higher strength and durability than natural diamond. But they have less hardness.

(iii) These are 20 times more costly than carbide but very long life, i.e., 50 times longer than tungsten carbide.

(iv) The tool inserts of thin layer of polycrystalline material is bounded to cemented carbide substrate. The substrate provides shock resistance whereas polycrystalline material provides very high wear resistance and cutting edge strength.

(v) There are used for cutting hardened steel and high temperature alloys at very high speeds.

13.4.10 Comparative Hot Hardness of Tool Materials

The variation of hardness of a few tool materials with temperature is shown in Fig. 13.1. The superiority of carbide and ceramics can easily be noted.

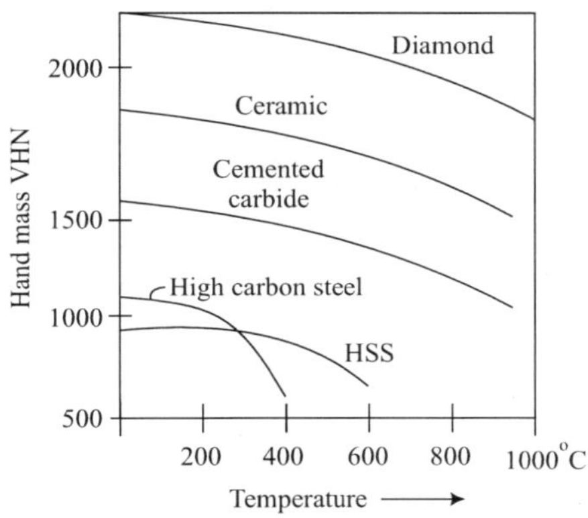

Fig. 13.1 Hardness Profile of Cutting Tool Materials.

QUESTION BANK NO. 13

1. Name various cutting tool materials. Briefly describe one important tool material emphasizing its characteristics and applications.
2. What are the desirable properties of cutting tool materials? Explain in brief.
3. Briefly describe the properties of high-speed steel as tool material. Name various types of high-speed steels.
4. Why is cemented carbide considered an useful tool material?
5. What is the significance of hot hardness of a tool material? Discuss its relation with wear resistance and toughness.

14

WEAR AND LIFE OF CUTTING TOOLS

14.1 INTRODUCTION

A cutting tool initially with sharp cutting edges and smooth flank is subjected to adverse conditions of operation during machining;

(i) High cutting forces are concentrated over a small contact over on the rake face and the flank.

(ii) The chip slides over the rake face.

(iii) The machined surface rubs over the flank surface.

(iv) The contact surfaces are subjected to high temperature.

(v) Tool is subjected to mechanical and thermal shocks during entering and exit from the cut.

Even HSS, carbides and other hard materials gradually wear out and may even fracture. This necessitates:

(i) Change of tool/grinding of tool.

(ii) Stepping of machine during retracting, changing and returning into cutting position.

Tool wear and tool life (time between two successive tool changes) are of great importance in the theory and practice of metal cutting. It has important economic consideration. A short tool life is uneconomical because tool-grinding and tool replacements costs are high.

Considerable efforts have been made over the years to develop new and improved materials with better tool life. It is very important to understand the nature of tool wear and other forms of tool failure for the development of new and improved materials. The life of tool can be brought to an end in various ways which may be classifed into two main groups.

1. Progressive tool wear machanisms.
2. Tool failure or damage bringing the tool to a premature end.

14.2 PROGRESSIVE TOOL WEAR MECHANISMS

The fundamental nature of the mechanisms of wear can be very different under different conditions. These wear mechanisms of a hard tool material in contact with a softer work material sliding past at a fairly high speed are discussed below:

14.2.1 Shearing at High Temprature

There are high temperatures in the cutting region. The strength and shear yield strength of hard tool material decrease at high temperature. The chip made of softer work material gets work hardened. The hardened chip can exert frictional stress on the tool and causes is shearing. This effect of plastic yielding and shearing increases with increase of temperature in the cutting region.

14.2.2 Diffusion Wear

Solid-state diffusion occurs where atoms in a metallic crystal lattice move from a region of high atomic concentration to one of low concentration. This process is a function of temprature and rate of diffusion increases exponenitally with increase in temperature. In metal cutting where intimate contact between the workpiece and tool material occurs and high temperature exists, diffusion can occur. Atoms may move from tool material to the work material. This process causes a weakening of the surface sturcture of the tool and is called diffusion wear.

14.2.3 Adhesion Wear (Attrition Wear)

The adhesion wear is caused by the failure of welded asperity junction between the two metals. In metal cutting, junctions between the chip and tool material are formed due to friction between the two. These junctions are fractured and small fragments of tool material are torn out and carried away on the underside of chip or on the new workpiece surface. In metal cutting new material surfaces free from oxide films and other contamination are continuously produced. This helps to form welded asperity junctions. There are good conditions for adhesive wear in metal cutting. When adhesion wear process continues for some time, the surface of tool material is nibbed away and becomes uneven.

14.2.4 Abrasion Wear

The hard particles on the underside of the chip pass over the tool face and remove tool material by mechanical action. This is called abrasion wear.

The hard particles required for abrasion wear can be:

(i) highly strain hardened fragments of an unstable built-up edge.

(ii) fragments of the hard tool material removed by adhesion wear,

(iii) hard constituents in work material.

(iv) oxide scales on the work surface.

These particles plough grooves into the surface of hard tool material.

14.2.5 Fatigue Wear

The fatigue wear is caused by variable thermal stresses or by cyclic mechanical stresses. The tool material is subjected to variable thermal stress by changing the temperature due to cutting fluid, chip breakage and variable dimensions of tool.

There is interlocking of asperities of tool surface and chip surface. The frictional stress can produce compressive stress on one side of asperity and tensile stress on the other side. Due to relative motion between chip and tool surface, there will by cyclic stress. This causes surface cracks resulting in crumbling of hard tool material. The phenomenon is called fatigue wear.

14.2.6 Electrochemical Effect

An e.m.f. is generated by seebeck effect when two junctions of a loop formed by two dissimilar materials are kept at different temperatures. The chip-tool interface acts as the hot junction of two dissimilar tool and work materials, a thermoelectric e.m.f. is set up. The current can assist the wear process at the rake face by aiding the diffusion of carbon ions from the carbide tool to the flowing chip.

14.2.7 Oxidation Effect

The chip and newly machined surface form abrasive oxides by reacting with atmospheric oxygen. These abrasive oxides form grooves and notches at the rake face and flank of the tool due to relative sliding motion between the tool and rusted surface. It has been found that this oxidation effect can be eliminated by using a jet of neutral gases (nitrogen, argon, etc.) into cutting zone.

14.2.8 Chemical Decomposition

There can be dissolution of bond between the binder and hard constituents of carbide tool. Localized chemical reactions take place. The weakened particles formed by chemical reaction can be easily torn away by the asperities at the underneath of chip or on newly machined surface.

14.3 TOOL FAILURE MECHANISMS

The damage caused by wear can be rectified by indexing and regrinding of tools. There is failure or damage of cutting tool resulting in premature end of tool life. The failure of cutting tool takes place in the following ways:

1. Wear on tool face and formation of a crater due to chip flowing along the face.
2. Wear on flank and formation of wear land due to rubbing action of newly machined surface.

3. Chipping of cutting edge of tool or insert due to excessive local stresses.

These damages cannot be rectified and the tool has to be thrown away.

14.3.1 Crater Wear

In metal cutting, the highest temperatures occur along some length of tool face. At high cutting speeds, these temperatures can be of the order of 1000°C or more. There is thermal softening and HSS tools wear very rapidly. In carbide tools, solid-state diffusion at these temperatures can cause rapid wear.

A crater is formed on the tool face taking the shape of chip under side. This factor determines the life of the cutting tool. The cratering becomes very severe and the tool edge is weakened and fractured.

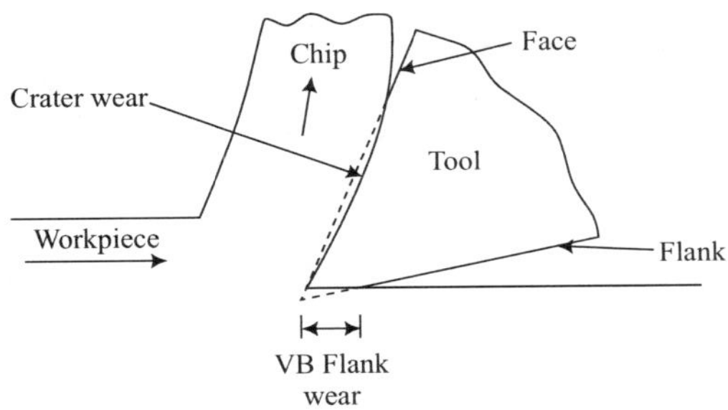

Fig. 14.1 Regions of tool wear.

14.3.2 Flank Wear

The flank wear of a cutting tool is caused by friction between the newly machined workpiece surface and contact area on the tool flank. The flank wear land is parallel to the resultant cutting direction due to rigidily of workpiece. The amount of wear is taken equal to the width of wear land. This can be measured with the help of a toolmaker's microscope.

A typical wear curve for a cutting tool is shown in Fig. 14.2.

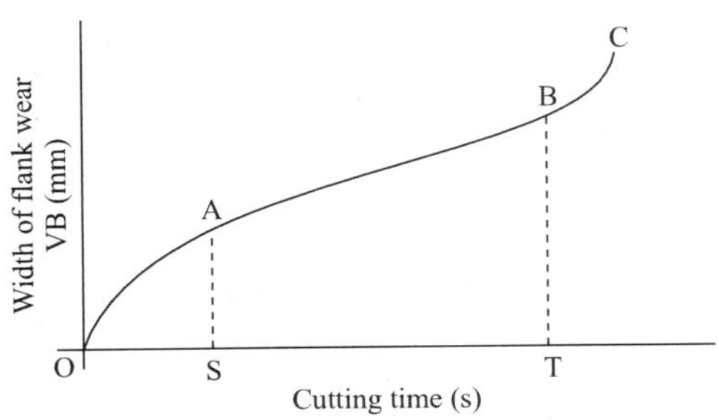

Fig. 14.2 Developmant of flank wear with time.

The curve can be divided into three regions.

1. Initial rapid wear (O-A): The sharp cutting edge is quickly broken down and a finite wear land is established.
2. Constant rate wear region (A-B): The wear progresses at a uniform rate.
3. Rapid wear region (B-C): The wear occurs at a gradually increasing rate. It is advisable to regrind or dispose of the tool before wear enters the region BC, where rapid breakdown occurs.

14.3.3 Chipping of cutting edge

Breaking away of small chips from the cutting edge of a tool or an insent takes place due to:

- Impact loads by intermittent cutting.
- Excessive plastic deformation due to large cutting forces.
- Transient thermal stresses due to sharp temperature gradients
- Localized cooling due to improper application of cutting fluid.
- Chatter from improper cutting conditions.
- Excessice crater and flank wear.

The cutting edge appears jagged and there are cavities or depressions in the wear land. If tool is not carefully ground or the insert is improperly brazed, fracture of tool may take place.

14.4 WEAR AND CHIPPING CHARACTERISTICS OF IMPORTANT TOOL MATERIALS

Different tool materials exhibit different characteristics towards wear and chipping of cutting edge.

14.4.1 High Speed Steel

(i) Low cutting speeds: The cutting temperatures are less than 700°C.

The adhesion and abrasion are main mode of wear. The adhesive wear is more rapid due to interrupted cuts and chatter.

(ii) High cutting speeds: The tool temperature above 700°C. HSS loses strength and deforms plastically under cutting loads crater wear is prominent due to diffusion at high temperatures.

14.4.2 Tungsten Carbide-Cobalt Tool Material (WC-Co)

HSS is strong in tension and can withstand localized tensile stresses caused by uneven chip flow and therefore, is strong against adhesion wear. However, it is less hard and weak against abrasion wear.

The behaviour of tungsten carbide tool is opposite in nature to HSS. At small cutting speeds and feed rates, the main wear machanism is adhesion and abrasion to some extent. At high cutting speed when temperature exceeds 850°C in cutting region the crater wear through diffusion is most prominent. WC-Co is liable to greater chipping and fracture failure due to low toughness as compared to HSS. The compressive deformation of cutting edge of a carbide tool limits the cutting speed, feed rate and strength of the material to be cut. For greater performance, carbide grades with lower cobalt content should be used.

14.4.3 Special Grades of Carbide Tool Material (WC-TiC-TaC-Co)

These cubic carbides have high compressive strength and hot hardness. These are more resistant to crater and flank wear and can be used at cutting speed two to three times used for WC=Co tools.

14.4.4 Cemented Oxide Tools (Al_2O_3)

Oxide tools have better high temperature strength and hardness properties as compared to carbides. They have negligible diffusion wear even at cutting speeds above 650 m/min. They are liable to premature failure due to chipping because of low toughness. A negative rake (more cutting edge thickness) helps to improve tool life.

14.4.5 Ceramic Tools

Ceramic tools have high toughness and high resistance to chipping. The problem of chipping can be further solved by the use of tools with negative rake and rigid machine tools. These are prone to crater wear and flank wear but much less than carbide tool.

14.4.6 Diamond

Diamond shows excellent resistance to wear. The sharp cutting edges of diamond chip easily due to lack of sufficient toughness.

Diamond is not used for machining steels, nickel and high strength alloys due to its rapid wear on account of degenerative transformation to graphite. This degeneration to graphite happens due to interaction with iron and atmosphere at temperatures above 730°C.

14.5 ANALYTICAL MODELS OF TOOL WEAR

The main tool wear processes can be expressed by mathematical equations.

14.5.1 Abrasion Wear

The volumetric rate of abrasive wear per unit time depends upon the number of abrasive particles in the chip and newly machined surface in contact with tool, rate of sliding and normal pressure.

The volumetric wear rate due is abrasion,

$$W_{ab} = k_1\, N.V.T$$

where, k_1 = constant of proportionality and is a function of normal presure, p

N = No. of wear particles at any given time and depends upon area of contact.

V = Relative velocity of sliding.

t = Time of abrasion

If p, n and V are constant

$$W_{ab} = k_{ab}.t$$

where, $$k_{ab} = k_1\, N.V$$

14.5.2 Adhesion Wear

(1) **Flank wear:** The volumetric flank wear during steady phase can be expressed as:

$$W = k_2 \frac{pl}{\sigma_0} \qquad ...(1)$$

where, k_2 = Constant of proportionality

p = Normal load

σ_0 = Compressive yield strength of work material (weaker material)

l = Sliding distance.

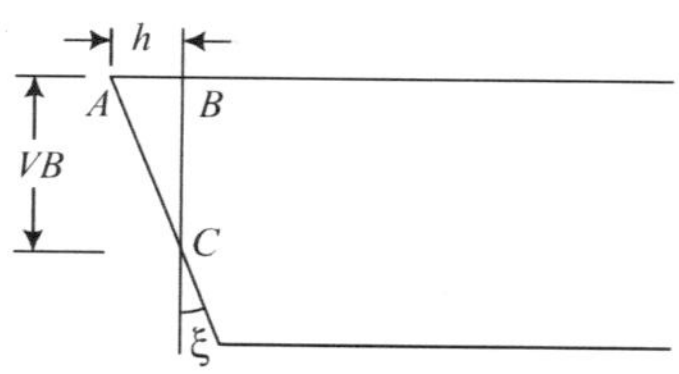

Fig. 14.3 Flank wear volume.

The constant k_2 is a function of mean number of contacting asperities per unit length and mean height of the wear particles. The number of contacting asperities increases and their mean height decreases with increase of normal load p. Therefore, k_2 may be assumed independent of p.

The volumetric wear may be expressed as:

$$W \cong \frac{1}{2} VB\, b\, h$$

$$\cong \frac{1}{2} b\, h\, VB^2 + \tan \xi \qquad ...(2)$$

where, b = width of cut

= width of wear mark or scar in a direction perpendicular to the plane.

VB = wear land on the flank

ξ = relief angle

$$W = \frac{1}{2} V_B bh = k_2 \frac{\rho l}{\sigma_0} \qquad ...(3)$$

Because rate of flank wear is constant during steady phase, therefore W is proportional to l. This is possible only if normal load p between the flank and newly machined surface is also proportional to VB.

$$\therefore \qquad P = k_3\, b\, VB$$

From equation (3)

$$VB = \frac{2 k_2 k_3 l}{\sigma_0 \tan \xi}$$

$$VB = k_{ad} . l$$

where, k_{ad} = constant depending upon tool-work pair.

(2) **Crater wear:** The wear volume,

$$W = \frac{1}{2} bR^2 (\theta - \sin \theta)$$

$$= \frac{1}{2} bR^2 \left[\theta - \left\{ \theta - \frac{\theta^3}{6} \right\} + ... \right]$$

$$= bR^2 \frac{\theta^3}{12} \qquad ...(1)$$

where, R = radius of curvature of crater

θ = included angle (rad)

$<< l$.

Assuming that the rate of wear is proportional to contact area

$$\therefore \qquad \frac{dW}{dE} = k_4\, b\, R\, \theta \qquad ...(2)$$

where, k_4 = constant of proportionality.

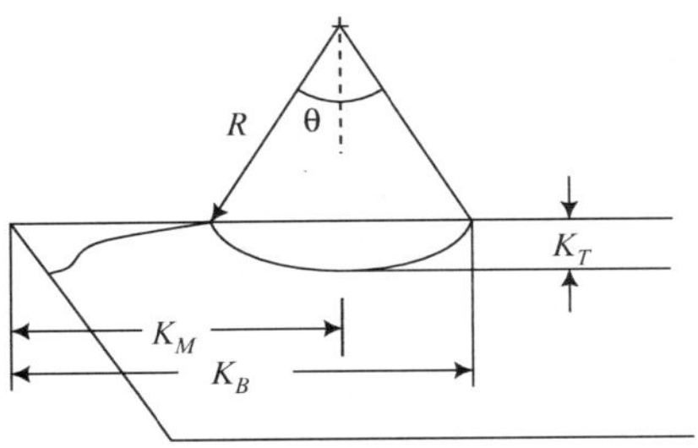

Fig. 14.4 Crater wear volume.

$$\therefore \qquad \frac{d}{dt}\left(\frac{1}{12}bR^2\theta^3\right) = k_4\, b\, R\, \theta \qquad ...(3)$$

R and θ are functions of time

Differentiating LHS of equation (3) and after simplification,

$$\frac{R\theta^2}{6}.\frac{dR}{dt} + \frac{R^2\theta}{4}.\frac{d\theta}{dt} = k_4\, R \qquad ...(4)$$

Multiplying both sides of equation (4) by $8R^{\frac{-2}{3}}$ and rearranging,

$$\frac{d}{dt}\left(\theta^2 R^{\frac{4}{3}}\right) = 8k_4 R^{\frac{1}{3}}$$

$$\therefore \qquad \theta^2 R^{\frac{4}{3}} = 8k_4 \int R^{\frac{1}{3}}\, dt + C \qquad ...(5)$$

C = constant of integrator
= 0.

The radius of curvature of the crater decreases exponentially with time (from experience)

$$\therefore \qquad R = at^n \qquad ...(6)$$

where, a and n are positive constants

Substituting equation (6) into equation (5)

$$\theta^2 a^{\frac{4}{3}} t^{\frac{-4n}{3}} = 8k_4 \int a^{\frac{1}{3}} t^{\frac{-n}{3}}\, dt$$

$$\therefore \qquad \theta^2 = 8k_4 \times \frac{t^{1+n}}{a\left(1 - \frac{n}{3}\right)} \qquad ...(7)$$

The depth of crater,

$$K_T = R\left(1 - \cos\frac{\theta}{2}\right)$$

$$\approx R\left[1 - \left(1 - \frac{\theta^2}{8}\right)\right]$$

$$= a\, k_{ad}\, t/8 \qquad ...(8)$$

where , $k_{ad} = 8k_4 \times \dfrac{1}{1 - \dfrac{k}{3}}$

The depth of crater wear increases linearly with time.

14.5.3 Diffusion Wear

(1) **Flank face:** Diffusion of constituents of tool material occurs at the contact land at the flank and also contact length at the chip-tool interface.

Fick's second law states that the rate of change of concentration, *dc*/*dt* of the diffusing constitunt is the proportional to the mass input in the *y*-direction. In case of cutting tool, the direction of diffusion is across the wear land at the flank. Mathematically.

$$\frac{dc}{dt} = D\frac{\partial^2 c}{\partial y^2} \qquad ...(1)$$

where, D = diffusion cofficient

$$D = D_o\, e^{\frac{-Q}{RT}} \qquad ...(2)$$

where, D_o = a constant

Q = Activation energy for the diffusion process.

R = Universal gas constant

T = Absolute temperature of flank

T is assumed constant over the flank leanth h_f, and diffusion cofficient D depends upon T.

The solution of the above equation gives the value of concentration c of the diffusing constituent in terms of t.

$$c = c_o\left[1 - erf\frac{y}{2\sqrt{Dt}}\right] \qquad ...(3)$$

where, c_o = concentration of diffusing constituent at the interface where $y = 0$

$$erf\frac{y}{2\sqrt{Dt}} = \text{error function}$$

$$= \frac{2}{\sqrt{\pi}}\int_0^{\frac{y}{2\sqrt{Dt}}} e^{-a^2}\, da$$

The diffusing flux, i.e., the diffusion mass per unit area per unit time in the *y*-direction as per Fick's Law,

$$J_y = -D\left(\frac{\partial c}{\partial y}\right)_{y=0} \times \tau_i$$

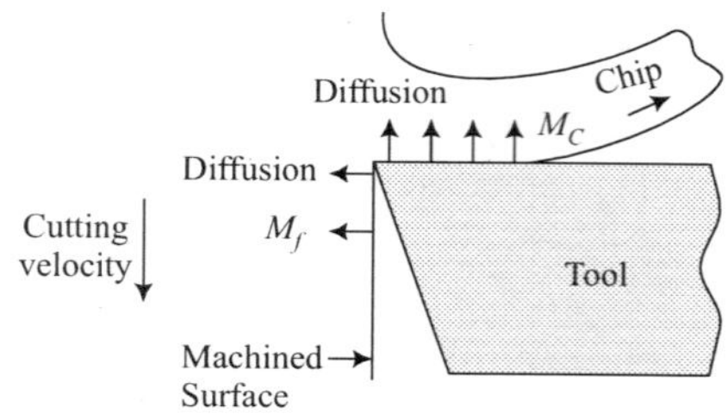

Fig. 14.5 Diffusion wear.

Differentating equation (3)

$$\left(\frac{\partial c}{\partial y}\right)_{y=0} = -2\frac{Co}{\sqrt{\pi D t}}$$

$$\therefore \qquad (J_y)_{y=0} = -2Co\sqrt{\frac{D}{\pi t}}$$

$\therefore$ diffusion mass across the flank face in time τ_f,

$$M_f = 2Co\; h_f b\sqrt{\frac{D\tau_f}{\pi}} \qquad \text{...(4)}$$

Substituting the value of D from equation (2)

$$M_f = 2Co\; hf b\sqrt{\frac{D\tau_f}{\pi}} \exp\left[1-\frac{Q}{2R\tau_f}\right]$$

The amount of diffused mass will directly depend upon flank wear land, h_f, the width of the cut b, time τ_f and flank face temperature, T_f.

(2) **Rake face:** The diffused mass across the chip-tool interface.

$$M_c = 2Co\; l_c\, b\sqrt{\frac{D_o\tau_e}{\pi}} \exp\left[1-\frac{Q}{2RT_c}\right]$$

where, T_c = Average temperature of rake face
l_c = length of chip-tool interface
τ_c = time of diffusion process.

The above relation for M_c is approximate due to the following reasons.

(i) temperature T_c over the rake face varies significantly over contact points.

(ii) M_c does not include loss of tool material post diffusional adhesion wear.

14.6 TOOL-LIFE CRITERIA

Tool life is the time for which a cutting tool/edge can be used without regrinding (HSS tool) or replacement (carbide and ceramic inserts). The bluntness of cutting edge if used beyond useful life results in the following:

1. Increase in cutting forces
2. Increase in tool temperature
3. Dimensional accuracay of component suffers
4. Surface finish of component suffers
5. Increase of rejections
6. Flank wear and crater wear increase exponentially
7. Total failure of the tool
8. Damage to the component and machine

Therefore, there should be certain tool-life criterion or tool failure criterion which should specify the maximum wear of the tool that can be tolerated before regrinding or changing it. There are two groups of tool-life criteria.

1. Direct
2. Indirect

14.6.1 Direct Criteria

It consists of direct measurement of tool wear or visual examination of the cutting edge and specifying limiting values.

(i) Width of wear land at the flank.
(ii) Maximum depth of crater.
(iii) Chipping and crack formation.

(1) *Width of wear land at the flank*

The wear of flank of the cutting tool is not uniform along the cutting edge. It is necessay to specify the locations and degree of the wear allowable before deciding on regrinding or disposal of the tool. The flank wear is greatest at the extremities of active cutting edge. The width of the flank wear land at the tool corner (zone C) is designated as VC. The width of the wear land at the opposite land of active zone called notch zone is designated as VN. In the central portion (zone B) the wear land is usually uniform. The average wear land width is designated as VB and maximum wear-land width is designated VB_{max}.

The criteria recommended by the ISO to define the effective tool life are as follows:

High-speed steel, ceramic tools and carbide tools

$$VB = 0.3 \text{ mm} \qquad VB_{max} = 0.6 \text{ mm}$$

(2) *Maximum Depth of crater*

The parameters of interest are:

- Max depth KT
- Location of max crater = KM

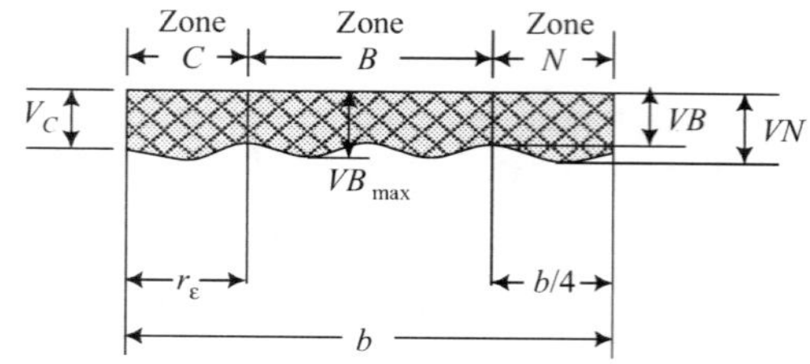

Fig. 14.6 View on major flank.

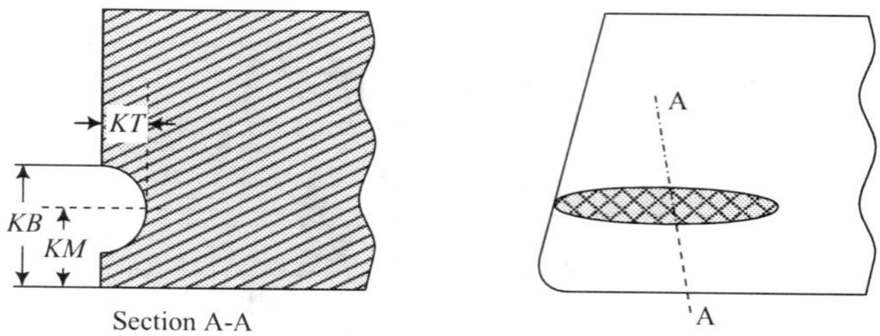

Fig. 14.7 Crater wear

ISO recommends the following value of crater depth as tool failure criteria,

$$KT = 0.06 + 0.3\,f$$

where, f = feed.

$$\frac{KT}{KM} = 0.2 \text{ to } 0.4$$

(3) *Chipping and crater formation*

Fine cracks may develop near the cutting edge of the tool or it way chip off shortly after the tool is put into operation. The reasons can be:

- Faulty sharpening
- Improper tool geometry
- Wrong selection of tool material
- Wrong selection of cutting conditions

If the life is very short and the tool has to be changed. The following steps are necessary to avoid such situation.

(i) Select a tough tool material
(ii) Use more high material
(iii) Use stiffer tool holder
(iv) Increase the side cutting edge angle
(v) Reduce the rake angle or even use negative rake
(vi) Reduce relief angles
(vii) Decrease depth of cut and feed.

14.6.2 Indirect Criteria

These are based on the effects produced by tool wear and chiping.

(i) *Surface Roughness*

The surface roughness of a workpiece increases with the uneven wear of the tool flank. The surface finish can be continuously monitored with the help of a portable surface analyzer and compared with its limiting value. Whenever the value of the surface finish deviates from limiting value, the tool is reground or insert replaced.

(ii) *Accuracy of dimensions*

The dimensional accuracy of a workpiece deteriorates as the tool gets worn out. The tool is said to have failed when the dimensional accuracy falls below a limiting value.

(iii) *Cutting forces*

A tool dynamometer can be used to monitor changes in cutting forces or rate of change of power consumption. When the increase in cutting force or power consumption exceeds a limiting value, the tool life is assumed to be exhausted.

(iv) *Rate of metal removal*

The metal removal rate by a cutting tool depends on the cutting speed and the depth of cut. The actual machining time and rate of metal removal directly depend upon the progress of flank wear. The limiting metal removal rate can be related to wear land of flank and can be used as tool failure criteria.

(v) *Tool failure*

The tool is said to have completely failed when there is appreciable rubbing between the machined surface and tool flank or cutting to edge has chipped off.

14.7 TOOL LIFE RELATIONSHIPS

Tool life is the cutting time required to reach a tool life criteria as discussed above. For a given work material, tool material and tool shape, the main factor affecting the tool life for a given machining operation is the cutting speed. Other machining parameters have only secondary effect.

14.7.1 Taylor Equation

Taylor gave the following empirical relation,

$$\frac{V}{V_r} = \left(\frac{t_r}{t}\right)^n$$

where, V = cutting speed

V_r = reference cutting speed given a known tool life t_r.

t = tool life

n = constant

The basic equation of Taylor can be written as:

$$Vt^n = c \qquad ...(1)$$

where, n = constant and is called tool life exponent

c = Taylor constant.

The constants n and c depend upon the tool and work materials, feed and depth of cut, type of coolant, tool geometry, etc. The value of n for different tool materials is:

Tool material	Value of n
1. High speed steel	0.15 – 0.20
2. Cemented carbide	0.30 – 0.49
3. Ceramics	0.50 – 0.70

The value of Taylor tool life constant c for different tool-work material combinations is given in Table 14.1.

Taylor'equation can be written as:

$$ln\ V + n\ ln\ t = ln\ c$$

This can be plotted as given in Fig. 14.8. From the graph n is 0.25 and reference speed V_r is taken as 1 m/s and t_r is 100 sec. The equation (1) can be written as:

$$Vt^{0.25} = 100^{0.25} = 3.16$$

This equation can be used for optimization of machining processes. This relationship between cutting speed and tool life holds good for a wide range of materials and cutting conditions. However, at very high cutting speeds, different values of n may be applicable.

Table 14.1 Value of Taylor tool life constant, c.

Work material	*Tool material*		
	HSS	*Carbide*	*Ceramic*
Carbon steel	40-100	100-160	2500
Cast iron	30-60	100-150	9000
Titanium	10-20	100-150	–
Nickle alloys	5-8	10-40	–
Niobium	40-60	150-200	–
Tungsten	120-160	400-600	–
Stainless steel	20-35	120-200	–
Molybdenum	–	150-250	–
Monel metal	30-50	–	–

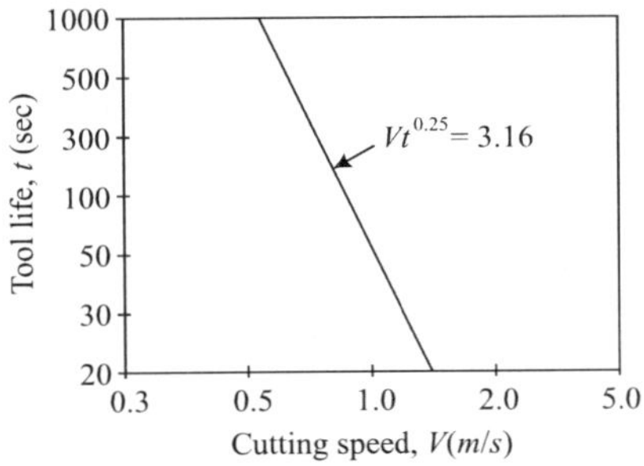

Fig 14.8 Relationship between tool life and cutting speed.

Taylor generalized the basic equation to include the effects of feed f and depth of cut d. Taylor's extended equations are:

$$t = \frac{c}{V^{\frac{1}{n}} . f^{\frac{1}{n_1}} . a_c^{\frac{1}{n_2}}} \qquad ...(2)$$

$$V = \frac{c}{t^n \, f^{n_1} \, d^{n_2} \left(\frac{BHN}{200}\right)^{n_3}}$$

where, n, n_1, n_2, n_3 are constants which must be established empirically through well-designed set of cutting tests.

14.7.2 Colding's Equation

Colding assumed that tool life is directly related to tool temperature. The tool temperature is controlled by chip equivalent q. For a lathe tool, the chip equivalent

$$q = \frac{\text{Engaged cutting edge length}}{\text{Plane area of cut}}$$

$$Vt^n = c_2(q)^m$$

where, n and m are exponents and c_2 is life constant.

Example 14.1 Determine the tool life of HSS tool for turning a mild steel bar at a cutting speed of 40 m/min. The tool life equation is:

$$Vt^{0.2} = 80$$

If tool life is 60 min, find the cutting speed.

Solution:

(i) $V = 40$ m/mm

$\therefore \quad 40t^{0.2} = 80$

$t = 32$ min.

(ii) For t = 60 min,

$$V(60)^{0.2} = 80$$

$$\therefore \quad V = 35.27 \text{ m/min}$$

Example 14.2 If Taylor's tool life exponent $n = 0.50$ and constant $c = 400$, what will be the percentage increase in tool life when cutting speed is reduced by half?

Solution: The Taylor's equation is

$$Vt^n = c$$

$$V_1 (t_1)^{0.5} = 400$$

$$V_2 (t_2)^{0.5} = 400$$

$$\frac{V_1}{V_2}\left(\frac{t_1}{t_2}\right)^{0.5} = 1$$

$$\therefore \quad \frac{t_1}{t_2} = (2)^2 = 4$$

$$\therefore \quad \frac{t_1 - t_2}{t_1} = \frac{4t_2 - t_2}{4t_2} \times 100 = \mathbf{75\%}$$

Example 14.3 Mild steel bars of 50 mm dimater have to be machined by HSS tools over a length of 160 mm. The cutting conditions are as follows:

Feed,	$f = 0.2$ mm/rev
Speed,	$N = 230$ rpm
Depth of cut,	$d = 1.5$ mm

Determine how many components can be machined before regrinding the tool if Taylor's tool life extended equation is

$$Vt^{0.2} f^{0.3} d^{0.12} = 50$$

Solution: Cutting speed, $V = \dfrac{\pi DN}{1000} = \dfrac{\pi \times 50 \times 230}{1000} = 36.13$ m/min

Substituting the values in tool life equation

$$36.13t^{0.2}(0.2)^{0.3} (1.5)^{0.12} = 50$$

$$\therefore \quad t = 44.5 \text{ min}$$

Time required for turning one job $= \dfrac{l}{fN} = \dfrac{160}{0.2 \times 230} = 3.48$ mm

$\therefore$ No. of components $= \dfrac{44.5}{9.78} = 12.78 = \mathbf{12}$ **Ans.**

It is not advisable to change the tool in the middle of the cut.

14.8 EFFECT OF PROCESS PARAMETERS ON TOOL LIFE

The wear mechanisms and tool life will depend upon tool material, work material, tool geometry, machining parameters and other cutting conditions.

14.8.1 Properties of Tool Materials

The following tool material properties help to increase tool life.

1. High hot-hardness is a very desirable property especially for high speed machining when high temperatures are occurring in the cutting region and on rake and flank faces. This gives resistance to plastic deformation, adhesion and abrasion mechanisms.
2. Toughness is desirable for taking heavy cuts, rough and interrupted cutting. Toughness helps to resist sudden loads during above machining conditions.
3. Wear resistance.
4. Lack of chemical affinity between tool and work materials.
5. High thermophysical properties such as thermal conductivity, specific heat and low coefficient of thermal expansion can resist thermal stress and thermal shocks.

14.8.2 Tool Coatings

A hard wear resistant coating is applied on a basic tool material with high toughness. This technology enhances the tool life considerably. The coatings are applied in single layer or multi-layers.

Examples: HSS tools are coated with titanium nitride; Tungsten carbide tools are coated with a TiG, Al_2O_3 TiN.

Two layer coatings are $T_i + T_iN$ or $T_iC + Al_2O_3$.

14.8.3 Work Material

The following properties of work material help to increase tool life.

1. Low strength and hardness. The cutting forces, cutting temperatures and abrasion wear are reduced.
2. Absence of surface scale, sand, slag and other hard and abrasive inclusions.
3. Inclusions of lead acts as boundary lubricant. Presence of sulphur acts as stress raiser thereby reducing cutting forces and temperature.
4. The cutting forces temperature and abrasion wear are reduced if work material does not have tendency to work harden.
5. In cast iron, tool life increases due to absence of iron carbide and presence of free graphite.
6. In carbon steel, spheroidal pearlite helps to increase tool life whereas lamellar pearlite adversely affects tool life.

14.8.4 Speed, Feed and Depth of Cut

The increase of these cutting parameters have direct effect on the tool life.

(a) *Cutting speed*

Tool temperature increases with cutting speed. The strength and hardness of tool material falls. This helps in abrasive, adhesive and diffusional wear. The cumulative effect is an exponential decrease in tool life.

(b) *Feed*

Increase of feed results in increase of cutting force per unit area at rake face as well as flank. There is increase of temperature and hence all types of wear increase. There are more chances of chipping of tool edge.

(c) *Depth of cut*

The chip-tool contact area increases with the depth of cut. There is small increase of tool temperature and small increase of abrasive, adhesive and diffusional tool wear.

All the alove cutting conditions increase cutting forces. The chipping and fatigue wear tend to increase due to mechanical shock.

14.8.5 Tool Geometry

All the geometrical dimensions of the tool, i.e., rake angles, cutting edge angle, relief angles and nose radius affect the tool life to varying extent.

(a) *Rake angles*

The effective rake angle (α_e or α_v) can be increased by increasing the side rake angle or back rake angle or both. The cutting force, temperature and tool wear decrease with the increase of rake angle. Therefore, tool life improves. On the other hand, with larger rake angle, the cutting edge becomes sharp. Its mechanical strength decreases resulting in chipping of tool edge. Therefore, there is an optimum value of rake angle for every tool-work material combination. In Fig. 14.9 the optimum rake angle is 14° when cutting high-strength steel with high-speed steel tool. Recommended normal rake angle for roughing operation with different tool-work material combinations are shown in Table 14.2.

(b) *Relief angles*

Large tool life is obtained when the relief angle is increased as the width of flank wear land is reduced. But the mechanical strength of cutting edge is reduced and is liable to chip. For maximum tool life, there is optimum relief angle. See Fig. 14.10

(c) *Side cutting edge angle*

When the side cutting edge angle is increased, the following effects are produced:

(i) Chip equivalent (q) increases

(ii) Area of chip-tool contact increases

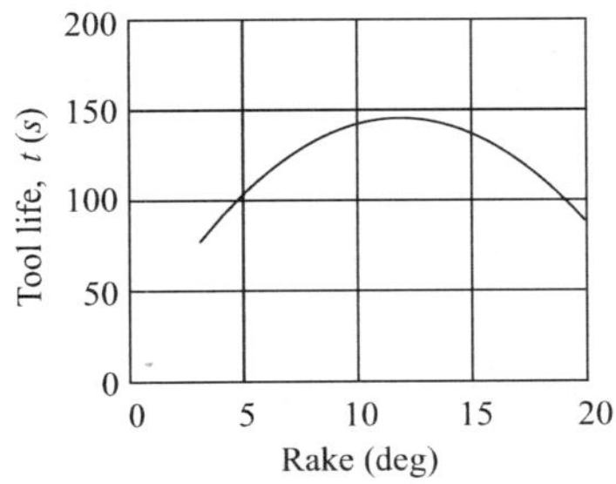

Fig. 14.9 Effect of tool rake on tool life.

Table 14.2 Recommended normal rake angle (degrees) for roughing operation.

Work material	*Tool material*	
	HSS	*Carbide*
1. Cast iron, cast brass	0	0
2. Brass and bronze	8	3.5
3. Soft brass and HSS	14	3.5
4. Mild steel	27	3.5
5. Light alloys	40	13

(iii) Cutting forces don't change much

(iv) Temperature decreases

(v) Rate of wear decreases

(vi) Tool life increases

(vii) Chances of chipping decreases

(d) *Nose radius*

The chip equivalent (q) and tool life increase with increase of tool radius.

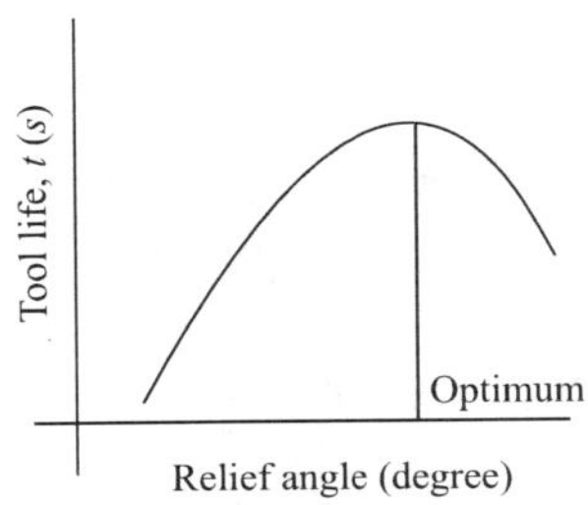

Fig. 14.10 Effect of relief angle on tool life.

14.8.6 Cutting Fluid

The cutting fluid when supplied copiously to HSS tools which have low hot hardness can enhance the tool life appreciably as follows:

(i) Cooling the chip and workpiece

(ii) Reducing frictional stress at tool-work and tool-chip interfaces

(iii) Decreasing temperatures in cutting region

(iv) Tool life increases.

On tools with high value of hot hardness like carbides and oxides, the cutting fluid has negligible effect on life.

14.8.7 Machine-tool Rigidity

The machine, tool and workpiece form a system of nodal stiffness and vibration damping. The dynamic rigidity of the system should be high to ensure low vibration or chatter. The chatter can cause fatigue failure and complete destruction of tool. Therefore, machine should be designed properly. The clamping of workpiece and tool should be rigid.

14.8.8 Intermittent Cutting

Every time a tool enters or exits from cutting, there will be impact loading of cutting edge. If the tool material is hard and brittle, the tool edge may chip and tool life is shortened. For such machaning, tools with high toughness should be used.

14.8.9 Built-up Edge

A built-up edge is formed in certain tool-workpiece combination under some operating conditions. The built-up edge is in work hardened state and sticks to the tool.This built-up edge starts acting as a cutting edge and increases the normal rake angle of tool. The cutting forces, temperatures and wear are reduced.

When the built-up edge grows in size, it becomes unstable and break-up occurs. The fragments get rolled on to the rake face and flank face and severely abrade the respective surfaces. Net effect of built-up edge is to increase flank wear and decrease crater wear.

14.9 MEASUREMENT OF TOOL LIFE

Tool life is the time elapsed between two successive grindings of a cutting tool. Tool life may be measured in the following ways:

1. Number of pieces machined between two sharpenings
2. The actual time the tool is in contact with the workpiece
3. Total time of operation
4. Equivalent cutting speed
5. Volume of material removed between tool sharpenings.

In practice, it is more profitable to assess the tool life in terms of the volume of metal removed because the wear is related to the area of the chip flowing over the tool surface.

The volume of metal removed from the workpiece between tool sharpening for a given depth of cut, feed and cutting speed can be estimated as follows:

$$\text{Cross-sectional area of chip} = df\ (\text{mm}^2)$$

where, d = depth of cut (mm)

f = feed (mm/rev)

$$\text{Length of chip in one revolution} = \pi D\ (\text{mm})$$

where, D = diameter of workpiece (mm)

$$\text{Volume of metal removed/rev} = \pi Dfd\ (\text{mm}^3)$$

$$\text{Volume of metal removed during tool life} = \pi DfdNT\ (\text{mm}^3)$$

where, N = speed of rotation of workpiece (rpm)

t = tool life (min)

$$\text{Cutting speed, } V = \frac{\pi DN}{1000}\ (\text{m/min})$$

$$\therefore \qquad \text{Tool life} = 1000\, fd\, Vt\ (\text{mm}^3/\text{min})$$

14.10 TOOL LIFE TESTS

Tool life data is required for the following reasons:

1. For optimization of cutting conditions, i.e., cutting speed, feed and depth of cut.
2. For optimum tool geometry
3. For performance evaluation of tool material, cutting fluid, work material and machine tool.

Large amount of data is available on tool life but is insufficient for all cases of work material, tool material, machines, etc. Tool life tests have to be carried out for specific cases. The tool life tests can be grouped as:

1. *Conventional tests:* Where data is selected from handbook and exponential constants are worked out by varying one machining parameter at a time. The work material, tool material, tool geometry and cutting fluids are tentatively selected. The tool life is measured on the basis of width of flank wear land or maximum depth of crater as a function of either cutting speed or depth of cut or feed.
2. *Accelerated tests:* The following tests are carried out to estimate the tool life quickly

   (i) *Extrapolation of steady wear rate*

   This test is carried out for a limited time and rate of wear is extrapolated from steady wear rate data.

(ii) *High speed tests*

The tests are carried out at higher cutting speeds so that tool life is shortened. Taylor's constants are found out from test data and same are applied to working speed to find out tool life.

(iii) *Variable speed tests*

Multiple machining passes are made on the sample to save cost. Taylor's tool life relationships are used. Variable cutting speed tests are carried out for facing, multi-pass turning and taper turning.

14.11 MACHINABILITY

The term machinability is applied to workmaterial to describe the machining properties. The "ease" with which a given material may be machined with a cutting tool is machinability. Machinability depends upon:

1. Chemical composition of workpiece
2. Micro-structure
3. Mechanical properties
4. Physical properties
5. Cutting conditions

In general, high hardness of work material will give poor machinability because power consumption and temperature and hence tool-wear rates will be high. However, some materials tend to adhere to the working surfaces of cutting tool resulting in high friction, high cutting forces and high tool wear rates. Even soft pure iron has as low machinability as very tough alloy steels because of its tendency to adhere to cutting tool. Free machining metals such as sulphur, lead and tellusium are added to steels and non-ferrous metals to impure their machinability. However, these metals cannot be added to high temperature metals.

14.11.1 Machinability Criteria

The machinability may be evaluated by considering the following criteria.

1. Tool life
2. Power consumption
3. Surface finish
4. Form and size of chips
5. Temperature in cutting region
6. Metal removal rate
7. Rate of cutting under a standard force

The main factor to be selected for evaluation of machinability will depend upon the type of operation and machining requirements. Tool wear and surface finish criteria are most important for finishing processes whereas tool wear and power consumption are important factors for roughing operations.

14.11.2 Tool Life

This factor is most important for machining economies. If a work material creates more rapid tool wear, the life of the tool will be short. Machining cost will involve increased cost of tools and greater production time lost during tool changes.

Machinability is not a basic standard but a relative one. A convenient machinability criterion based on tool life is called specific cutting speed. It is the cutting speed corresponding to a predetermined tool life.

Example: If the tool life during a turning operation under standardized conditions of feed, depth of cut, tool material and geometry is found to be 60 min at a cutting speed of 100 m/min. The specific cutting speed for 60 minute tool life.

$$V_{60} = 100\text{m/mm}$$

The machinability rating or relative machinability of test material

$$M_R = \frac{V_{60}^t}{V_{60}^s} \times 100(\%)$$

where, V_{60}^t = specific cutting speed by test material for 60-minute tool life.

V_{60}^s = specific cutting speed of a standard material for 60-minute tool life.

A free-cutting steel which is machined relatively easy has a machinability rating of 100% which is fixed arbitrarily. This standard steel has the following composition.

$$C = 0.13 \text{ max}$$

$$Mn = 0.06 \text{ to } 1.10$$

$$S = 0.08 \text{ to } 0.03$$

Machinability ratings based on specific cutting speed for different work material have been worked out.

14.11.3 Power Comsumption

The specific cutting energy or specific cutting pressure or specific cutting power is taken as index of machinability.The specific cutting energy is taken is kW/cc/mm. The machinability index is inversely proportional to specific cutting energy. The specific cutting energy for different engineering materials and for important machining operations are available in literature.

14.11.4 Surface Finish

In cases where components are rejected on the basis of poor surface finish the surface finish has to be used as machinability index. A more easily machinable work material will take better surface finish under specific machining conditions.

14.11.5 Miscellaneous Criteria

The machinability of different work materials can be assessed on the basic of following factors:

1. Temperature developed in cutting region.
2. Penetration test of a dulling operation, i.e., depth of a hole cut by a standard drill rotating at standard speed and downward thrust in a given time.
3. Sawing tesk i.e., depth of a cut produced by a power handsaw on a specified bar under standard sawing conditions of speed and downward pressure.
4. Machinability index has been corelated with a combination of physical and thermal properties, hardness i.e., thermal diffusivity, ductility, specific heat, etc.

The term machinability has little meaning except in loose quantitative sense. However, as seen above, many attempts have been made to obtain a quantitative measure of machinability i.e., machinability index or number. A certain group of materials may be arranged in one order of machinability on a tool wear basis but will be in different order if surface finish or power consumption criteria are applied. Therefore, these results give some guidance. For example, these results can be used to measure the variation of some machining property of materials having same specifications.

QUESTION BANK NO. 14

1. What is meant by machinability? Explain the methods of representing machinability.
2. What is meant by tool life? State Taylor's equation for tool life and explain the symbols used.
3. Discuss the different tool wear mechanisms.
4. Explain the following type of tool damages:
   (i) Flank wear
   (ii) Crater wear
   (iii) Chipping of cutting edge
5. Compare the performance of HSS, carbides and oxides againest flank wear, crater wear and chipping.
6. Explain the main type of tool life criterion.
7. Explain the effect of various process parameters on tool life.
8. Dicuss the causes of variability of constant n and c of the tool life equation.

15

THE GRINDING WHEEL

Introduction: Abrasives; Bonds & bonding process, Grain type, Size and wheel structure; Wheel shapes and sizes; Designation of grinding wheel. Selection of grinding wheels; Mounting of grinding wheel, Dressing and truing of grinding wheel; Testing of grinding wheel, Question Bank.

15.1 INTRODUCTION

A grinding wheel consists of a large number of abrasive particles held together by a suitable material. The abrasive particles are called *grains* and the holding material is called *bond*. It has cutting action like a multi-point tool (like a milling cutter). The cutting points are irregularly shaped and randomly distributed over the active face of the wheel. The cutting speeds in grinding are much higher than other machining processes. The wheel may be in one piece or made up of segments of abrasive blocks built up into a solid wheel. It is mounted on an adapter suitable for the grinding operation.

15.2 ABRASIVES

The abrasives used for grinding and polishing should have uniform physical proportion for manufacturing grinding wheels. These should have high hardness, toughness and resistance to fracture. There are two types of abrasives:

1. *Natural* substances such as sandstones, quartz, emery, corundum and diamond. The natural abrasives have impurities which impair uniformity of physical properties. These abrasives are not very popular for manufacturing of grinding wheels.
2. *Artificial* substances include silicon carbide and aluminium oxide.
   (i) *Silicon carbide* (*SiC*) The composition of raw materials for manufacturing silicon carbide abrasive is as follows:

			Function
1.	Silicon sand	: 56 parts	Provide Si
2.	Powdered coke	: 34 parts	Provides *C*
3.	Salt	: 12 parts	Fusion
4.	Sawdust	: 12 part	Porosity

Rectangular electric resistance furnaces are used to heat the mixture. The gases are allowed to escape through joints of the bricks of the furnace.

These wheels are designated by the letter '*S*'. There are two types of silicon carbide abrasives.

Green grit : 77% Si C

Black grit : 95% Si C

These abrasives have hardness lower than diamond and aluminium oxide abrasives. These are used as low strength materials for grinding of cemented carbides, stone and ceramic materials, gray cast iron, brass, bronze, copper, aluminium, vulcanized rubber, etc.

(ii) *Aluminium oxide* (Al_2O_3): Mineral bauxite, ground coke, iron borings are mixed together and heated in electric arc furnace. The mineral bauxite is a hydrated aluminium oxide clay containing silica, iron oxide, titanium oxide, etc.

Aluminium oxide is hard, tough and has high fracture resistance. It is used for grinding carbon steels, high speed steels, wrought iron, annealed malleable iron, tough bronzes, etc. The wheels are designated by letter "*A*".

Following companies are manufacturing silicon carbide and aluminium oxide under their trade names.

Table 15.1 Trade names of abrasives.

Sl no.	*Manufacturing company*	*Trade Name*	
		Silicon carbide 'S'	*Aluminium oxide 'A'*
1.	The Carborundum Co	Carborundum	Aloxide
2.	The Norton Co.	Crystol	Alundum
3.	Maklin Co.	Silicon carbide	Aluminium oxide
4.	Abrasive Co.	Electrolon	Borolon

15.3 BONDS AND BONDING PROCESSES

Adhesive substances are used to hold abrasive grains together in the form of grinding wheels or sharpening stones. These bonds may be used either to hold silicon carbide or aluminium carbide. The different bonding materials are used to produce different products as shown in Table 15.2.

Table 15.2 Different Bonding Material and Products.

Sl no.	*Bond*	*Product*
1.	Vitrified bond	vitrified grinding wheels
2.	Silicate bond	silicate wheels
3.	Shellac bond	elastic wheels
4.	Resinoid bond	resinoid wheels
5.	Rubber bond	vulcanized wheels
6.	Oxychloride bond	oxychloride wheels

The six bonding processes are as follows.

1. **Vitrified bonding process:** Vitrified wheels are manufactured by bonding clay and abrasive grains. The clay and abrasive grains are mixed together thoroughly with sufficient water to get a uniform mixture. The fluid mixture is poured into moulds and left to dry. The dried material is cut and trimmed to suitable shape and size. The cut pieces are burnt in a kiln. The clay fuses (vitrifies) and forms a porcelain which is a glassy substance to surround and connect the abrasive grains. The abrasives are annealed at high baking temperature.

   More than 75% of grinding wheels are manufactured by this process and vitrified wheel is designated by letter "*V*".

   1. These wheels have good strength
   2. These are porous and used for high stock removal with cool cutting.
   3. These are affected by heat, cold water and acids.
   4. These are sensitive to impact.
   5. These have low bending strength.

2. **Silicate bonding process:** Silicate of soda or water glass is mixed with abrasive grain and poured in moulds to dry. The moulded shapes are backed in a furnace at 260°C for ten days. A silicate bonded wheel is designated by letter "*S*".

   The silicate bond is waterproof, grains are not annealed and are easily released. These wheels are used for grinding edged tools and other operations without a coolant.

3. **Shellac bonding process (Elastic bonded wheels):** The abrasive grain and shellac are mixed in heated containers, rolled or pressed in heated mould. The moulded shapes are backed at 150°C for a few hours. A shellac bonded wheel is designated by letter "*E*".

   This bond has considerable strength and elasticity. These wheels are used for light cuts. It is used for cool cutting of thin sections of hardened steel. It is also used for finishing of chilled iron, cast iron, steel rollers, hardened steel cams, aluminium pistons and abrasive cutting of thin sections.

4. **Resinoid bonding process:** Synthetic resins (Bakelite, Redmanol) are mixed with abrasive grains, placed in moulds and heated at 200°C. The resin are set and hold the abrasive grains in wheel form. The resinoid bond wheel is designated by the letter '*B*'.

   These wheels are strong and are used for high speed grinding and removal of stock rapidly. These are used for grinding cams and rolls requiring high finish.

5. **Rubber bonding process:** Abrasive resins, rubber and sulphur are mixed, rolled into sheets. The wheels are punched out of sheet on a punch press and vulcanized. A rubber bonded wheel is designated by letter "*R*".

Rubber bonded wheels are very resilient, dense but less heat resistant. These are strong and tough and very thin wheel can be made. These are used for producing good finish.

6. **Oxychloxide bonding process:** Abrasive grains are mixed with oxide and chloride of magnesium. The bonding process is similar to that used for vitrified bonded wheel. These wheels are designated by the letter "*O*". Wheels and wheel segments used for dirc-grinding operation are manufactured by oxychloride bonding process. The bond ensures a cool cutting action and is used for dry grinding.

15.4 GRAIN TYPE, SIZE AND WHEEL STRUCTURE

15.4.1 Grits

The size of the abrasive grains is indicated by a grain or grit size. Grinding is a true cutting operation and the grain size also indicates the size of the cutting teeth. Grain size is the mesh size, i.e., the number of meshes per linear inch (25.4 mm) of a sieve through which the grains pass when graded after crushing. The common grain sizes and types used for the manufacture of grinding wheels are given in Table 15.3.

Table 15.3 Common grain size and type.

Sl. No	*Grain type*	*Grain size*						
1.	Coarse	10	12	14	16	20	24	–
2.	Medium	30	36	46	54	60	–	–
3.	Fine	80	100	120	150	180	–	–
4.	Very fine	220	240	280	320	400	500	600

The grain size used in a grinding wheel depends on the amount of metal to be removed, the desired finish and hardness of work material.

Example: Coarse wheels are used for fast material removal. Fine wheels are used for soft ductile materials. These should also be used for hard and brittle work-material.

15.4.2 Grade of Grinding Wheels

The grade of grinding wheels does not indicate hardness of abrasive grain but the tenacity and strength of bond to hold the cutting edges or grains in place. The softest grade starts from letter "*A*" and hardest grade ends with letter "*Z*". The soft or hard grade indicates the strength of the bond to resist the break away of grains. A hard wheel holds the grains more securely. The grade of grinding wheels are given in Table 15.4.

Table 15.4 Grade of grinding wheels.

Sl. no.	*Wheel grade*	*Designation*									
1.	Soft	A	B	C	D	E	F	G	H	–	–
2.	Medium	I	J	K	L	M	N	O	P	–	–
3.	Hard	Q	R	S	T	U	V	W	X	Y	Z

Hard wheels should be used for soft material and soft wheel for hard materials.

The grade of grinding wheel is selected on the basis of hardness of work material, the arc of contact, speed of the wheel and workpiece and grinding conditions.

15.4.3 Structure of Grinding Wheels

The abrasive grains are distributed through the bond and the structure of spacing is measured by the cutting edges per unit area of the wheel face. This indicates the number and size of void spaces between the grains. The structure can be open or dense. The structure numbers are given in Table 15.5.

Table 15.5 Structure of grinding wheels.

Sl. No.	*Type*	*Number*							
1.	Dense	1	2	3	4	5	6	7	8
2.	Open	9	10	11	12	13	14	15	Higher

Open structure is useful for heavy cuts on soft, tough and ductile materials whereas dense structure should be used for hard and brittle materials. The selection of structure depends upon chip clearance to be provided, hardness of work material, condition of grinding operation and finish required.

15.5 WHEEL SHAPES AND SIZES

Grinding wheels are manufactured in wide variety of shapes and sizes to suit the type of grinding operation and the grinding machine. The main shapes of wheels are.

1. **Straight-side grinding wheels:** These are used for cylindrical, centreless and surface grinding operations. The size of the wheel is designed by diameter and width of the wheel face. The selection of size will depend upon class of work, size and power of the grinding machine. The shapes of the wheel are also designated by a number.
2. **Tapered face grinding wheel:** This shape of wheel is used for grinding of the threads and gears, etc.
3. **Cylinder wheel:** The end face of the wheel is used for grinding a flat surface.

4. **Cup wheel:** This type of wheel is used to produce flat surface by traversing the workpiece past the face of the wheel. The flaring cup wheel is used in the tool room.
5. **Dish wheel:** The small thickness of wheel is used to grind narrow surfaces in tool room.

   Saucer wheel is used for sharpening the teeth of circular and band saws.
6. **Segmented wheels:** These are used on vertical spindle rotary and reciprocating table surface grinders and way grinders.
7. **Mounted wheels:** These are small wheels (less than 50 mm diameter) mounted permanently on a mandrel or spindle by cementing. The mounted wheels are used very carefully to avoid springing of spindle due to excessive pressure.

   The wheels have different types of ends.

The different shapes of grinding wheels are given in Table 15.5 along with numerical shape number. The main dimensions of a grinding wheel are: outside diameter, bore diameter and wheel width.

15.6 DESIGNATION OF GRINDING WHEELS

Grinding wheels are manufactured with a wide range of characteristics. In order to aid in selection of a proper wheel for a given application, ISO standard marking system has been developed. Each wheel is designated by seven alphanumeric symbols in succession as follows:

1. Type of abrasive
2. Nature of abrasive
3. Grain size
4. Wheel grade
5. Structure
6. Nature of bond
7. Manufacturer's record

ISO standard marking system for grinding wheels using aluminium oxide or silicon carbide abrasives is shown in Table 15.6.

15.7 SELECTION OF GRINDING WHEELS

There are national and international standards available as recommendations or guide for the selection of grinding wheels for different applications. The following ISO guidelines can be used for selecting a grinding wheel with aluminium oxide or silicon carbide grains.

1. Aluminium oxide for steels and silicon carbide for carbides and nonferrous metals.

Table 15.5 Different shapes of grinding wheels.

Sl. no.	*Type no.*	*Designation*	*Sketch*
1.	1	Straight side	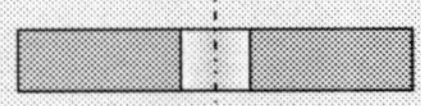
2.	2	Cylindrical	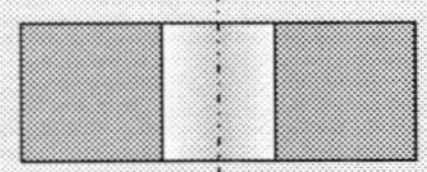
3.	4	Tapered face	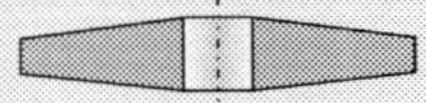
4.	5	Straight side with one side recess	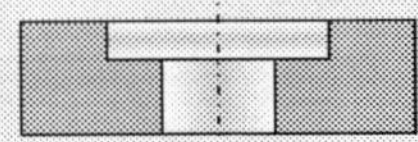
5.	6	Straight cup	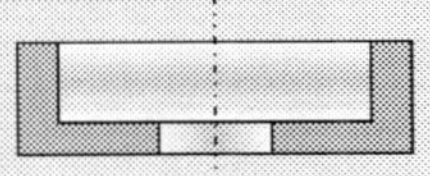
6.	7	Straight side with both side recesses	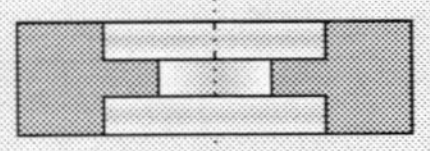
7.	11	Flaring cup	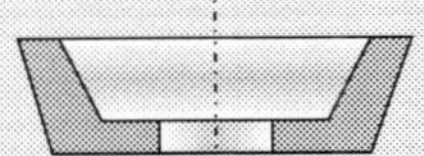
8.	12	Dish	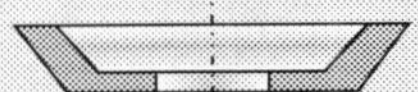
9.	13	Saucer	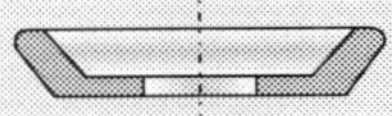
10.	–	Segmented	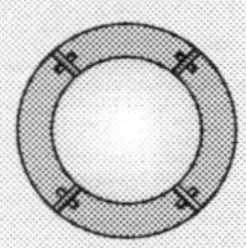
11.	–	Mounted	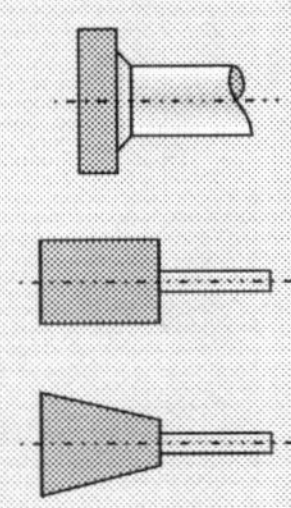

Table 15.6 ISO Standard Marking System for Aluminium, Oxide and Silicon Carbide Grinding Wheels*.

Sequence	1	2	3	4	5	6
Type of abrasive	Nature of Abrasive	Grain Size	Grade	Structure	Nature of Bond	Manufacturer's Record
51	A	36	L	5	V	23
Manufacturer's symbol indicating exact type of abrasive (optional)	Aluminum Oxide—A Silicon Carbide—C					Manufacturer's private marking to indentify wheel (optional)

Grain Size:

Coarse	Medium	Fine	Very Fine
10	30	70	220
12	36	80	240
14	46	90	280
16	54	100	320
20	60	120	400
24		150	500
		180	600

Structure:

Dense to open	
0	8
1	9
2	10
3	11
4	12
5	13
6	14
7	Etc.

Nature of Bond:

V—vitrified
S—silicate
R—rubber
RF—rubber reinforced
B—resinoid
BF—resinoid reinforced
E—shellace
Mg—magnesia

Grade Scale

Soft	Medium	Hard

ABCDEFGHIJKLMNOPQRSTUVWXYZ

* The sample grinding wheel designation shown is 51A36L5V23.

2. Hard-grade wheel for soft materials and soft-grade wheel for hard materials.
3. Large grit for soft and ductile materials and a small grit for hard and brittle materials.
4. Small grit for a good finish and large grit for maximum metal removal rate.
5. Resinoid, rubber, or shellac bond for a good finish and a vitrified bond for maximum metal removal rate.
6. Do not use vitrified bond for surface speeds greater than 32 m/s.

The Indian standard, 1S: 1249-1958 gives recommendations for selection of grinding wheels for different applications. The selection factors can be grouped as follows.

A. **Constant factors**
   1. Material to be removed
   2. Amount of stock to be removed
   3. Area of contact
   4. Type of grinding machine

B. **Variable factors**
   1. Wheel speed
   2. Work speed
   3. Condition of machine
   4. Personal factor.

15.7.1 Material to be Ground

This influences the selection of

(i) Abrasive
(ii) Grain size
(iii) Grade
(iv) Structure
(v) Bond

The general guidelines are:

1. Aluminium oxide for high strength materials and silicon carbide for lon strength materials.
2. Hard wheel for soft materials and soft wheel for hard materials
3. Fine grain for hard and brittle materials and coarse grain for soft and ductile material.
4. Closer spacing for hard and brittle materials and wide spacing for soft and ductile materials.
5. The bond selection as per manufacturer's recommendations.

The guidelines on selection of grit and grade for different classes of work are given in Table 15.7.

Table 15.7 Grit and grade for different classes of work.

Sl. no.	Class of work	Grit range	Grade range
1.	Fettling, snagging	12-30	Q-T
2.	Tool grinding	30-80	M-Q
3.	General rough work (off hand)	14-30	Q-S
4.	Cylindrical grinding	36-120	J-N
5.	Carterless and crankshaft grinding	46-80	J-N
6.	Internal grinding	46-60	M-N
7.	Tool and cutter grinding	46-60	I-M
8.	Surface grinding (segments)	20-36	G-M
9.	Surface grinding (cylinders or cups)	20-36	G-K
10.	Surface grinding (straight wheels)	46-60	H-K

15.7.2 Metal Removal rate

Metal removal rate involves accuracy and surface finish.

1. Coarse grain for fast cutting and fine grain for fine finish.
2. Wide spacing for fast metal removal rate and close spacing for fine finish grinding.
3. Resinoid, rubber and shellac bond for high surface finish.

15.7.3 Area of Contact

The area or arc of contact influences the selection of:

(i) grit size
(ii) grade
(iii) structure

1. For small area of contact, fine grain and close spacing.
2. For large area of contact, coarse grain and wide spacing.

15.7.4 Grinding Machine Type

The grade of wheel depends upon the type of grinding machine

1. Heavy and rigid machines: soft wheels and light machines flexible wheels.
2. Precision machines affect grade of wheels.

15.7.5 Wheel Speed

The selection of grade and bond depends upon the wheel speed. The wheel speeds for different types of grinding wheels are given in Table 15.8.

Table 15.8 Recommended wheel speed for different types of grinding.

Sl. No.	*Type of grinding*	*Surface speed (m/min)*
	(a) Vitrified bonded wheels	
1.	Cylindrical	1500-2000
2.	Surface	1200-1500
3.	Internal	600-1800
4.	Tool and cutter	1500-2000
5.	Centre less snagging	1500-1800
	(b) Resinoid bonded wheels	
6.	Snagging	2000-3000

In general, surface speed for vitrified bonded wheel is specified as 2000 m/min and for rubber, shellac or resinoid bonded wheel higher speeds.

15.7.6 Work Speed

The hardness of wheel is dictated by relative speed of wheel with respect to speed of workpiece. Higher the speed of workpiece with respect to wheel speed, the harder should be the wheel. The relative speed will change with the wear of wheel diameter. This is adjusted by changing the speed of the workpiece. The surface speed of cylindrical workpieces are given in Table 15.9 for different materials.

Table 15.9 Work speed for different materials.

Sl. No.	*Work material*	*Surface speed (m/min)*	
		Rough cut	*Finish cut*
1.	Cast iron	60	120
2.	Aluminium	30	60
3.	Soft steel	9	15
4.	Hard steel	20	30

15.7.7 Condition of Grinding Machine

The grade of wheel depends upon the condition of grinding machine. If the machine is worn, spindle is loose in bearings and foundations are shaky, harder wheels should be used as compared to machine in good operating condition.

15.7.8 Personal Factors

The experience and skill of workman is very important especially for off-hand grinding and grinding cost. The selection of wheel is influenced by the workman.

15.8 MOUNTING OF GRINDING WHEEL

Proper mounting of grinding wheel on the spindle is very critical because of high cutting speeds used in grinding. The wheels are subject to high centrefugal force.

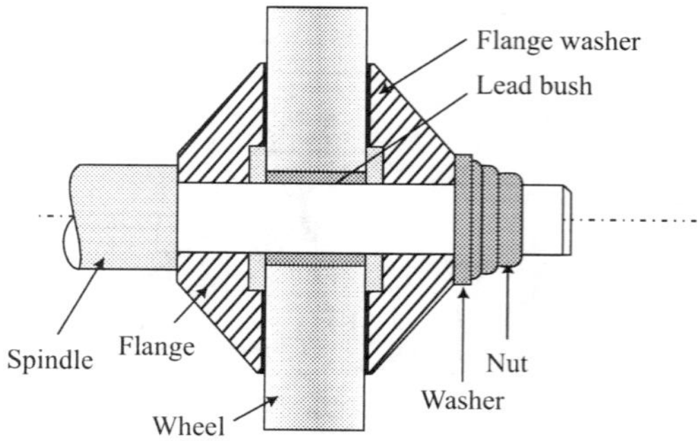

Fig. 15.1 Mounting of wheel.

The wheel is mounted on a spindle as per Fig. 15.1. The following precautions must be observed in grinding wheel mounting.

1. Visually inspect the grinding wheel against damage in transit and storage.
2. Ringing Test: Put the wheel on an arbour and strike with a hammer. A clear ringing sound from a vitrified silicon wheel shows that there is no crack. This test cannot be carried out on shellac, resinoid or rubber bonded wheels which do not ring against hammering.
3. The abrasive wheels should be fitted easily on spindles without use of force.
4. The lead bush should not project beyond the sides of the wheel.
5. The mounting flanges should be of same diameter and should be large enough to avoid bending of wheel.
6. The flanges should settle flatly and uniformly on the wheel.
7. The flanges should have recess to avoid pressure on the wheel near the bore.
8. Flange washers made of cardboard, leather or rubber of thickness less than 1.5 mm should be inserted between the wheel and flanges.
9. The inner flange should be keyed to the spindle and outer flange should be free so that it can provide a uniform pressure on the compressible washer.
10. The nut should not be tightened excessively to avoid damage of wheel due to excessive pressure.
11. The wheel gaurd must be used before running the wheel.
12. After starting the wheel, it should be run idly for 10 to 15 minutes.
13. If the wheel does not run properly, it should be dressed and trued before starting any grinding work.

15.9 DRESSING AND TRUING OF GRINDING WHEELS

Two important tasks are carried out after proper mounting and balancing of wheel. Unbalanced wheels can put undue load on grinding machine and quality of work can be poor.

1. Small wheels are balanced by milling a short recess on the inside of flanges and filling with lead.
2. Large wheels are balanced by putting on balancing machines and moving weight around a recess on the flange.
3. If wheels cannot be balanced due to wear even after truing and dressing, it should be discarded.

15.9.1 Dressing

It is the process of sharpening the wheel to expose active sharp grits. For from grinding, the wheel surface should be shaped to the required shape for the particluar operation. The wheels become dull and clogged during production. Many methods are used for redressing a grinding wheel.

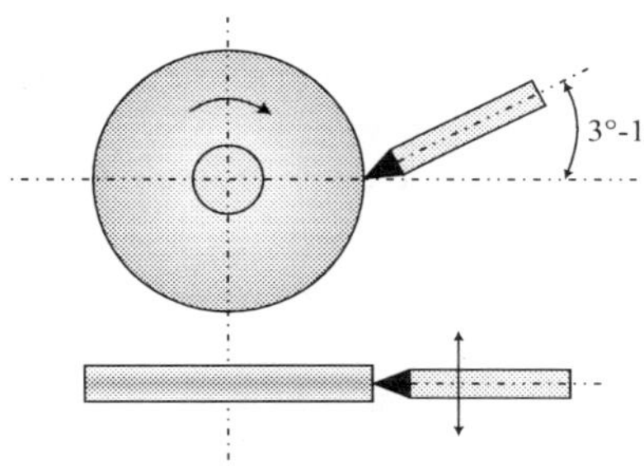

Fig. 15.2 Single point dressing of grinding wheel.

1. **Single point dressing:** This is the most common method. The dressing tool consists of a diamond tip mounted onto a steel holder. The wheel is rotated and diamond pointed dresser is moved across the surface of the wheel as shown. The dressing tool is mounted at an angle to the wheel surface. The feed rate of dressing point across the wheel surface affects the cutting action of the wheel. High feed rate produces a coarsely dressed surface used for rough grinding. Low feed rates produce a fintely dressed surface used for finish grinding.
2. **NC redressing:** For from grinding, the wheel surface is accurately shaped by moving the single point dressing tool under automatic control including the use of numerical control (NC).
3. **Crush dressing:** A dressing roller made of hardened steel, tungsten carbide or other hard material is pressed against the surface of rotating grinding wheel to be redressed. The dressing roller rotates independently or through contact with wheel. The grits at the wheel surface are crushed and dressed. This is a very fast method and useful for large quantity products. The shaped rollers required for form grinding are very expensive.

4. **Diamond dressing rolls:** The process is similar to crush dressing and wheel material is removed by the crushing action of diamond grits on the dressing roll. Shaped diamond dressing rolls are more expensive but generate very accurate dressed surface.
5. **Electro-discharge machinery (EDM):** An electrode corresponding to the surface of the grinding wheel is used. The electric discharge erodes away the matrix material. The current density is controlled to produce coarse or fine surfaces. This method is useful for metallic bonded wheels.

15.9.2 Truing

Truing is the process of elimination or minimization of radial and axial run out of the wheel surface. For most grinding wheel applications, truing and dressing are carried out simultaneously. For metallic bonded wheels, the wheel is mounted very accurately and truing is done separately by minimizing the run out.

Truing is also the process of changing the shape of wheel for form grinding. Therefore, truing of wheel includes working the wheel concentric with the bore and shape the face contour for form grinding. Truing and dressing are done with the same tools including.

1. **Diamond tool:** The feed rate is restricted to 0.02 mm to avoid grooves cut in the wheel.
2. **Crush roll:** This is a more popular method for form wheels. Crushing produces a wheel with many sharp pointed grains. The process of truing is very fast and the wheel runs cool.

15.10 TESTING OF GRINDING WHEELS

A grinding wheel is composed of abrasive particles bound together with a bonding material. The following characteristics of a grinding wheel must be determined to evaluate its performance.

1. Measurement of active grains per unit area of wheel surface.
2. Hardness of grinding wheel
3. Force required to break the bond between grains.

15.10.1 Determination of the Density of Active Grains

The number of active grains per unit area of wheel surface can be found out by the following methods.

(i) **Method of Baker, Marshal and Shaw:** A glass plate is covered with carbon black. The grinding wheel is rolled over. The pattern of carbon black is enlarged and photographed. The number of marks per unit area can be counted from the photograph.

(ii) **Grisbrook method:** The surface of grinding wheel is projected on a projection microscope. The number of cutting points passing a line on the projection screen are counted by rotating the grinding wheel. The number of active grains found by this method agrees with the number of chips produced during one revolution of the wheel in grinding.

(iii) **Oscilloscope screen:** A thermocouple is located on the surface of workpiece with insulated wire. A thermocouple junction is formed between wire and workpiece when the active grain of rotating wheel passes over. A pulse is obtained which can be displayed on an oscilloscope screen and counted.

15.10.2 Wheel Hardness

The wheel hardness as designated in the standard coding system can be measured by the following methods:

(i) **Depth of penetration:** A spade-type drive is used. A hole is drilled in the wheel with a constant force. The depth of penetration gives the wheel hardness.

(ii) **Jet erosion:** An air-abrasive jet is used to break the bond. The depth of penetration of the jet erosion in a standard period of time is used as equivalent wheel hardness.

(iii) **Resonant frequency:** An isolated wheel is struck with a rubber hammer. The resonant frequency is measured which is calibrated in wheel hardness.

15.10.3 Strength of Bond

The force required for breaking the bond of a wheel is found out by crushing test devised by Colwell.

A sintered-metal carbide or a diamond chisel is passed over the wheel surface so that a layer of grains are torn off from the bond. The tearing force is measured with the help of a dynamometer. This is the measure of the strength of the bond.

QUESTION BANK NO. 15

1. Outline the nature and characteristics of abrasives used in grinding wheels.
2. What is the function of bonds in grinding wheels? Indicate bonding materials and name their corresponding grinding wheel.
3. Describe grit, grade and structure of a grinding wheel.
4. Sketch various shapes of grinding wheels. Give their applications.
5. How are grinding wheels designated as per ISO?
6. How is a grinding wheel selected? Discuss various factors that influence the selection.
7. Describe various precautions to be observed in mounting of a grinding wheel. How these wheels are balanced?

8. Explain various methods of dressing and truing of grinding wheels.
9. What is the need for testing of grinding wheels? How these tests are carried out?
10. What do you understand by GRIT-STRUCTURE and HARDNESS of BOND of a grinding wheel? What effect do they have on usage of wheels?
11. What decides the hardness of the grinding wheel? Distinguish between dressing and truing of grinding wheel.

16

DESIGN PRINCIPLES OF CUTTING TOOLS

Introduction; Design of form tools, Flat form tools, Circular form tools, Design of milling cutters, Question Bank.

16.1 INTRODUCTION

Selection of proper tool geometry and tool material greatly affects the economics of machining of components. Design of cutting tool is very important to achieve required size, shape, accuracy and surface finish and integrity of the component. The tool geometry determines the magnitude and directions of forces acting on the tool, type of chip formation, tool temperature, tool life in addition to above factors i.e., dimensional accuracy, surface finish and integrity of machined surface.

Optimization of tool geometry can result in reduction of cutting costs by 10 to 50%. The tool life can be increased tenfold simply by proper selection of side cutting edge angle, side relief angle and nose radius. For a given cutting operation and workpiece material, the tool geometry must be otptimized for enhancing productivity. The tool material also affects the optimization of tool geometry. The various tool materials have already been discussed in Chapter 13. Geometry of simple point tools has been discussed in detail in Chapter 12. Design of special tools is discussed below.

16.2 DESIGN OF FORM TOOLS

Form tools are used on capstan, turnet and automatic lathes. Proper rake angle and clearance angle have to be provided to generate desired profile on the workpiece and create better machining conditions.

There are four types of form tools:

1. Flat form tools
2. Circular form tools

3. Dovetail form tools
4. Milling cutter.

16.2.1 Flat Form Tools

A typical flat form tool is shown in Fig. 16.1.

Fig. 16.1 Flat form tool.

A flat form tool used for *V*-groove turning is shown in Fig. 16.2. It has the following angles.

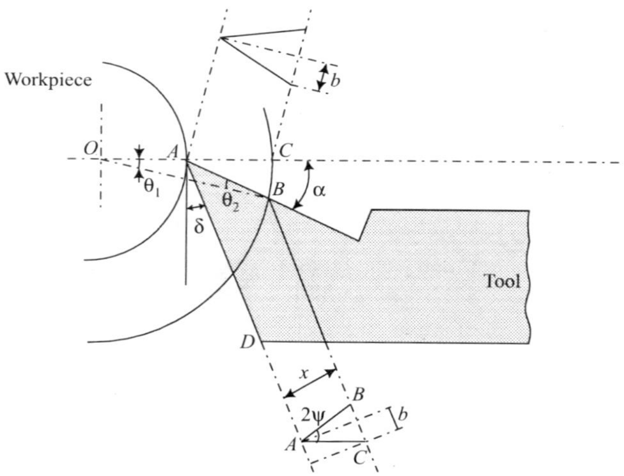

Fig. 16.2 Geometry of turning a *V*-groove.

$$\text{Rake angle} = \alpha$$

$$\text{Clearance angle} = \delta$$

The dimensions of the circular *v*-groove are as follows

Width of groove on outer periphery = $2b$

Radial depth of groove = h

Semi-groove angle = λ

The semi groove angle,

$$\lambda = \tan^{-1}\frac{b}{h}$$

Now $\quad OB = R$ = outer radius of job

$OA = r$ = radius at the groove bottom

$$\angle AOB = \theta_1$$

$$\angle ABO = \theta_2$$

In $\Delta\, ABO$ using Lami's theorem,

$$\frac{R}{\sin(180° - \alpha)} = \frac{AB}{\sin\theta_1} = \frac{OA}{\sin\theta_2}$$

$$\therefore \quad AB = \frac{R\sin\theta_1}{\sin(180° - \alpha)}$$

$$\sin\theta_2 = \frac{OA}{R}\sin(180° - \alpha)$$

$$= \frac{r}{R}\sin\alpha$$

Also $\quad \alpha = \theta_1 + \theta_2$

θ_1 and AB can be calculated

$$X = AB\cos(\alpha + \delta)$$

where $\alpha + \delta$ = angle between the plane normal to AD and face plane.

Width of rake face at point $B = 2b$

$$\therefore \quad \tan\psi = \frac{b}{x} = \frac{b}{AB\cos(\alpha + \delta)}$$

$$= \frac{b\sin(180° - \alpha)}{R\sin\theta_1\cos(\alpha + \delta)}$$

where ψ = angle between two faces of the tool in a plane normal to flank edge AD.

Example 16.1: A V-groove of angle 90° and depth 10 mm is turned on a shaft of 50 mm diameter with a form tool. The tool has a rake angle of 15° and end clearance angle of 10°. Calculate the angle to be ground on the tool in a plane normal to end flank angle.

Solution: The V-groove angle = 90°

$$\therefore \quad \lambda = 45°$$

$$\tan\lambda = \tan 45° = 1 = \frac{b}{h}$$

$$\therefore \quad h = b = 10 \text{ mm.}$$

The angle to be ground on the tool normal to plane of end edge of tool,

$$\psi = \tan^{-1} \frac{b \sin \alpha}{R \sin \theta_1 \cos (\alpha + \delta)}$$

Now $\sin \theta_2 = \frac{OA}{R} \sin (180° - \alpha)$

$$= \frac{40}{50} \sin \alpha = \frac{40}{50} \sin 15°$$

$$= 0.207.$$

$\therefore$ $\theta_2 = 11.946°$

$$\theta_1 = \alpha - \theta_2 = 15 - 11.946 = 3.05°$$

$$\tan \psi = \frac{10 \sin 15°}{50 \sin 3.05° \cos (15 + 10°)}$$

$$= 1.090$$

$\therefore$ $\psi = 47.4°$

Angle of V-groove $= 2\psi =$ **94.8°** **Ans.**

16.2.2 Circular Form Tools

Circular form tools can be of the following types.

1. Circular form tool with zero rake angle
2. Circular form tool with rake angle, α.

Let the tool centre be fixed at a height y above the centre of workpiece. The relief angle δ can be found out from the relation,

$$\delta = \tan^{-1} \frac{y}{R_t}$$

where, R_t = tool radius

1. **Circular form tool with zero rake angle:** A circular form tool with zero rake angle is used to turn a V-groove:

   The arrangement is shown in Fig. 16.3

   $O_1A = R_w$ = Radius of workpiece,

   $O_2B = R_t$ = Radius of tool

   $BA = h$ = Depth of V-groove.

   $2b$ = Width of V-groove

   $\therefore$ $O_1B = R_w - h$ = Root radius of groove.

   $$\tan \psi = \frac{b}{h}$$

   where ψ = Included angle of V-groove in radial plane

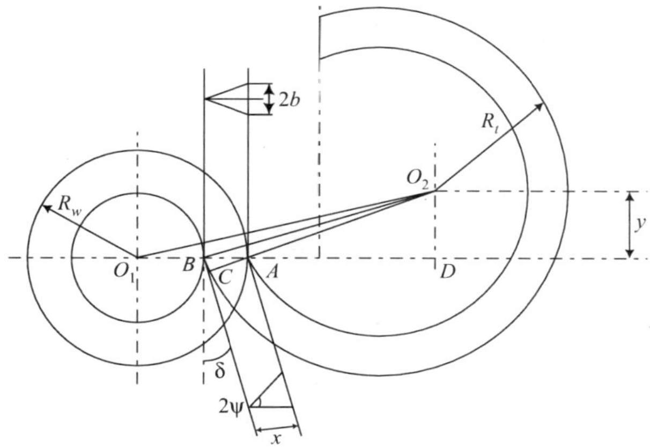

Fig. 16.3 Turning with circular Form Tool with zero Rake Angle.

Form geometry of Fig. 6.3.

$$AD = R_t \cos \delta - h$$

$$AO_2 = \sqrt{AD^2 + Y^2}$$

$$AC = X = R_t - AO_2$$

$$= R_t \sqrt{[(R_t \cos \delta - h)^2 + y^2]}$$

$$2\psi' = 2 + \tan^{-1}\frac{b}{x}$$

where, $2\psi'$ = Angle between the two faces of the tool

$\therefore$ $$2\psi' = \tan^{-1}\frac{b}{R_t - \sqrt{[(R_t \cos \delta - h)^2 + y^2]}}$$

Example 16.2: A circular form tool with zero rake angle and maximum radius of 60 mm is used to cut a V-groove on a circular workpiece of 80 mm radius. The V-groove has an included angle of 60° and a depth of 15 mm. The clearance angle desired is 7°. Calculate the angle ground between the two faces of the tool in radial plane.

Solution: The given data:

$$h = 15 \text{ mm}$$

$$2\psi = 60°$$

$$R_w = 8 \text{ mm}$$

$$R_t = 60 \text{ mm.}$$

Width of groove $2b = 2h \tan \psi$

$\therefore$ $$b = 15 \tan 30° = 8.66 \text{ mm.}$$

Let us fix $y = 7.5$ mm

$$\therefore \quad \delta = \sin^{-1}\frac{y}{R_t} = \sin^{-1}\frac{7.5}{60}$$

$$= 7.18. \text{ Acceptable}$$

$$X = AC = R_t - AO_2$$

$$= R_t - \sqrt{[(R_t \cos\delta - h)^2 + y^2]}$$

$$= 60 - \sqrt{[(60 \cos 7.18° - 15)^2 + (7.5)^2]}$$

$$= 14.84 \text{ mm}$$

$$\tan\psi' = \frac{b}{x}$$

$$\therefore \quad \psi' = \mathbf{60.53°.} \quad \textbf{Ans.}$$

2. **Circular form tool with rake angle:** The arrangment of turning a V-groove on a circular workpiece of radius R_w with the help of a circular form tool of radius R_t is shown in Fig. 16.4.

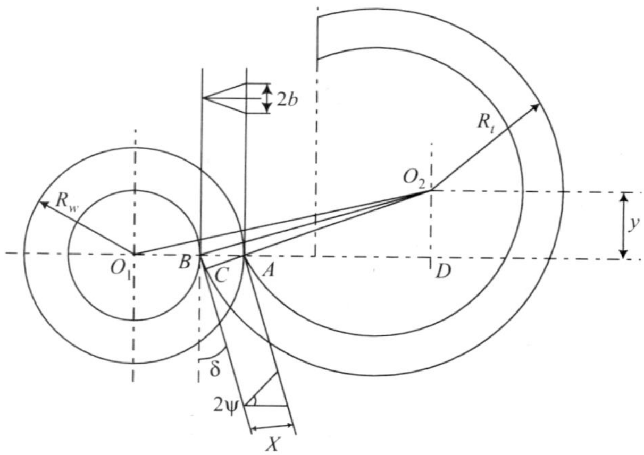

Fig. 16.4 Turning with circular Form Tool with rake angle.

Included angle of V-groove = 2ψ
Depth of groove = $BE = h$
Rake angle ground on tool face = $\angle DBA = \alpha$

Clearance angle ground on tool = $\delta = \sin^{-1}\frac{y}{R_L}$

$$\therefore \quad \cos\alpha = \frac{BE}{AB}, \quad \text{and}$$

$$\tan \alpha = \frac{AE}{BE}$$

The clearance angle $\delta = \angle O_2BE$

$$\therefore \quad \angle O_2BA = \alpha + \delta$$

$$\angle O_2AF = \lambda$$

$$\therefore \quad \tan \lambda = \frac{Y + FD}{AF}$$

$$= \frac{Y + EA}{(R_t \cos \delta - h)}$$

$$= \frac{Y + h \tan \alpha}{R_t \cos \delta - h}$$

where, $\quad Y = R_t \sin \delta$

$$\angle BAC = \angle O_2AG = \lambda + \alpha$$

ΔBAC is right angle triangle

$$\therefore \quad AC = X = ABC \cos (\lambda + \alpha)$$

$$= \frac{h \cos (\lambda + \alpha)}{\cos \alpha}$$

Example 16.3: In example 16.2 if rake angle is 10° ground on rake face, calculate the angle ground on the circular tool.

Solution: Form example 16.2

$$R_w = 80 \text{ mm}$$
$$R_t = 60 \text{ mm}$$
$$y = 7.5 \text{ mm}$$
$$b = 8.66 \text{ mm}$$
$$\delta = 7.18°$$
$$h = 15 \text{ mm}$$
$$\alpha = 10° \text{ (given)}$$

$$\tan \lambda = \frac{Y + h \tan \alpha}{R_t \cos \delta - h} = \frac{7.5 + 15 \tan 10°}{60 \cos 7.18° - 15}$$

$$\therefore \quad \lambda = 12.8°$$

$$X = \frac{h \cos (\lambda + \alpha)}{\cos \alpha}$$

$$= \frac{15 \cos (12.8° + 10°)}{\cos 10°} = 14.04 \text{ mm}$$

$$2\psi' = 2 \tan^{-1} \frac{b}{x} = 2 \tan^{-1} \frac{8.66}{14.04}$$

$$= \mathbf{63.33°} \quad \textbf{Ans.}$$

16.3 DESIGN OF MILLING CUTTERS

There is a wide variety of milling cutters to cope with machining of many shapes and sizes. The main groups of milling cutters are:

1. Profile cutters 2. Form relieved cutters

Standard types of profile milling cutters are:

(i) Plane milling cutters
(ii) End milling cutters
(iii) Face milling cutters
(iv) Slitting or sawing cutters:

The milling cutters can be designed to machine any form. The general form relieved cutters are:

(i) Gear cutters (ii) Shaped slotting cutters

16.3.1 Slab or Plain Milling Cutter

The tooth geometry of a slab milling cutter is shown in Fig. 16.5. It has straight teeth on the circumference and without side relief on ends. The teeth have relief angle on the flank surface followed by primary clearance angle and secondary clearance angle.

In order to reduce force fluctuations during cutting, helical teeth are used.

The selection of slab milling cutter should be as follows:

1. The cutter width should be more than width of the workpiece to achieve complete machining in one cut.
2. In order to minimize torque, use cutter diameter as small as possible. This will avoid twisting of arbor.

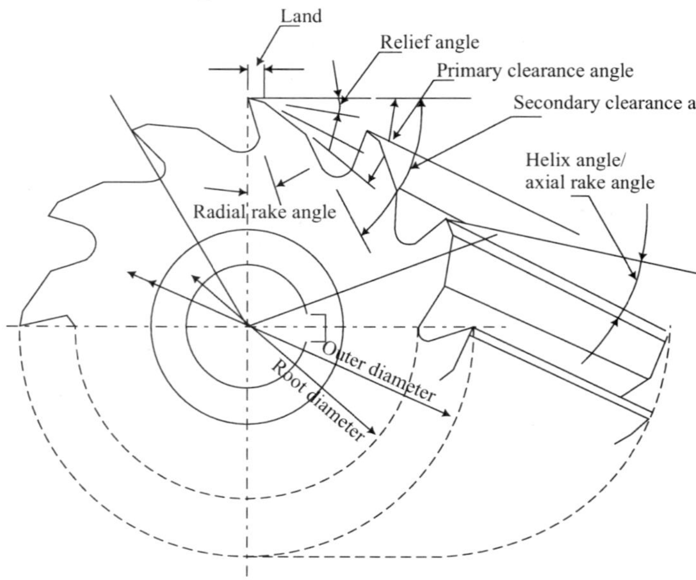

Fig 16.5 Slab milling cutter.

16.3.2 End Milling Cutter

End milling cutter has cutting edges both on periphery as well as end face. Therefore, it can machine horizontal as well as vertical surfaces. Two types of milling cutters with tooth profile are shown in Fig. 16.6. The performance of end milling cutters depends upon the following elements.

1. Proper choice of tool material and heat treatment to have high toughness, high hardness (67- 69 RC).
2. Flute helix angle of 25° needs lowest torque and longitudinal feed force.
3. Radial rake angle or normal rake angle. A combination of 15° rake angle and 30° helix angle gives the best results.
4. **Relief and clearance angles:** The primary relief angle may range from 20° to 45° followed by a secondary relief angle. The primary relief angle increases with the diameter of the cutter.

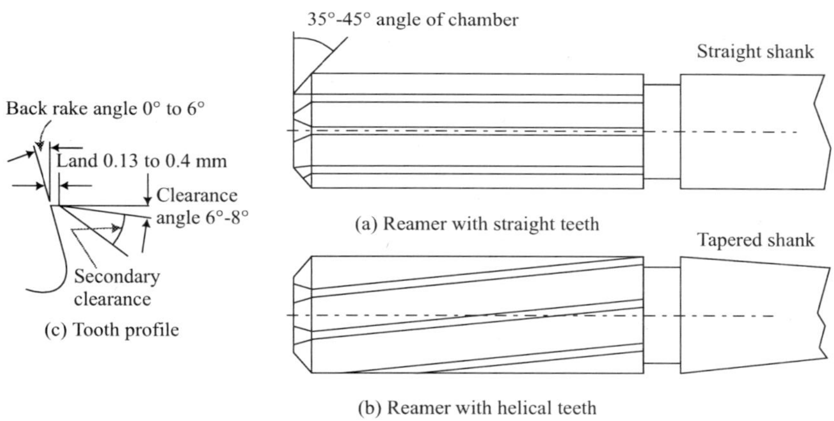

Fig 16.6 End milling cutter.

16.3.3 Face Milling Cutters

The face milling cutters have large deameter, i.e., more than 120 mm and are used for machining of plane surface of large forgings and castings. The tool materials are:

1. HSS for solid tools.
2. Inserted tool bits of tungsten carbide.
3. Polycrystalline diameter tool bits for machining highly abrasive materials.

The geometry of face milling cutters are shown in Fig. 16.7(a) and (b). The important geometrical dimensions are.

1. **Radial rake angle:** Both positive and negative rake angles can be provided
2. **Axial Rake Angle**

3. **Form or chamfer**
4. **Clearance angle**
5. **Tooth pitch:** Coarse pitch cutters should be used for heavy cuts, fine pitch cutters for low depth of cut.

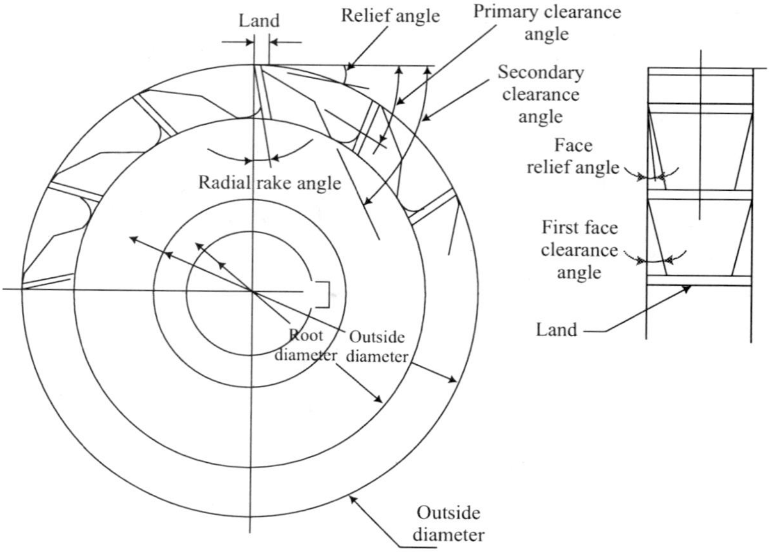

Fig. 16.7(a) Face Milling Cutter.

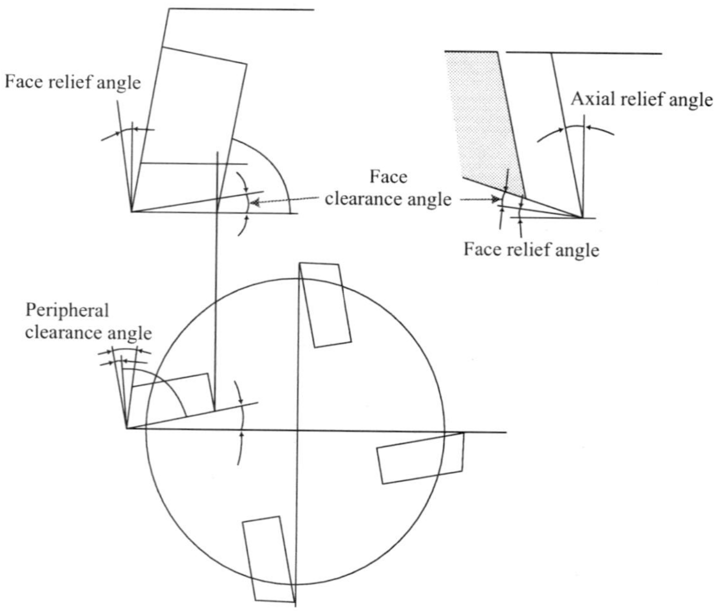

Fig. 16.7(b) A Face Milling Cutter.

16.3.4 Slitting or Sawing Cutter

These cutters are used for slitting of metal. The geometry of such a cutter is shown in Fig. 16.8. The geometry has the following parameters:

1. **Rake Angle:** The following rake angles are used
   (i) Steels: 10° to 80°
   (ii) Brass and non-ferrous alloys: 16° to 24°
2. **Clearance angle**
   Ferrous alloys: 7° to 10°
   Non-ferrous alloys: 10° to 12°
3. **Concavity:** $1\frac{1}{2}^{\circ}$.
4. **Pitch:**
   Fine tooth for hand materials. Coarse tooth or medium tooth for soft materials.
5. **Other dimensions to be specified:** Bore diameter, outer diameter and material.

16.3.5 Form Relieved Milling Cutter

These cutters are used for milling of gear teeth and other curved surfaces. The special features fo these cutters are:

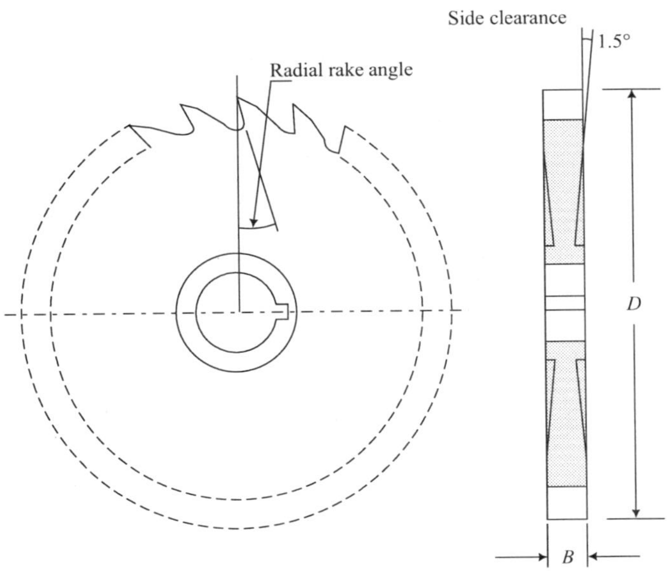

Fig 16.8 Slitting cutter.

1. **Resharpening:** The grinding is done on flat rake face of the teeth and no grinding on flank face. The cutter diameter gets reduced but form of teeth remains unchanged.
2. **Flank clearance:** Logarithmic and Archimedean spiral is used for the curve of the teeth to ensure constant flank clearance. Archimedean spiral is mostly used for easy machining of cutter.

QUESTION BANK NO. 16

1. What are the main types of milling cutters?
2. Classify different form tools. How is the geometry of a flat form tool worked out?
3. What is the difference between a circular form tool without rake and with rake angle? How are their geometrical dimensions fixed when used for turning a *V*-groove?

Part III

Machine Tools

17

DESIGN PRINCIPLES OF MACHINE TOOLS

Introduction; Performance requirements of machine tools; Design requirements of machine tools; Testing of machine tools, Question Bank.

17.1 INTRODUCTION

There are a large variety of machine tools such as general purpose machines (lathes, milling machines, drilling machines, etc), special purpose machines used for mass production (camshaft grinding machine, piston turning lathe), NC machines, etc. However, there are common features in their design e.g., static and dynamic stiffness of machine structure, operational motions (cutting and feed), gear boxes, controls and accuracy considerations, etc. The various componenets in all machine tools can be identified as beds, columns, spindles, bearings, sideways, etc. Basically, a machine tool is required to provide proper support to the workpiece and cutting tool. It also provides motion to one or both of them in order to generate the required shape on the workpiece.The form generated depends upon the type of machice tool.

A machine tool is designed to perform certain primary functions. It should be possible to use it for other secondary functions. This capability is called its flexibility. The flexibility of a machine tool is increased by the use of secondary functional attachments. Similarly, special attachments are added to improve its productivity.

17.2 PERFORMANCE REQUIREMENTS OF MACHINE TOOLS

General performance requirements of machine tools are listed below.

1. **Safety** against all possible accidents and gaurd against damage to operator and other machines and structures.
2. **Automatic operation:** The machine should be operated easily without fatigue of the operator. The operations such as loading of workpiece,

unloading of component, moment of tool with respect to workpice, feed control, tool resetting dimension measurements during process, etc., should be made easy by the use of quick-acting clamping derives, copying attachments, sensors and controls, etc. It should be possible to make some or all of the operations automatic within the constraints of initial cost of the machine.

3. **Accuracy and finish:** The components machined should have required accuracy of shape and dimensions, surface finish and surface integrity. The processing accuracy of machine tool depends upon its static stiffness, dynamic rigidity to withstand the weight of workpiece and cutting forces. The various components of machine tool should be accurately machined and assembled. The spindle axes and guideways must be accurately aligned.
4. **Productivily:** The productivity and production capacity of the machine tool should be high. These can be measured as:
   - number of workpieces produced per unit time
   - amount of material removed per unit time, or
   - surface area finished per unit time.
5. **Reliability:** The machine tool should give uninterrupted operation without breakdown for long service life. All non-replaceable components of the machine tool should have reliable operations.
6. **Maintainability:** The construction of machine tool should be simple for ease of service. The availability and cost of spare parts will affect the maintainability of the machine tool.
7. **Compactness:** The machine tool should be compact in design and occupy minimum floor space. High compactness can be achieved through proper design and construction.

   Compactness can be measured as:
   - weight per unit power (kg/kW), or
   - volume per unit power (m^3/kW).
8. **Simple construction:** The machine tool should be simple in construction. A complex machine with many components has poor productivity in manufacture and reliability in operation. Simplicity of design based on standard components enhances manufacturing simplicity.

   The productivity can be measured by the ratio of number of standardised components to total components of the machine tool.
9. **Machine costs:** The overall cost of machining will depend upon the initial machine cost and annual operating and maintenance costs. The machine tool should have minimum overall cost within technical performance requirements.

10. **Appearance:** Principles of industrial design should be used to improve the appearance of machine tool with pleasing shape and colour of parts and covers. However, the visual appearance should not adversely affect the above requirements especially operational ease, productivity and maintenance of machine tool.

17.3 DESIGN REQUIREMENTS OF MACHINE TOOLS

The various requirements for the design of machine tools are listed below.

1. **Machining requirements:** Process planning should identify the following requirements of machine tool under design.
   (i) Cutting operations to be carried out on machine tool.
   (ii) Size, shapes and materials of workpiece to be machined.
   (iii) Desired rate of production.
   (iv) Dimensional accuracy and surface finish of components to be achieved.
2. **Selection of main type of machine tool:** The machine tools in general may be classified as
   (i) General purpose machine tool
   (ii) Numerically controlled machine tool
   (iii) Special purpose machine tool.

Any machining operation, i.e., turning operation can be performed on general purpose lathe (centre lathe, turret lathe capstan lathe, single or multi-spindle automatic lathe), or NC lathe or a special-purpose lathe.

The machining cost consists of:

(i) fixed costs which depend upon the initial costs of the machine, space and overheads, and

(ii) variable costs including cost of labour, electicity, coolant, lubricating oil, etc.

The fixed costs are independent of quantity of production. The variable costs are directly propertional to quantity of production.

A graph can be plotted between quantity of production rate for different type of machine tools and total cost of production. One such graph is plotted in Fig. 17.1.

The fixed cost are as follows:

OA: General machine

OC: NC machines

OE: Special purpose machine.

These costs are in ascending order as $OE > OC > OA$. The variable costs are represented by slopes of lines as follows:

AB: General purpose machine

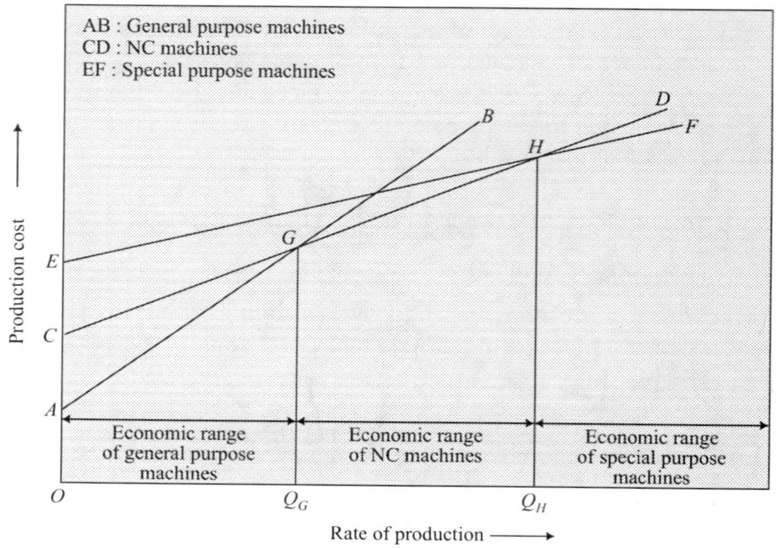

Fig. 17.1 Production cost vs. Rate of production for different machine tools.

CD: NC machines

EF: Special purpose machines.

For quantities of production, $Q < Q_G$, general purpose machines are economical. For Q between Q_G and Q_H, NC machines are most economical. When quantity $Q > Q_H$, the special purpose machines result in lowest production cost. The selection of the type of machine tool to be designed will depend upon the production rate.

3. **Selection of cutting parameters:** The selection of optimum speed, feed and depth of cut will depend upon the tool material, workpiece material, surface finish and dimensional accuracy required. These are available from exprimental data.

   The spindle and table speeds can be calculated from the dimensional data of workpieces to be machined.

Example

V_{max} = maximum cutting speed from the workpieces to be machined on the machine tool (m/min)

V_{min} = minimum cutting speed for the workpieces to be machined on the machine tool (m/min)

D_{max} = diameter of largest workpice to be machined on the machine tool (m)

D_{min} = diameter of smallest workpiece to be machined on the machine tool (m)

The spindle speeds of machine tool under design,

$$N_{max} = \frac{1000\, V_{max}}{\pi\, D_{min}}$$

$$N_{min} = \frac{1000\, V_{min}}{\pi\, D_{max}}$$

The maximum feed rate will depend upon the recommended value for the roughing cut for the hardest tool-soft workpiece combination. The minimum feed rate will depend upon the finest finishing cut.

4. **Calculation of power requirements:** The drive power for cutting motion depends upon the cutting force and cutting speed. The cutting forces depend upon work material, size of cut, tool geometry, cutting speed and tool material to some extent. Specific cutting power per unit chip section (N/mm^2) is available from experimental data for different work materials.

The power needed for feed drive depends upon feed force and feed rate.

Specific power,

$$F_h = K_s fd(1 + k_r + k_w)\ [N]$$

where, k_s = specific cutting resistance for a work material

d = depth of cut

f = feed rate

k_r = allowance for rake angle

k_w = allowance for tool wear

The power rating of cutting drive,

$$P_{rc} = \frac{F_h V_{max}}{60{,}000\eta_f}\ [kW]$$

where, V_{max} = cutting speed (m/min)

η_f = efficiency of kinematic chain

Feed for lathe operation,

$$Q = kF_v + \mu(F_h + F_r + W_d)$$

where, F_v = resultant force in the feed direction

F_k = resultant force in the cutting direction

F_r = resultant for the radial dierction

k = factor for overturning moment

= 1.1 to 1.4.

μ = coefficient of friction between carriage and guideways

W_d = weight of parts being traversed.

The radial component,

$$F_r \approx 0.3F_h$$

The feed component depends upon tool angles.

$$F_v = 0.15F_h \text{ to } 0.5F_h.$$

The power rating of feed drive,

$$P_{rf} = \frac{Qf_{max}N_{max}}{60 \times 10^6 \times \eta_f} \text{ [kW]}$$

where, f_{max} = maximum feed rate (mm/rev)

N_{max} = maximum rotational speed of spindle (rpm)

η = efficiency of kinematic chain associated with feed drive.

Example 17.1(a): The specific cutting resistance for steel is 500 N/mm². Determine the cutting force and feed force for rough turning of steel workpiece of 100 mm diameter using depth of cut as 1.5 mm/rev. Assum the following data.

Allowance for rake angle = 0.1

Allowance for wear = 0.15

$$\frac{F_V}{F_h} = 0.25$$

$$\frac{F_r}{F_h} = 0.30$$

Coefficient of friction $\mu = 0.2$

Weight of traversing parts = 500 N.

Correction factor for overturning moment = 1.2

(b) If the maximum cutting speed is 315 m/min and maximum feed rate is 0.5 mm/rev, find the power rating of cutting and feed drive. Take depth of cut as 2.5 mm and mechanical efficiency of kinematic chain both for cutting and feed motions as 80%.

Solution (a): The data given is

$$k_s = 500 \text{ N/mm}^2$$

$$d = 1.5 \text{ mm}$$

$$f = 0.30 \text{ mm/rev}$$

$$k_r = 0.1$$

$$k_w = 0.15$$

$$\frac{F_V}{F_h} = 0.25$$

$$\frac{F_r}{F_h} = 0.30$$

$$\mu = 0.02$$

$$W_d = 500 \text{ N}$$

$$k = 1.2$$

The cuting force,

$$F_h = k_s f d(1 + k_r + k_w)$$

$$= 500 \times 0.3 \times 1.5)(1 + 0.1 + 0.15) = 281 \text{ N}$$

$$F_v = 0.25 \times 281 = 70 \text{ N}$$

$$F_r = 0.3 \times 281 = 84 \text{ N.}$$

The feed force,

$$Q = k.F_v + \mu(F_h + F_r + W_d)$$

$$= 1.2 \times 70 + 0.20(281 + 84 + 500)$$

$$= 101 \text{ N.}$$

(b) The maxium values of resultant force.

$$F_h = 500 \times 0.5 \times 2.5 \times (1 + 0.1 + 0.15)$$

$$= 781 \text{ N.}$$

$$F_v = 0.25 \times 781 = 195 \text{ N}$$

$$F_r = 0.3 \times 781 = 234 \text{ N}$$

$$Q = kF_v + \mu(F_h + F_r + W_a)$$

$$= 1.2 \times 195 + 0.2(781 + 234 + 500)$$

$$= 264\text{N}$$

The power rating for cutting

$$P_{rc} = \frac{F_h V_{max}}{60{,}000 \times \eta_f} \text{ [kW]}$$

$$= \frac{781 \times 315}{60{,}000 \times 0.8} = 5 \text{ kW.}$$

$$N_{max} = \frac{1000\ V_{max}}{\pi D} = \frac{1000 \times 315}{\pi \times \frac{100}{1000}} = 1000 \text{ rpm}$$

The power rating for feed drive

$$P_{rf} = \frac{Q f N_{max}}{60 \times 10^6 \eta_f} \text{ [kW]}$$

$$P_{rf} = \frac{264 \times 0.5 \times 1000}{60 \times 10^6 \times 0.8} = 36.66 \times 10^{-3} \text{ kW}$$

$$= 36.66 \text{ W.}$$

5. **Detailed design of components:** The main components of a machine tool are:
   (i) Drive system for cutting and feed motion
   (ii) Spindle
   (iii) Spindle bearing
   (iv) Structure
   (v) Guideways
   (vi) Controls
   (vii) Other components and devices

These are designed based on general principles of machine design.

17.4 TESTING OF MACHINE TOOLS

The following types of tests are conducted on machine tools after erection. These test are carried out as acceptance tests at the following stages,

1. Purchase of new machine tools
2. Purchase of reconditioned machine tools
3. Repairs or maintenance of existing machine tools

Two types of tests are carried out to assess the process capability and manufacturing accuracy of machine tools.

1. Geometrical alignment tests
2. Cutting performance tests
3. Special tests

The measuring and testing instruments are: dial gauges, mandrels, straight edges and squares, spirit level, micrometers, autocollimator, measuring microscopes, telescopes, etc.

17.4.1 Geometrical Alignment Test

These tests are carried out to check the manufacturing accuracy of components and sub-assemplies on the fully assembled machine tool.

1. **Leveling and installation:** The levels of four corners of the machine bed are checked with the help of spirit level. This will ensure the machine bed lies in a horizontal plane. This will provide datum for other tests.
2. **Slide ways, tables and bed plate:** The surface furnish, flatness, straightness and parallelism of slideways, tables and bed plates are checked.

3. **Spindles:** The trueness of rotation, axial slip, alignment with axis of other spindles, parallelism/prependicularity of the main spindle with slideways are measured.

17.4.2 Cutting Performance Tests

Performance test is carried out by machining a component. The dimensional accuracy and surface finish are measured. The machining of various components is carried out on different specimen by varying cutting conditions (speed, feed, depth of cut, etc.). Measurements made are checked whether those are within acceptable limits.

17.4.3 Special Tests

Special tests may be carried out on machine tools:

(i) Power consumption under different machining parameters

(ii) Mechanical efficiency of kinematic drive

(iii) The input and output speeds of gear boxes

(iv) Noise test for high speed machines

(v) Vibration tests to measure amplitude of vibration between tool and workpiece under different machining conditions

(vi) Static stiffness tests

17.4.4 Test Charts

These are the charts and tables which include allowable tolerances and standard of geometrical and performance tests supplied by manufacturers of machine tools. Permissible errors for vairous tests are also supplied. The recommended methods of tests and measurements are included.

QUESTIONS BANK NO. 17

1. Discuss the performance requirements of modern machine tools.
2. List the main design requirements of machine tools.
3. How is a machine tool selected?
4. How are speeds, feeds and power requirement selected for a proposed machine tool?
5. Discuss the important acceptance tests carried out on machine tools.
6. Write brief notes on

   (i) Test charts

   (ii) Special tests.

18

MACHINE TOOL VIBRATIONS

Introduction; Types of machine tool vibrations; Forced vibrations; Self-excited vibrations (chatter); Control of chatter;

18.1 INTRODUCTION

The workpiece, cutting tool and machine form a structure system with complicated dynamic characteristics. The cutting tool applies force on the workpiece during machining. The reactive forces from the workpiece are transmitted to the machine. The cutting tool receives its cutting forces from the machine. Under certain operating conditions, the structural system may experience vibrations. The presence of vibration results in poor surface finish, cutting-edge damage and irritating noise. It is, therefore, very important to study the causes and control all types of free and forced vibrations due to interaction between the cutting process and the machine tool structure.

Machine tools are complex structures consisting of beds, columns, gears, shafts, bearings, slides, cutting tools, workpieces, etc. These are distributed mass system with infinite number of mass points and therefore infinite degrees of freedom. The cutting forces can be resolved into steady or constant component and time-dependent dynamic component. The steady component of cutting forces along with dead loads can cause static deflections in the elastic workpiece tool-machine system. These deflections disturb orientation and motion of tool relative to workpiece. The point of application of cutting load changes with time and hence the deflection. The machine tool must have *static stiffness* to resist the constant loads of the system.

The dynamic component of forces cause dynamic deflection or vibration in the machine-tool-workpiece system which produce dimensional, shape and finish imperfections and endanger the life of cutting tool and machine by fatigue and impact loading. The machine tool must, therefore, be designed to be *dynamically rigid*. In addition, it should be *dynamically stable* and amplitude of vibration should go on decreasing and system must return to its equilibrium position in short time. The machining performance will suffer if the machine tool has insufficient static stiffness, dynamic rigidity and stability.

18.2 TYPES OF MACHINE TOOL VIBRATIONS

The machine tool vibrations are classified as:

1. Free vibrations or transient vibrations or random vibrations.
2. Forced vibrations.
3. Self-excited vibrations or machine tool chatter.

18.2.1 Free Vibration

This is caused by impulsive or shock loading on the machine tool. The sources of impulse are:

(i) The tool strikes a hard grain in the workpiece
(ii) Initial engagement of cutting tool
(iii) Sudden entry of tool unto workpiece
(iv) Rapid reversals of reciprocating masses such as machine tables
(v) Impulses transferred from machine foundation.

A relatively large force acts for a short time on the elastic structure of machine tool. The structure is deflected and oscillates in its natural modes of vibration. The damping present in the structure causes the decay of vibrations in a fairly rapid time.

18.2.2 Forced Vibrations

These are caused by period force within the system. The sources of such forces are:

(i) Unbalanced rotating masses
(ii) Intermittent engagement of multi-tool cutters (milling)
(iii) Transmitted from nearby machinery through foundation
(iv) Mounting defects
(v) Imperfect bearings
(vi) Misalignment of machine elements.

The machine tool will oscillate at the frequency of forcing force. If this frequency of oscillations corresponds to one of natural frequencies of the structure, there will be resonance of vibration.

18.2.3 Self-excited Vibration

This is due to dynamic instability of the cutting process and commonly known as *machine tool chatter*. If large tool-work engagements are attempted, oscillations suddenly build up in the structure. The structure oscillates in one of its natural modes of vibration. The metal removal rates have to be reduced. The operator has to stop the machine to change the cutting conditions; i.e., reduction of feed rate or depth of cut.

The vibrations occur due to interaction between the cutting process and the machine tool structure.

18.3 FORCED VIBRATIONS

When a machine tool structure is subjected to a periodic force, it will vibrate at the forcing frequency. There will be resonance when angular frequency of the external harmonic force equals the natural angular frequency of the structure. The machine tool-workpiece system behaves like a damped mass-spring system. In order to minimize the amplitude of vibration of the system, the damping should be large and natural frequency of the system should be significantly less than the frequency of the disturbing force.

The cyclic variations of the cutting forces cause forced vibrations in machine tools. In side or face milling, there will be forced vibrations. The frequency of forced vibration will be equal to the product of the tool rotational frequency and number of teeth on the cutter. The frequency of spindle torque variation during a stab-milling operation will depend upon the number of teeth. The frequency of torque variations increases and the peak torque decreases with the increase of number of teeth.

18.3.1 Harmonic Acceptance of System

The machine tool should be designed so that the natural frequencies of the various parts of the structure do not approach the forcing frequencies. The harmonics of the periodic forces should also be avoided to safeguard against resonance.

A harmonic response locus or harmonic acceptance of the system is the response curve plotted in polar form. These frequency response curves are useful for analysis of complex structures with many degrees of freedom and also analysis of machine tool chatter.

18.3.2 Complex Structure

A machine tool is a three-dimensional structure consisting of a number of masses, springs and damping elements connected together. When excited with a periodic force, the structure will resonate with large amplitude vibrations. The frequency response curve will have many peaks corresponding to each natural frequency. Each resonance will have a mode of vibration or mode shape that the structure adopts. The modes of vibration can be plotted in the form of frequency-amplitude ratio and frequency-phase angle as shown in Fig. 18.1.

In a horizontal-milling machine, the important modes of vibration are those which causes relative displacement of tool with respect to workpiece. This will deteriorate the surface finish of workpiece.

In complex structures, there is *cross-frequency response* as the directions of applications of forces and displacements are not same. In simple structure with single-degree-of-freedom, there is *direct response* as the direction of displacement is same as point of application of force.

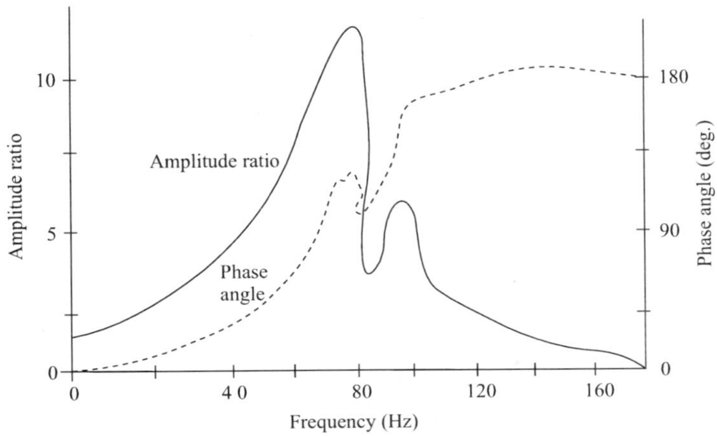

Fig. 18.1 Frequency response of a machine tool structure.

18.4 SELF- EXCITED VIBRATIONS (CHATTER)

18.4.1 Interaction of Cutting Process and Machine Structure

The basic cause of chatter or self-excited vibration is the interaction of the cutting process and the machine tool structure. An idealized orthogonal cutting system is shown in Fig. 18.2. The system consists of a cutting tool and workpiece connected by an elastic structure. A cutting force F_r is generated between the tool and workpiece acting at an angle θ_c to the cut surface. The value of F_c depends upon tool-workpiece engagement and depth of cut. The cutting force causes elastic strain of the structure. There is relative displacement of tool and workpiece, x. The underformed chip thickness changes. The machine tool behaves as a closed-loop feedback system as shown in Fig. 18.3. A hard spot in the work material will cause deflection of structure resulting in change of undeformed chip thickness. This causes change of cutting force. The machine vibrates in one of its natural mode.

There are two effects which can cause instability of the closed-loop cutting process structural system.

1. Regenerative effect
2. Mode-coupling effect.

The energy is transferred into the structure to sustain the vibration.

1. **Regenerative instability:** A disturbance in the cutting process generates a wavy surface on the workpiece. In the next pass of the cutting tool, the undeformed chip thickness will depend upon the present tool motion as well as wavy surface already cut during previous pass. Vibration is created by addition of the two tool motions. The vibration cycle will depend upon the phase difference between the two waves.

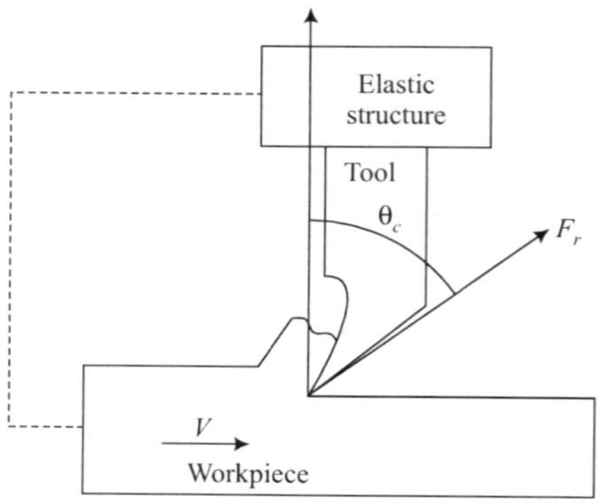

Fig. 18.2 Idealized orthogonal cutting system.

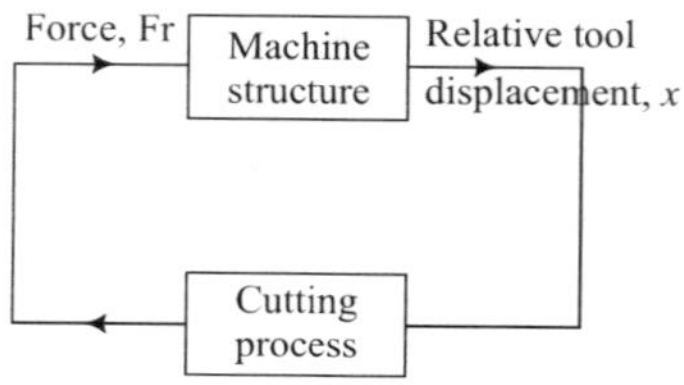

Fig. 18.3 Closed-loop Feedback system.

2. **Mode-coupling Instability:** In machining operation like thread cutting, the successive tool passes do not overlap. In one of the passes of the tool relative to workpiece, there may be coupled modes of vibration because of characteristics of the machine structure. The free vibrations of the structure causes the cutting tool to follow an elliptical path relative to workpiece. The cutting force varies and more work may be put into the system during return cycle of elliptical path. The increased input of energy can cause higher amplitude of vibrations.
3. **Stability charts:** The stability chart of a machine tool is the graphical representation between width of cut and relational speed of tool or workpiece. The instability portions are in the form of lobes and represent chatter. A judicious selection of cutting speed can increase the chatter-free width of cut.

18.5 CONTROL OF CHATTER

The self-excited vibrations or chatter can be controlled by the following methods.

18.5.1 Design of New Machine Tools

Existing models of machine tools are tested for chatter behaviour. The following changes are made and their effectiveness is assessed by finite element method or lumped-parameter method. The dynamic characteristics of the machine can be improved by the following methods

1. Mass redistribution
2. Change of size and shape of some structural members to strengthen them
3. Viscous damping at some places like spindles.
4. Frictional damping at some joints.

18.5.2 Existing Machining Tools

Chatter can be prevented in an existing machine tool by the following methods.

1. **Change of cutting parameters:** Decrease the feed rate, depth of cut and cutting speed.
2. **Change of tool geometry:** Increase of rake angle and method of clamping of workpiece
3. **Change of characteristics of vibratory system:** The following methods can be employed:
   (i) Use tuned undamped vibration absorbers to counteract forced vibration with constant frequency.

   Example: Electromagnetic imbalance of motion.

   (ii) Use of stiffner between table and over-arm of a horizontal milling machine, reduction of overhang of the tool in lathe, use of tighter clamping of workpiece, use of steady for long slender workpieces, etc.

   (iii) Introduction of vibration absorbers in the vibratory system especially in boring, milling and turning operations.
4. **Modification of regenerative effect**
   (i) The regenerative instability can be destroyed by the use of milling cutters of irregular tooth pitch for slab milling or different helix angles on successive teeth.

   (ii) The regenerative instability can be destroyed by the use of continuously variable spindle speed under programme control.

QUESTION BANK NO. 18

1. Discuss the importance of machine tool vibration analysis.
2. Explain different types of machine tool vibrations and their causes.
3. Explain different methods of control of chatter.
4. Write a detailed note on self-excited vibrations.
5. What is the importance of forced vibration in machine tools?

19

NUMERICALLY CONTROLLED MACHINE TOOLS

Introduction, Main components of NC system; NC machines; Types of NC machines; CNC and DNC system; Position and motion control in NC machine; Measuring system for NC machines; Question Bank.

19.1 INTRODUCTION

Numerically controlled machine tools have been developed for high productivity, low production cost and highly precise machining. Numerical control is the most sophisticated automation for controlling machine tools, equipment and processes. The complete NC machine tool system consists of three basic elements.

1. The programme of instructions
2. The controller unit
3. The machine tool.

The programme of instructions which is a mixture of alphabetical codes and numerical data are fed into the controller unit. This programme of set of operations is converted into electrical signals by the controller unit that control the various actions of machine tool. The instructions are carried automatically by the machine tool in predetermined sequence with pre-set or self-adjusted speed, feed depth of cut, etc. without human intervention.

Avoidance of human intervention, omission of conventional tooling and fixturing and quick-charge capability of NC system are the primary advantages of NC machine tools. Other advantages of NC machine tools over conventional machine tools are.

1. Optimization of machining parameters as per cutting tool life and quality of components.
2. Possibility of machining components not possible with conventional machining system.
3. Quick and accurate inspection and detection of error in design and machining.

19.2 MAIN COMPONENTS OF NC SYSTEM

The main components of a traditional NC machine are described below.

19.2.1 Programme of Instructions

The programme of instructions is called part programme. It is a detailed set of directions fed to a NC machine for machining a component. It is a mixture of alphabetical codes and numerical data which is interpreted by the NC controller. The programmes were fed to the controller using a paper tape reader. Presently, the programme is fed electronically from an on-line programming system remote from the machine.

The programme is prepared by a part pragrammer. It is a specialized process planning exercise. The preparation of NC part pragramme is discussed in chapter 22.

The programme instructions are fed to the controller to execute the following functions by the machine tool.

1. Movement of machine slides to generate the required surface on the workpiece
2. Control of displacements, velocities and accelerations of various machine tool components.
3. Control of various auxiliary function of machine tool.

   For example: Coolant on and off; automatic tool change; turret indexing, etc.

19.2.2 Machine Tool Controller

The machine tool controller accepts the part programme of instructions and converts them into electrical signals. These signals control various actions of a machine tool. The controller forms part of a feedback control system for the motion of the machine tool slideways.

The machine tool control system performs a number of functions as shown in Fig. 19.1.

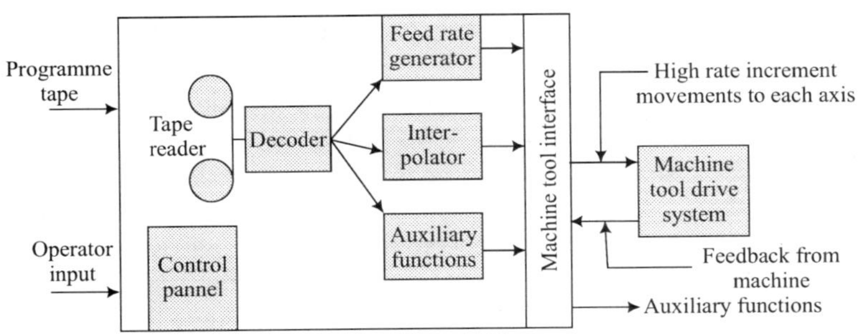

Fig. 19.1 Functions of any NC controller.

Input instructions are provided through programmes tape which are read by tape reader and decoded by the NC controller. In addition, instructions can also be provided by the operator manually from a control panel. The control panel also displays data for the operator.

The data is fed into a data input buffer. The information is stored as a block. Each block contains one complete set of instructions for the next step in machining sequence. Each block is terminated by an end of block character. The instructions in the block are decoded. The information is stored in the controller for execution.

The main functions of a controller are:

1. Interpolation
2. Feed rate generation
3. Auxiliary functions which are machine dependent

In order to machine surfaces of complex shape at constant feed rate, signals are given to various slides and spindles. The individual motions are integrated to produce the required shape. The interpolator breaks down (interpolates) the complex curves into small individual increments for each controlled motion of machine tool. Controller also interfaces with various machine units like drive motors, transducers and other controls of the machine tool.

19.2.3 Machine Tool

The machine tool responds to the electrical signals from the controller and executes various slide motions and spindle rotations to machine a component. NC machine tool is a general purpose machine with table, spindle and other hardwares. It is fitted with drive motors and controls to operate the spindle and slides automatically and to switch on or off the various auxiliary functions of machines. Transducers are fitted to feed back data on the positions and velocity of slideways to the controller.

19.2.4 Computers in Numerical Control

NC controllers use control systems with all the functions handled by hardware, logic circuits, switches, relays, etc. Adaptation of these controllers to the applications where more axes of control are required involve addition of more hardware.

Rapid development and increased availability of low cost, high-speed computing systems have influenced the development and operation of NC system. A micro-computer system has taken over the system and the functions of NC system are provided by software pragrammes. These systems are called *computer numerical control system (CNC).* The programme is stored within the CNC controller. The various functions of the controller, interpolation, decoding, etc. are carried by software rather than hardware. CNC systems are more flexible than NC system.

19.3 NC MACHINES

The majority of NC machines are basically similar in configuration to general purpose manual machines. The NC machines range in complexity from a simple drilling machine to NC machining centres.

1. **Single spindle drilling machine:** It is the simplest NC machine. These are programmed on three axes. *X*-axis controls the table movement to the right and left of the column. *Y*-axis controls the table movement towards or away from the column. *Z*-axis controls the up and down movement of spindle.
2. **NC lathe:** Lathe is the most productive machine tool to manufacture round components. Lathes are programmed on two axes. *X*-axis controls the cross motion of the cutting tool and *Z*-axis controls the carriage travel towards and away from the head stock.
3. **Turning centres:** These are NC lathes with turrets of tools to increase the flexibility of operation. The turning centres are capable of machining complex turned parts. These are more accurate and productive than centre lathes and has a higher spindle rate. The turret is of disc-type and can accommodate 12 tools and can produce a wide range of components without change of tool. Programming is for two axes. *X*-axis is for movement of cross slide and *Z*-axis for saddle movement towards or away from head-stock. Turning centres with increased capacity tool changes are also making a strong appearance in modern factories.
4. **Milling machine:** NC milling machine is the most versatile machine tool used in manufacturing industry. These are programmed on three axes. *X*-axis controls the table movement left or right. *Y*-axis controls the table movement towards and away from the column. *Z*-axis controls the vertical movement of the knee or spindle.
5. **NC turn-milling machine:** The availability of secondary milling spindles on turning machines has led to the development of hybrid operation called turn-milling. The workpiece is slowly rotated with the main spindle which provides the feed motions. The rotating milling tool is used for cutting. A highly versatile turn-milling centre can perform a wide variety of operations in one setup.
6. **NC Machining centres:** The most versatile complex NC machine tools are called NC machining centres. These machines incorporate multi-axis control and are usually horizontal or vertical spindle milling machines. These are provided with high-capacity tool magazines and automatic tool changers. The high-capacity tool magazines increase the flexibility of the machines, by enabling complex parts with a wide range of features without any change of tools. A large amount of machining can be carried out in one workpiece setup and a variety of operations including milling, drilling, tapping and boring can be performed.

Use of indexing table helps machining on several orientations of workpiece. The specifications of a Horizontal Machining Centre model MC 1000 Manufactured by Bharat Fritz Werner are given in Table 19.1.

Table 19.1 Specifications of Horizontal Machining Centre, MC 1000.

	Characteristic	*Unit*	*Range*
	Spindle		
1.	Drive motor (44% ED)	kW	32
2.	Speed.	rpm	20-3600
3.	Maximum spindle torque	N-m	1150
4.	Tool shank taper (DIN 69871)	–	A50
5.	Front bearing diameter	mm	100
	Traverse		
6.	Table (x-axis)	mm	1600
7.	Spindle (y-axis)	mm	1250
8.	Column (z-axis)	mm	1100
	Axis drive		
9.	Feed rate	mm/min	1-10,000
10.	Rapid traverse x, y, z	m/min	15
	Index table (CNC rotary tabled)		
11.	Pallet table size	mm × mm	1000 × 1000
12.	Load capacity	kg	2000
13.	Max. job swing	mm	1600
14.	Index increments	deg.	1(0.001)
	Automatic tool changer		
15.	No. of tools	–	60(72)
16.	Max. tool diameter	mm	120
17.	Max. tool length	mm	550
18.	Max tool weight	kg	20(35)
	Accuracy		
19.	Positional accuracy	mm	0.015
20.	Total connected load	kVA	60

19.4 TYPES OF NC MACHINES

There are three basic types of NC motions.

Type of NC motions

- Positional (point-to-point)
- Paraxial (straight line)
- Continuous path (contouring)

19.4.1 Point-to-point (PTP) or Positional Control

The machine tool elements (tools, table, etc) are moved to programmed location and the machining operations are performed after the motions are completed. The path or speed of movement between locations is unimportant. The coordinates of the end points of the motions are accurately controlled. This control is suitable for drill presses and boring machines where drilling, tapping or boring operations have to be performed at various locations on the workpiece. Example of a component machined by PTP system is shown in Fig. 19.2.

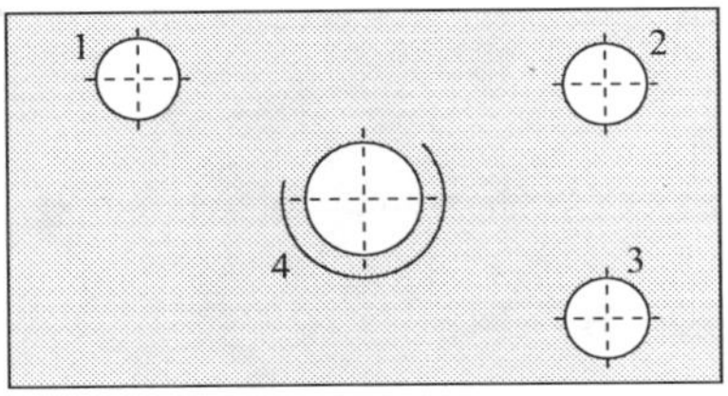

Fig. 19.2 Workpiece example of positional control.

19.4.2 Straight-line or Linear Control

The cutting tool is moved parallel to major axis of machine tool at a controlled rate suitable for machining. The tool can be moved in one direction only. Angular cuts in a workpiece are not possible.

Examples

1. Only rectangular configurations can be machined on a milling machine.
2. Surface parallel or perpendicular to spindle axis can be machined on a lathe.

These machines are also suitable for point-to-point control. The example of a workpiece machined by straight line control system is shown in Fig. 19.3.

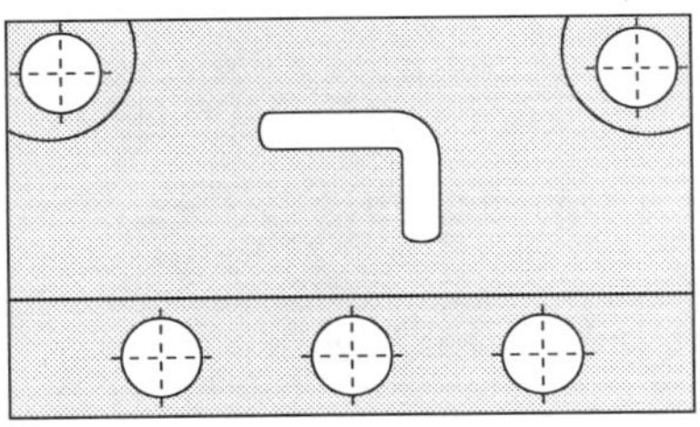

Fig. 19.3 Workpiece example of straight line control.

19.4.3 Continuous Path or Contouring Control

The motions of two or more machine axes are controlled simultaneously so that the position and velocity of the tool are changed simultaneously. The curves and surfaces can be machined at controlled feed rate. The interpolator determines the increments of individual controlled axes of the machines to produce the desired motion.

The control is also called continuous control or full axis of control. The majority of NC machines are produced with continuous control of all axes.

The example of workpiece for continuous controls are shown in Fig. 19.4.

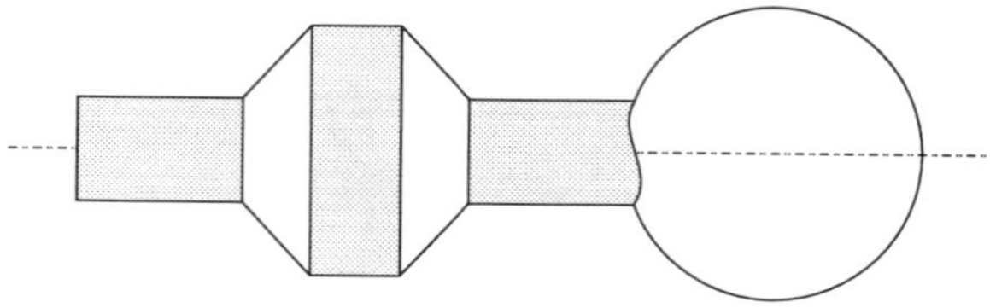

Fig. 19.4 Workpiece example of contouring control.

19.5 CNC AND DNC SYSTEMS

CNC denotes a numerical control system that uses a dedicated microprocessor as an integral part of MCU to execute the basic NC control function. A programme is loaded in MCU. The main components of CNC system is shown in Fig. 19.5.

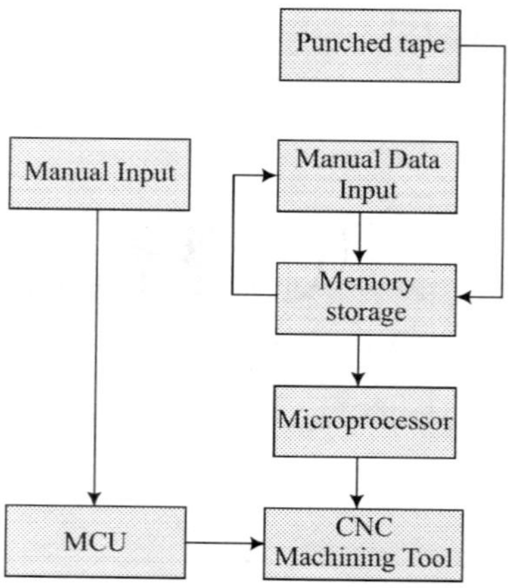

Fig. 19.5 CNC system.

Direct numerical control (DNC) system makes use of a mainframe computer to manipulate operations of a number of NC machines. DNC computer is located at a central place away from NC machines and manages information flow to a group of machines. The DNC computers have high processing power. The DNC system is shown in Fig. 19.6.

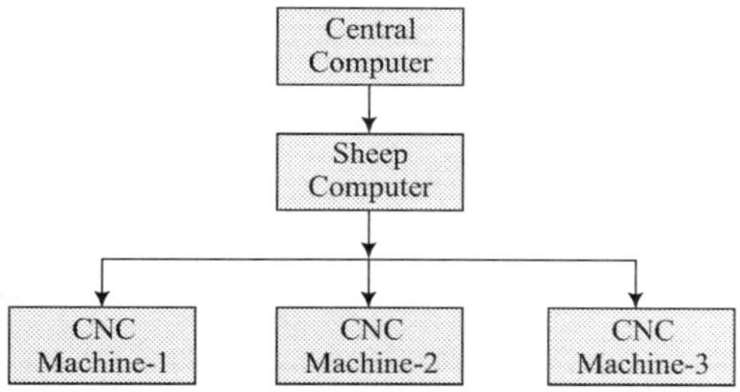

Fig. 19.6 DNC system.

19.6 POSITION AND MOTION CONTROL IN NC MACHINES

The location of NC tool at any point of time is controlled by ISO axes system. The system has been discussed in chapter 1.

The NC part programming consists of collection of data required to produce a component, calculation of a tool path along which the machine operation will be performed, and the arrangement of given and calculated data in a standard format which could be converted into an acceptable form for a give machine control unit (MCU). The NC part programming is discussed in chapter 22.

Open and Closed Loop

A group of devices are used to control the position of machine tool slides and spindle speed. These devices are electrical, hydraulic or pneumatic. There are two control systems:

1. Open loop system
2. Closed loop system

19.6.1 Open Loop System

This is a control system without feedback. It works as follows

1. The information stored in the tape is decoded by the tape reader.
2. Tape reader stores the information till the machine tool is ready to receive it.
3. Tape reader converts the information into electrical pulses or signals and are sent to control unit.

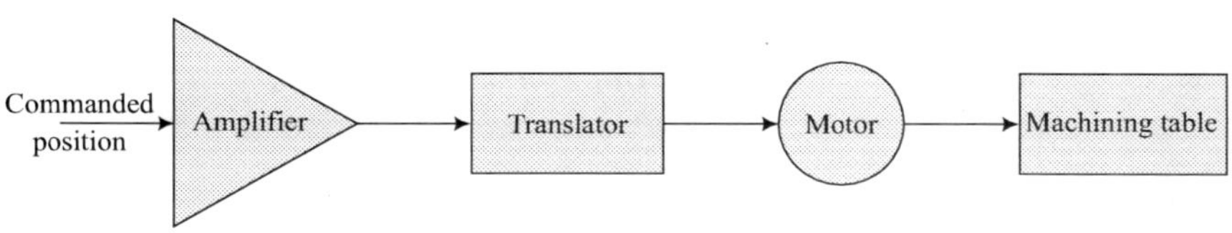

Fig. 19.7 Open loop system.

4. Control unit energizes the driving control unit which actuates D.C. motors to perform the desired function as per programme instructions in the tape.
5. Stepper motor rotates proportionally to the number of electrical signals received from servo control unit.
6. A precise lead screw is coupled to the motor and rotates. The machine table slides accordingly.
7. The table movement per turn of motor depends upon the pitch of the lead screw. Each pulse of signal rotates the motor by a fraction of revolution called stepping angle.

   Example: The lead screw has pitch of 4 threads per cm and 100 number of electrical signals rotates the motor by one revolution.

   The movement of machine table per pulse will be equal to $\frac{1 \times 10}{4 \times 100}$ = 0.025 mm. The position of machine table is controlled by number of pulses.
8. The motor controller sends back the signals indicating the motors have completed the motion. However, the stepper motor may have lost one or more pulses. The actual position of the table may not be the same as the desired position.
9. The open loop system is shown in Fig. 19.7. There is no provision for correction of deviation.
10. Point-to-point positioning system accepts open loop control system.

19.6.2 Closed Loop System

A closed loop system is similar to open loop system but a feedback system is added to the electrical circuit as shown is Fig. 19.8. Many types of feedback sensors are available for comparing the actual table movement with the desired table movement. For an error, the corrective signal is sent back to a D.C. servomotor (not stepper motor as used in open loop system). The D.C. servomotor makes necessary correction by readjusting the table to compensate the deviation.

In closed loop system, the accuracy is very high. The least count for slide of machine table is about 0.0025 mm.

The special drive motors used are A.C., D.C and hydraulic servos. Large N.C. machine tools use the powerful hydraulic servomotor. The speeds of driving motors are variable and depend upon the amount of electrical current passed. Closed loop systems are used for contouring NC system.

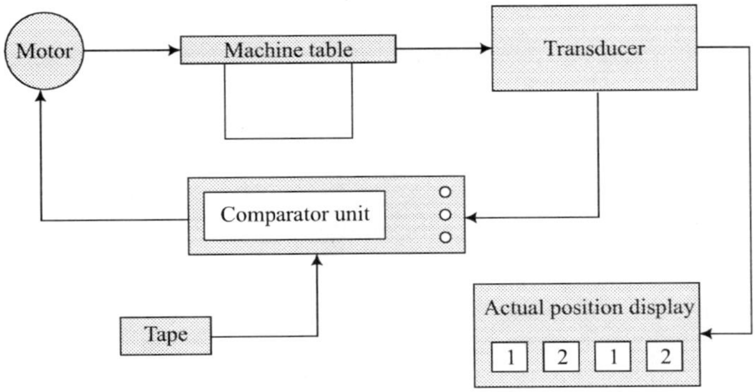

Fig. 19.8 Closed loop system.

19.7 MEASURING SYSTEMS FOR NC MACHINES

The position transducers are used as measuring system for N.C. machine tools. The transducers can be electric scales, electrooptical scales, magnetic scales, synchors, shaft digitizers, resolver's, laser types, linear or rotary transducers, etc.

The traducers can be categorized as:

1. Analogue control
2. Digital control

19.7.1 Analogue Control

The movement of machine table, rotation of spindle or movement of spindle are measured in terms of calibrated analog signals. These signals are continuous, modulated by events to provide the message and are proportional to continuous movement of machine parts. The accurate positioning is achieved by matching the continuous input voltage with the transducer output voltage. Analog control systems are ideal for point-to-point positioning systems.

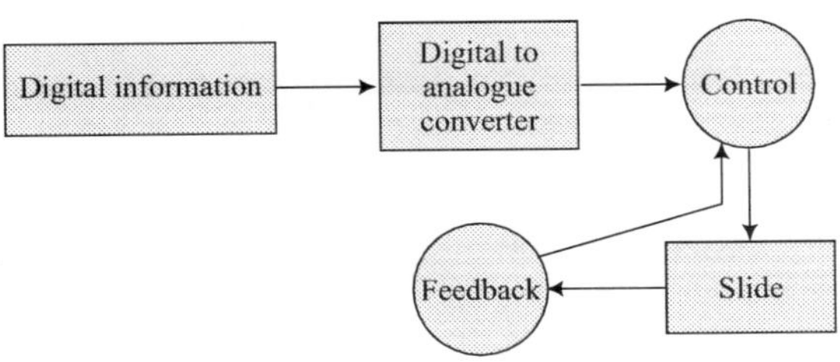

Fig. 19.9 Analog control system.

19.7.2 Digital Control System

In digital control system, the movement of worktable is a discrete quantity and in the form of a voltage pulse. Each pulse represents a basic length. Electrical pulses equivalent to the dimensional information are generated and used to drive the servomotor. The transducer, in turn, produces electrical pulses as the table moves and is compared with the input signal. Digital feedback systems are ideal for continuous path control. Theoretically, digital system is of a finite accuracy, whereas the analogue system can accept any value.

NC machine tools use both the control systems. The input signal is always digital in the form of dimensions given in the drawing. The output of NC system is always analog as the machine tool slides move continuously. The converter is used to change the information from one control system to another. The control is digital when a sequence of input is applied. The control is analogue when the input variable is continuous.

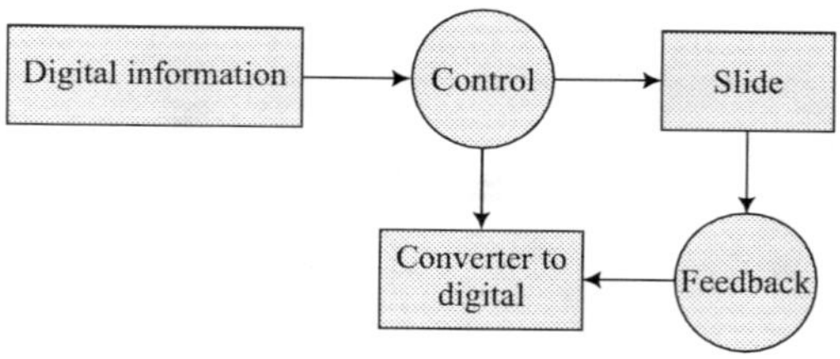

Fig. 19.10 Digital control system.

QUESTION BANK NO. 19

1. What is numerical control? What are the advantages of NC machine tool over conventional machine tools.
2. Describe important components of NC system.
3. What are different NC machine tools? Describe any two.
4. Explain different types of positioning control system.
5. Compare open loop system and closed loop system used for position and motion control in NC system.
6. Differentiate between NC, CNC and DNC systems.
7. Differentiate between analog and digital control system as measuring systems for NC machines.

Part IV

Automation

20

AUTOMATION IN MANUFACTURING SYSTEMS

20.1 INTRODUCTION

The efficiency of a manufacturing enterprise can be measured by:

1. Manufacturing cost
2. Productivity
3. Profit

The **manufacturing cost** includes cost of machining, assembly, testing, moving of components from machine to machine and temporary storage of components and products. This cost is affected by:

- efficiency of manufacturing process used,
- arrangement and integration of these processes,
- labour costs,
- quality of product design.

Productivity is the efficiency of labour utilization and is measured as average output per man-hour.

Example: Factory *X* employs 400 workers to produce 300 refrigerators per week. Factory *Y* employs 400 workers and produces 150 refrigerators per week. Therefore, labour productivity of factory *X* is twice that of factor *Y*.

Higher productivity of factory *X* may not lead to lower production cost as it may be using very expensive automated manufacturing equipment which may offset the saving in labour cost.

The **profit** made by the company will be affected by manufacturing cost as well as productivity.

Manufacturing cost, productivity and profit cannot be optimised simultaneously. The conditions for minimum cost, maximum productivity (minimum production time) and maximum profit rates are all different.

20.2 TYPES OF PRODUCTION SYSTEMS

The production systems can be classified on the basis of volume of production. The various categories of production system are:

(1) Mass production
(2) Large-batch production
(3) Small-batch production.

The cut-off limit of volume of production for various categories will depend on type of industry and country, however, it has been found that more than 75% of production of manufacturing industries is by batch production. Therefore, automation in batch manufacturing requires most attention.

The general *characteristics* of various production systems are as follows:

20.2.1 Mass or Continuous Production

(1) Large numbers of components of same configuration are produced.
(2) The demand is relatively constant and predictable.
(3) The manufacturing equipment is fully loaded continuously or for long time with the orders.
(4) Special purpose machines and tools are used to match the features of the components.
(5) High productivity can be achieved.
(6) High level of automation is justified economically.

20.2.2 Large-Batch Production

(1) Large quantities of similar components are produced in batches.
(2) Flexibility of operation of the machines is required to some extent.
(3) Special purpose machines and toolings can be justified. These machines may be associated with long set-up times.

20.2.3 Small-Batch Production

(1) A wide variety of components with fluctuating demand is produced in small batches.
(2) General purpose machines are used.
(3) The machines have to be frequently reset for batches of different components.

(4) There is considerable loss of productive time for resetting of machines.
(5) Automation must be flexible in operation and readily reprogrammable for different batches of components.

20.3 TYPES OF PLANT LAYOUT

The machines and facilities within department may be arranged along work flow or as per function of the machine. There are three basic types of layout:

(1) Process Layout or Functional Layout.
(2) Line Layout or Product Layout.
(3) Group or Cellular Layout. Combination of process and product layouts.

20.3.1 Process or Functional Layout

(1) Process layout has been traditionally used for batch production.
(2) Similar machines are arranged in one section. For example: turning section, milling section, drilling section, etc.
(3) Components have to be transported from one process-specialised section to another.
(4) Scheduling and control of production is complex due to poor workflow.
(5) Work-in-process is very high.
(6) Many supervisors are responsible for production.
(7) The layout should be flexible and suitable for fluctuating demands.
(8) Similar skills are grouped together in one section.

20.3.2 Product Layout or Line Layout

(1) Machines are arranged in sequence of operations.
(2) Parts flow from one machine to next during processing.
(3) Suitable for mass and large-batch production systems as there should be continuous flow of similar components.
(4) There is high level of dedicated automation.
(5) The work-in-process is low.
(6) Short and predictable production times.
(7) Simple production control.
(8) High capital investments in automatic flow lines.
(9) Considerable design and development costs.
(10) The layout is inflexible.
(11) Not suitable for variable product demand and frequent product design changes.

20.3.3 Group or Cellular Layout

(1) Suitable for small-batch production.
(2) Suitable for stable distribution of types of components.
(3) Suitable for companies manufacturing single items such as machine tools, hydraulic pumps, etc.
(4) Cells or groups of machines are formed to process families of related parts, requiring similar process in a variety of sequences.
(5) It gives better workflow and lower production time as compared to process layout.
(6) Both labour and machine productivity or utilization.
(7) Lower work-in-process.
(8) Easy production control.
(9) Increased labour mobility and flexibility. Number of machines are more than number of operators.
(10) The key machines are fully loaded requiring full time operator.
(11) Secondary machines may be used by fewer operators. The operators move from machine to machine.
(12) Modem flexible manufacturing system is based on this type of layout.
(13) Multi-product or group flow lines can be used for some families of components who have similar sequence of operations. Unidirectional flow lines can be used for the above operation.
(14) General purpose machines may be used along with mechanical conveyors for work handling.
(15) Although resetting of machines is required for a change of product, however due to similarities between components, the resetting can be kept at low level.

Typical different layouts are shown in Fig. 20.1.

20.3.4 Comparison of Different Types of Plant Layouts

The general characteristics of different types of plant layouts are summarised and compared in Table 20.1.

20.4 TYPES OF AUTOMATION

Automation is a technology to operate and control manufacturing systems by the use of mechanical, electronic, mechatronic, electrical, computer and information technology based systems. The scope of this technology includes the following:

(1) Industrial robots.
(2) Automatic machining centres.
(3) Automatic assembly machines.

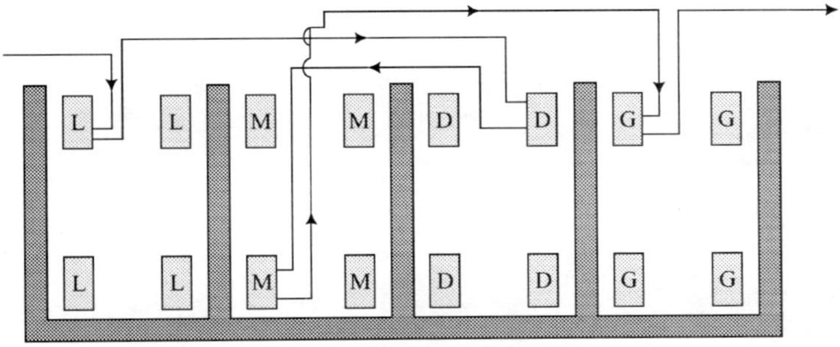

Fig. 20.1(a) Process Layout.

Key, L = Lathe, M = Milling Machine D = Drilling Machine G = Grinder

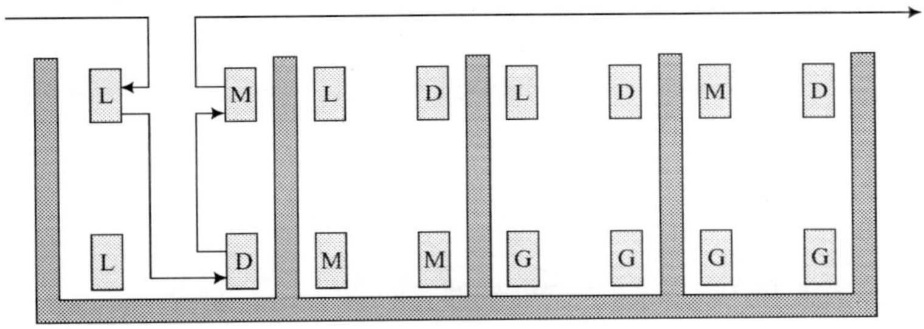

Fig. 20.1(b) Product Layout.

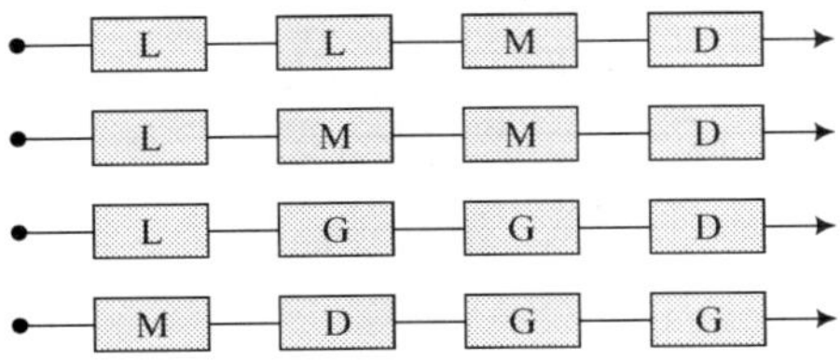

Fig. 20.1(c) Group Layout.

(4) Automatic material handling and storage systems.

(5) Automated drafting machines used along with CAD.

(6) Automated process planning, data collection, decision support systems and management information systems.

(7) Rapid prototyping.

(8) Feedback control systems.

(9) Nanotechnology and micro-machining.

(10) Automatic inspection systems for quality control.

(11) Material resource planning.

Table 20.1 Comparison of Different Plant Layouts.

Sl. No.	*Characteristics*	*Types of Layout*		
		Process	*Product*	*Group*
1.	Specialization	By process	By component type	By component type
2.	Material flow between machines	In batches	Continuous	Nearly continuous
3.	Output times of components	Long	Shortest	Short
4.	Stocks of work-in-process	High	Lowest	Low
5.	Responsibility for quality	Many foremen per part	One foreman per part	One foreman per part
6.	Responsibility for delivery time	Many foremen per part	One foreman per part	One foreman per part
7.	Investment in special tooling	High (one set per operation per part)	High (one set per line, per part)	Low (one set per tooling family)
8.	Investment in buildings	High	Lower	Lower
9.	Control of material flow	Complex	Simple	Simplest

(12) Supply chain management software.

(13) Real time systems.

(14) Life support systems used in hospitals.

(15) CAD/CAM.

The industries using automation technology include automobile, aircraft, metal working, electronic appliances and machine tool manufacturing.

Automation used in manufacturing can be divided into three basic types:

(1) Fixed automation

(2) Programmable automation

(3) Flexible automation.

20.4.1 Fixed Automation

This type of automation is dedicated to a specific task or a limited range of tasks. The sequence of machining, processing and assembly operations is fixed by the configuration of equipment.

Some features of fixed automation are:

(1) Cams and other mechanisms are used to control motions.

(2) High initial investment and long set-up times.

(3) Restricted to large-volume production for which high productivity can be achieved.

(4) The equipment is relatively inflexible and cannot be readily adapted to new tasks or products.
(5) Fixed or dedicated automation use transfer lines and automatic lathes.
(6) Mechanized automobile assembly line is an example of fixed automation.

20.4.2 Programmable Automation

The equipment can be readily programmed for new tasks and products. Some features of programmable automation are:

(1) Can be used for low-volume production.
(2) Flexibility in production equipment to deal with changes in product configuration.
(3) Low production rates.
(4) High investment in general purpose machines and equipment.
(5) Programmable automation use numerically controlled (NC) machines, computer numerically controlled (CNC) machines, robots, etc.
(6) Resetting times are long.
(7) Frequent changes in setting of machines, tools, jigs and fixtures.

Two basic types of programmable automation are:

(i) Programme sequence control
(ii) Numerical control.

(i) ***Programme sequence control:*** A sequence of motions or operations is controlled by timers, relays or programmable controllers. The actual motions are limited by preset mechanical stops and limit switches.

Example: Many Capstan and Turrent lathes use sequence control, programmed by microprocessor-based sequence controllers.

(ii) ***Numerical control:*** Numerical control (NC) is the most important form of programmable automation for manufacturing. A programme of instructions in a format specific to the controller of the machine is fed to the machine controller. The position, velocity and acceleration of the machine motions can be controlled. Other auxiliary functions such as tool changing, switching on or switching off of coolant supply, etc. are also controlled.

In the past, the programmes for NC machines consisted of numerical data only. However, programmes for modem controllers use a mixture of alphabetical codes and numerical information.

The main features of NC machine tools are:

(1) Very flexible in operation.
(2) Highly suited to small-batch manufacturing.

(3) A new programme of instructions for the change of job can be prepared quickly.

(4) Changeover to new job can be achieved quickly.

(5) The jigs and fixtures used are simple and their change over time is quick.

(6) Programming is carried out away from the machine tool and does not interfere with production time.

(7) Numerical control has been applied to a wide range of manufacturing equipment.

Examples: Metal-cutting machine tools, welding, sheet metal working, tube bending, flame cutting, etc.

The most widespread application of NC is metal cutting machine tools.

20.4.3 Flexible Automation

Flexible automation is an extension of programmable automation where resetting or changeover time is reduced or eliminated. The part programmes are prepared off-line and then transmitted to the machine computer.

The main features of flexible automation are:

(1) Continuous production of mixed products.

(2) No or minimum time lost in changeover and resetting of machines, tools, jigs and fixtures.

(3) Low and medium production rates.

(4) Flexibility to deal with product design and manufacturing variations.

(5) High investment in specialised custom-engineered equipment.

Different forms of automation in machine tools are:

(1) Transfer Machines.

(2) Automatic Machines.

(3) Numerically Controlled (NC) Machine Tools.

20.5 TRANSFER MACHINES

A transfer machine is a special purpose manufacturing system for continuous production of identical or very similar components in mass production of consumer goods. The production of a component requires multiplicity of operations such as milling, facing, boring, drilling, etc. Each operation is carried out on unit head machine. A transfer machine consists of several machining unit heads fastened together by conveying units. The whole system constitutes one large automatic installation. Therefore, a transfer machine is a combination of individual machine tools sequentially arranged and interlocked with controls and a transferring system. Workpieces are loaded at one end and completed components leave the transfer line at the other end.

Transfer machining leads to increased productivity, reduced number of machine tools, reduced production cost, better quality of products and saving in floor space. However, the capital cost of such machines is very high. It requires skilled personnel. The main industries using transfer machining is the automobile industry where large-scale production of cylinder heads and blocks, gear boxes, etc. can justify the large capital layout required for this form of manufacture.

20.5.1 Types of Transfer Machines

In transfer machines, the component is automatically transferred from one machining operation to the next by one of the following systems.

(1) In-line transfer system along a conveyor.

(2) Rotary transfer system by a circular indexing table.

(3) Drum machine.

The transfer machines are constructed in modular fashion. The machining heads for drilling, boring and so on are standard machine tools added to the system at appropriate stations.

20.5.2 In-line Transfer Machine

It consists of a straight conveyor or a central bed into the sides of which the machining heads are dowelled and bolted at convenient fixed pitch. The component is held on a special pallet or fixture. A rapidly moving return conveyor parallel to the main transfer line is used to return the empty pallets. The size of an in-line machine can be almost unlimited. These machines have been employed to perform all the machining operations, for example, on an automobile engine cylinder block. In addition to machining operations, these systems also incorporate stations for inspection and sometimes assembly operations. A plan view of in-line transfer machine is shown in Fig. 20.2.

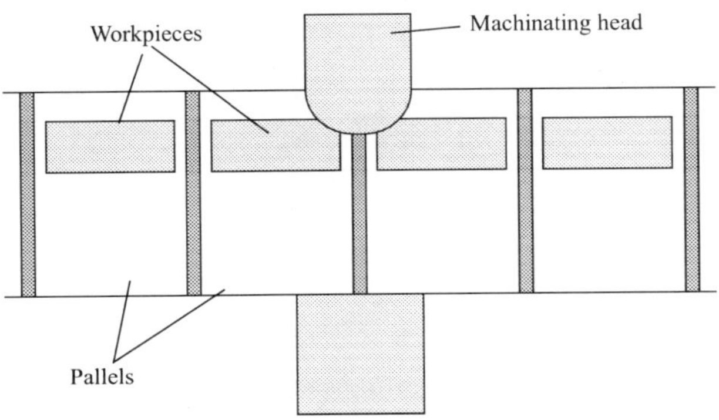

Fig. 20.2 In-line Transfer Machine.

20.5.3 Rotary Transfer Machine

When the space does not allow work to be conveyed in straight line, it may be more convenient to transfer work around a circular line. The workpieces are located on a circular table and are indexed around each successive machining station which are spaced as in-line transfer machine. The circular table top is supported on a high-quality bearing system at its centre with driving arrangements for its indexing. A rotary transfer machine is simpler than in-line transfer machine when number of machining stations is limited to say eight. This system is very successful for complete automatic assembly of a component where no metal removal is involved. The plan of a rotary indexing table transfer machine is shown in Fig. 20.3.

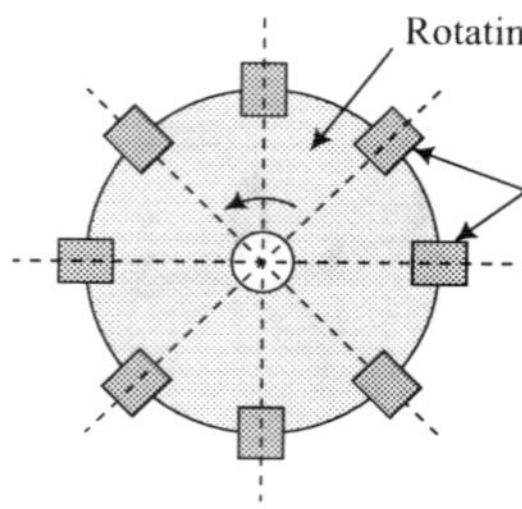

Fig. 20.3 Rotary Transfer Machine.

20.5.4 Drum Transfer Machine

The drum transfer machine is also a rotary transfer machine where the rotary table is replaced by a rotating drum along a horizontal axis. The workpieces are fastened to the outside surface or periphery of a drum. The workstations are radially positioned around the path at equal distances. The drum transfer machine has limited working capacity because of the limited space available in the system.

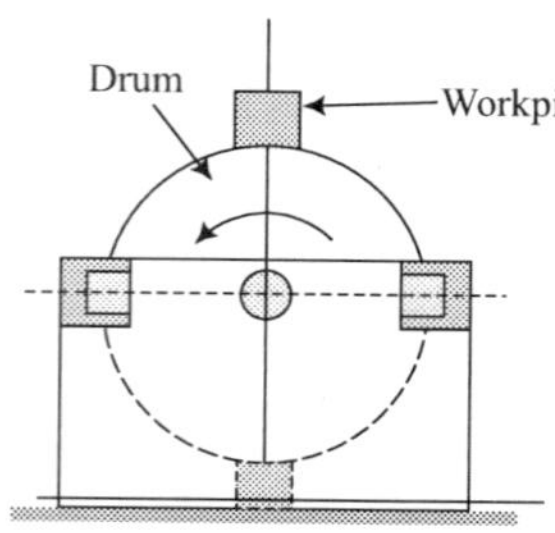

Fig. 20.4 Drum Transfer Machine.

20.5.5 Advantages and Disadvantages of Transfer Machines

The main *advantages* of use of transfer machines are:

(1) More output and lesser cost of production.

(2) A very few operations are needed.

(3) Close grouping of machines helps to save space.

(4) The life of cutting tools can be enhanced resulting in reduction of production cost of tool, time saving in tool replacement and resetting.

(5) Greater accuracy due to clamping of workpiece in fixture throughout.

(6) Fully automatic operation reduces time of production considerably.

(7) Heavy and odd shaped components can be handled more conveniently.

The following *disadvantages* limit the use of transfer machines.

(1) Heavy capital cost outlays.

(2) Complex control system.

(3) High overhauling and reshuffling costs.

(4) Breakdown of one machine results in stoppage of whole transfer line.

(5) Limited to mass production only.

20.6 AUTOMATIC MACHINES

Automatic machines are designed for transfer machines to meet the demand for mass production of components. Automatic lathes are most common as almost all machining operations can be carried out.

1. **Automatic Lathes:** There are single spindle automatic lathes as well as multi-spindle automatic lathes. The holding device for bar stock is collet and for other workpieces, chuck is used. In multi-spindle lathes, as many jobs as the number of spindles are machined simultaneously. There is huge increase in rate of production. Automatic lathes with only a few operations, turning and facing are used.

   Turret type automatic lathes with various slide motions can perform machining operations in a pre-set sequence with the help of cams and levers or hydraulic/pneumatic power.

   NC/CNC machines should be used if number of components required in a batch is not very large. These machines require very little time for reprogramming and change over for a new component. In case of automatic machines working with cams, changeover time is large as a new set of cams have to be used.

2. **Automatic Screw Cutting Machines:** These machines are used to produce screws and other similar components automatically from bar stock. These machines may be single-spindle or multi-spindle type. Cams and levers are activated by limit switches for automatic tool engagement and withdrawal, indexing of tool turret and feed of bar stock. These machines are economical for large production batches as machine setting time is large and costly.

3. **NC and CNC Lathes and Other Machine Tools:** Numerical Control (NC) and Computer Numerical Control (CNC) machine tools and lathes

are being used more widely for mass production of a variety of components. These are very flexible in operation and take very little time to switch over to a new component as compared to machine tools with hard automation.

NC machines operate on two types of control system.

(1) *Open loop control system:* The lead screw of the slide is activated by a stepper motor through a gear system. The computer feeds desired number of electrical impulses/oscillations to the stepper motor. The stepper motor rotates through an angle proportional to the signal. The rotation is transmitted to the lead screw directly or through a gear box. The speed of lead screw can be changed by varying the rate of pulses fed to the stepper motor. Thus, speed and distance moved by the slide are controlled.

(2) *Feedback control system:* The motion of the slide is measured and compared with the value fed to the computer. The error signal generated is fed to the drive of the slide. Electro-optical measuring system is popular for accurate measurement of slides. The slides can also be activated with the help of hydraulic power system.

Circulating balls are used between nut and lead screw to minimize the friction in the lead screw drive.

(3) *CNC chucker:* These machines are used for performing turning operations on large number of workpieces with length-to-diameter ratio less than one. These are nano-block type beds which support the head stock and slide. The NC/CNC machines are discussed in detail in chapter 21.

4. **Automatic grinders:** These are internal and external cylindrical grinding machines fitted with mechanisms for loading workpieces automatically and with automatic wheel dressing devices. These machines can continuously produce finished components.

20.6.1 Comparison of Various Automatic Systems

Selection of appropriate level of automatic for a particular manufacturing situation should be based on economics of production. The important factors are:

(1) Batch size

(2) Variety of components to be processed.

Manual machines are most suitable for a few components. NC machines are economical for medium batches and dedicated automation should be restricted to mass production. The more automated machines are not economical for very small batches because of high portion of production costs for set-up and programming.

A representative comparison of different levels of automation is given in Fig. 20.5.

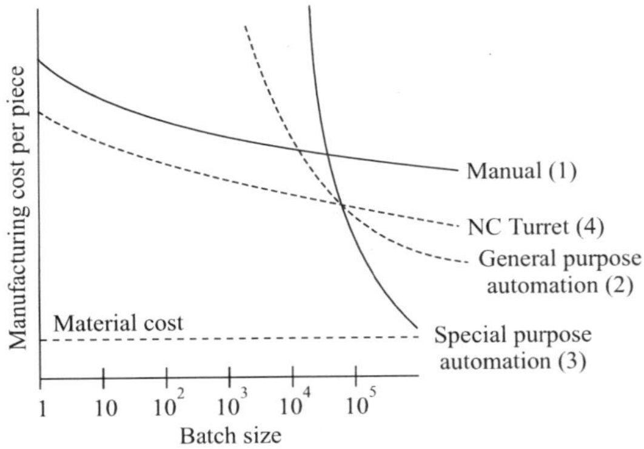

Fig. 20.5 Variation of manufacturing cost with batch size for different levels of automation.

(1) Manual production; use of a centre lathe
(2) General purpose automatics; a turret lathe
(3) Special purpose automatics; a transfer machine
(4) NC turret lathe

20.6.2 Automated Component Handling

Industrial robots are programmable handling devices which can be fitted with tooling for a variety of tasks such as welding, paint spraying, etc. A gripper or hand is used to handle workpieces. Many types of grippers have been developed to handle different types of items. For adapting a robot to a new task, the grippers have to be redesigned and robot must be reprogrammed.

The use of robots along with NC machines in a cellular layout is very attractive. The grippers are designed to handle a family of component parts without need of modification. Reprogramming of robots is very simple as compared to modification of grippers.

20.7 FLEXIBLE MANUFACTURING SYSTEMS

Flexible manufacturing systems (FMS) are computer-controlled interlinked multi-station machining complexes using group layout, NC machine tools and robotics. These are highly automated cells manufacturing families of components. The main features of FMS are:

(1) Interlinked NC workstation which operates on a limited range of work-pieces. General purpose NC machines or machining centres are most commonly used.
(2) Automatic guided vehicles (AGVs) and robots are used for automatic transportation, loading and unloading of tools and workpiece.

(3) The workpieces are mounted on pallets for transportation to overcome the problems associated with new set-up at each workstation.

(4) Centralized NC or DNC machines are used along with overall computer control of the system.

(5) There is little or no manual intervention and operation continues without interruption for long periods of time.

(6) There is inbuilt flexibility in the system to machine a variety of components without need to adjust machine setup or changing of tools.

20.7.1 Types of Flexible Manufacturing System

There are several variants of basic FMS concept:

1. ***Flexible manufacturing cells (FMCs):*** The main features are:
   (i) Stand-alone machine without interlinkage.
   (ii) Single-stage machining.
   (iii) Automated tool exchange.
   (iv) Machine tool with one buffer place.
   (v) Machine tool with workpiece storage.
   (vi) Wide range of parts (40 to 8000) in small batches from 15 to 500.
2. ***Flexible transfer lines (FTLs):*** The main features are:
   (i) Internal interlinking of NC machines with automatic machining transfer systems.
   (ii) Multi-stage machining but without flexible routing of workpieces.
   (iii) Cyclic transport of workpieces.
   (iv) Directed material flow.
   (v) Partial versatility of processing equipment with relatively short set-up times.
   (vi) The family of components is small (<20) and components must be similar to each other.
   (vii) The work cycles at each station must be quite well balanced.
   (viii) Production quantities should be large (1500 to 15000 per annum for each component).
3. ***Flexible manufacturing systems (FMSs):*** The main features of the system are:
   (i) External interlinking of NC machines by automatic workpiece transfer and handling.
   (ii) Flexible routing and automatic loading and unloading.
   (iii) Single and/or multi-stage machining.
   (iv) Non-cyclic transport of workpieces.
   (v) Automated material flow in a variety of sequences.

(vi) No manual setting up.

(vii) Machining time at each station can be different.

(viii) Number of different components that can be processed from 10 to 150.

20.7.2 Design of Flexible Manufacturing Systems

1. ***Machine tools***
   - Horizontal spindle machining centre.
   - Head-indexing machines for drilling, tapping.
   - Multi-spindle heads for processing several holes simultaneously.
2. ***Work handling***
   - Workpieces are mounted on standard pallets.
   - Work handling devices for transport of parts, pallets, tools around the system can be:

     Tow carts, automatic guided vehicles (AGVs), rail-carts, roller conveyors, industrial robots.
3. ***Layout:*** The layout depends upon the type of handling devices for transportation of workpieces from one workstation to another. Examples are:

   (i) In-line rail-cart layout with horizontal machining centres located at the side of the track.

   (ii) Simple loop layout using roller conveyors.

   (iii) Multiloop layouts with tow carts.

   (iv) Tree-type layouts with AGVs.
4. ***Tooling:*** There is a need for large number of tools of the order of 250 or more on each machine. Therefore, there should be means of automatically changing the tools from machine to machine. The following approaches can be followed:

   (i) Interchangeable tool magazine carousels.

   (ii) Automated tool transfer from fixed to moving carousels at each workstation.

   (iii) Automatic handling of tools and components individually together with supplying these together at each workstation.

   Additional capital expenditure and development costs are needed to adopt the above solutions.
5. ***Pallets and fixtures:*** The pallets are used to transport workpieces around FMSs and are precision items and costly. There is need for flexible fixturing which can be readily rebuilt and adopted to the range of workpieces to be machined.

20.8 AUTOMATED FACTORY

The sub-systems of a manufacturing system have been fully computerized. These sub-systems are:

(1) Product design.

(2) Process planning.

(3) Machine tool operation.

(4) Materials handling.

(5) Inventory control.

(6) Quality control.

(7) Sales.

Completely automated factory is being computered which will have information as the prime input regarding human needs and finished products. There would be no worker except for maintenance purposes. There will be a need for combination of software and hardware for product design, planning, control and operations. The factory will be completely managed by computer with the help of robots and AGVs. The factory building will have no windows, no lighting, no canteens, etc. Although complete automation in individual sections has been successfully achieved, a completely automatic factory has yet to come.

QUESTION BANK NO. 20

1. Discuss briefly various production systems.
2. Compare various characteristics of different plant layouts.
3. What is automation? Classify different types of automation.
4. Discuss various types of transfer machines.
5. Write a brief note on automatic machines.
6. How would you design a flexible manufacturing system?
7. Discuss various types of manufacturing system.
8. What are the main features of a FMS?

21

COMPUTER AIDED MANUFACTURING

Introduction: Product cycle: Scope of CAD/CAM/CIM: Process planning: Computer aided process planning (CAPP): Processing of NC programmes: Computerised information system: Question Bank.

21.1 INTRODUCTION

Computers are widely used in commercial and scientific fields. The scope of applications of computers is fast increasing due to new developments in computer systems. These developments include high computing power, high reliability, small size low cost, etc. The complicated and time consuming tasks can be rapidly completed by computers. There can be programmed to carry out complex calulations, process and store large amount of data with high speed and efficiency. Consequently, computers are used in the design and manufacture of component.

Computer-aided design (*CAD*) is the use of computer systems for creation, modification, analysis and optimization. The computer system is a combination of hardware and software.

Computer-aided manufacture (*CAM*) is the use of computer system to plan, manage and control the operation of a manufacturing plant.

Computer-integrated manufacture (*CIM*) is the integration of various activities of design, process-planning and manufacturing into a unified system so that "paperless" factory is achieved.

21.2 PRODUCT CYCLE

It is important to appreciate various stages in the design and manufacture of product before scope of CAD/CAM can be considered. The product cycle is shown in Fig. 21.1 which is built up of the following steps.

1. **Product concept:** A product is conceived on the basis of market and customer requirements. This product can be a modification of precious products.

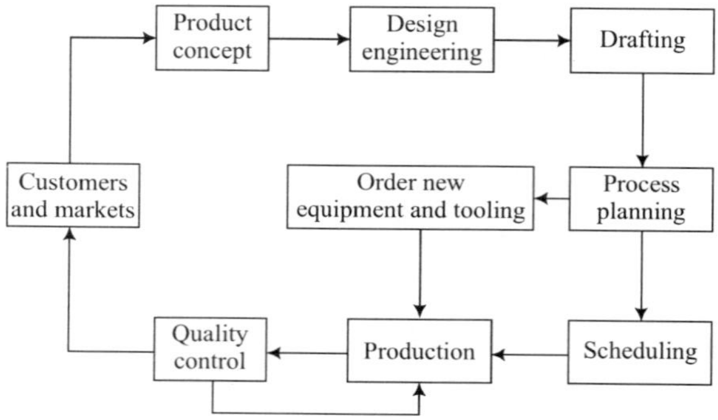

Fig. 21.1 Product cycle.

2. **Design engineering:** The product is designed in detail, including design analysis, geometric modelling and part lists are prepared.
3. **Drafting:** Drawings of assemblies, sub-assemblies, components alongwith bill of materials are prepared.
4. **Process planning:** Plans for production of various components and assemblies are prepared. This involves the selection of sequences of processes, machine tools, estimation of cycle times, determination of process parameters, i.e., feeds and speeds for machining operations. Process planning will include ordering of new equipment, tooling, digs and fixtures.
5. **Scheduling:** For sending the product for manufacture, scheduling and control of manufacturing has to be planned. The order and timing of each manufacturing step for each component and assembly is determined to meet an overall manufacturing schedule. Scheduling and production control involve large amount of data. Calculations have to be carried out for ordering quantities by subtracting stock levels from forecast of the number of items required during a particular period.
6. **Production and quality control:** The actual manufacturing and control of product quality has to be carried as per schedule.
7. **Customer and market:** The final products are delivered to the customers and markets.

21.3 SCOPE OF CAD/CAM/CIM

Computer-based procedures have been developed to facilitate each stage of production cycle of Fig. 21.1. Product cycle with computer-aided procedures is shown in Fig. 21.2.

1. **Computer-aided design:** The markets and customer requirement are used in the computer-aided design software. Geometric model of the product and its components are created in the computer.

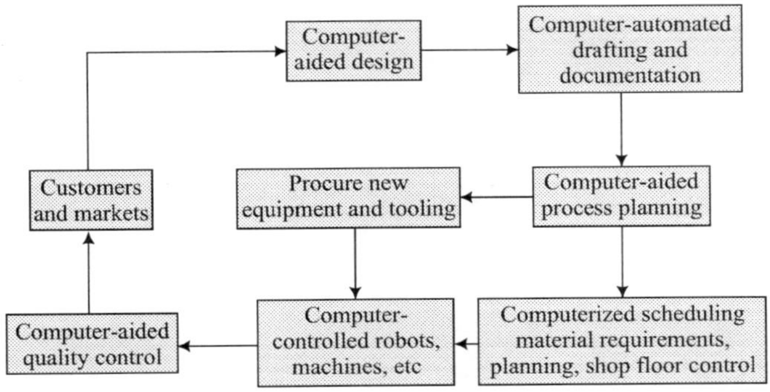

Fig. 21.2 Product cycle with computer aided procedures.

The models are analyzed using software packages for finite element stress analysis, mechanisms design, etc.

2. **Computer-automated drafting and documentation:** Drawing and part lists, if requnid, are produced with computer-aided drafting software.
3. **Computer-aided process planning:** Computer-aided process planning systems and NC programmes are prepared with the help of computer softwares to produce work plans, estimates, manufacturing instructions automatically from geometric description of the component and assemblies.
4. **Computerized scheduling, material requirement, planning, shop floor control:** Many commercial software packages are available for scheduling, inventory control, materials requirement planning (MRP) and shop floor control. Computers are used for control and monitoring of individual machines (CNC machines) at the shop-floor level.

21.3.1 Computer-integrated Manufacture (CIM)

The ultimate aim of CAM is to produce a "paperless" factory. This can be achieved by integration of various activities in the product cycle into a unified system in which data is transferred automatically from the present stage to the next. This is called computer-integrated manufacturing.

The design and process planning functions have to be carried out once in the product cycle, therefore, these functions can be integrated easily. The geometric data generated during design process forms the basic input of process planning to prepare manufacturing sequences and work plans. The design and process planning activities will form a common design and manufacturing data base.

The integrated design and manufacturing database is illustrated in Fig.21.3.

1. Geometric models of the products and components are created during design.
2. This data is accessed by NC progamming, process planning and robot programming.

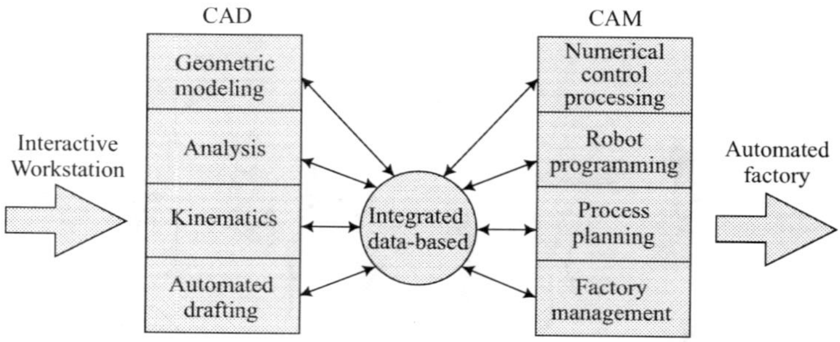

Fig. 21.3 Integrated Design and Manufacturing Database.

3. The programmes and work plans generated are added to the database.
4. The workplaces time estimates, bill of materials are accessed by production control and inventory control programmes and prepare the manufacturing schedules.

21.4 PROCESS PLANNING

The American Society of Manufacturing Engineers has defined process planning as "the systematic determination of the methods by which a product is to be manufactured, economically and competitively". Process plans are used:

1. To specify the sequence of production operations.
2. Order/arrange tool and facilities to transfer raw stock into finished parts.

Process planning is based on detailed knowledge of machining processes available and the production capabilities of the plant.

21.4.1 Stages in Design and Process Planning

The sequence of stages of design and process planning for manufacture of a product and is component parts is shown in Fig. 21.4. These stages can be grouped as follows:

1. **Stages 1 to 3: Product design:** The three stages are product conception, calculations, design and drafting. The output is the component and assembly drawings, specifications, part lists and bill of materials.
2. **Stages 4 to 9: Process planning:** The process planning consists of preparation of sequences of manufacturing operations, determination of processing parameters (feeds and speeds) for each operation based on drawings, part geometry, material and quantities required. Processing planning will involve in:
   (i) determination of operation sequences.
   (ii) selection of machines for each operation.
   (iii) detailed planning of each operation on each machine.

(iv) selection of work-loading methods.

(v) sequence of operations and tools.

(vi) determination of cutting data, i.e., feeds and speeds.

(vii) selection of cycle times.

3. **Stage 10 to 14: Manufacturing:** Manufacturing involves execution of detailed work plans prepared by process planning as per stage 4 to 9.

21.4.2 Scope of Computerization

1. **CAD system:** Computer-aided design packages are available to carry out design and drafting efficiently. In integrated computer-based system, the geometric modelling information and bill of materials (BOM) files are stored in a common database as shown in Fig. 21.3.
2. **Computer-aided process-planning:** Computer-aided process-planning systems have been developed to carry out operation from stage 4 to stage 9. Process planning is not an easy task to be computerized as it is not an activity that can be easily flow-charted. Many parameters have to be considered simultaneously and detailed decisions have to be taken. It is difficult to structure this process in an ordered manner. The computer-aided processing-planning system has to follow the planning logic of an experienced planner.

   However, process planning deals with lot of data and documents, standards, machinability data, machine specifications, tooling inventories, stock availability and existing process plans. Process planning is an information processing job and computers can handle it efficiently. The optimization of work plan may require many iterations. The use of computer enables alternatives to be readily explored. The computer-aided process planning can result in consistency in plans.
3. **Computer-aided machining:** Many computer-aided programming systems for NC machines have been developed. Determination of tool paths, automatic selection of tools and cutting conditions and driving of machine tools have been incorporated into programers.

21.5 COMPUTER-AIDED PROCES PLANNING (CAPP)

The plans for machining processes can be developed in two basic ways.

1. **Forward planning:** The initial decision is shape and size of basic workpiece. The sequence of operations is developed by removing material progressively in successive operations until final part geometry is achieved.
2. **Backward planning:** The starting point is final shape of the component. The operation sequence is developed by adding material removed by each operation in the reverse order until geometry of starting workpiece is achieved.

Table 21.1 Stages in Design and Process Planning.

Stage no.	*Description*	*Scope of computerization*
1.	Product conception	CAD system
2.	Calculations	
3.	Design and drafting	
4.	Machining squence	CAPP system
5.	Squence of machines	
6.	Clamping device	
7.	Sepuence of operation	
8.	Tool selection	
9.	Cutting data	
10.	Tool paths	NC Programming
11.	Tool location	
12.	Tool setting distance and collision checks	
13.	Post processing	
14.	Machining	

These are two computer-aided systems for work planning.

1. **Retrieval process-planning system:** This system utilizes the principles of group technology and work plans are based on families of similar parts.
2. **Generative processing-planning system:** The sequences of operations and process parameters are generated for a specific part.

21.5.1 Retrieval Process-planning System

The main features of the system are as follows.

1. This system is based on the principle of group technology. All parts having common or similar mechining operations are grouped together. However, shape of parts in a family may be similar or dissimilar.
2. Production schedule, tooling and machine set-up are worked out for the family and not individual component.
3. The system is very efficient as large production quantities in one family of parts shares common tooling and machine set-up.
4. High productivity equipment is economically feasible for large production quantities.
5. Only a few production sequences are needed for equal number of families. There is huge saving of time and paper work.
6. The machine tools are arranged as per cellular layout.
7. The data for various families such as sequence of operations, tools, jigs and fixtures is stored on magnetic tapes. This information can be retrieved for production of a new component.

8. The classification number of the part is to be manufactured is fed to the computer and information in the form of print-outs can be retrieved. The information will contain sequence of operations, machine group number, cutting speed, depth of cut, feed rates for individuals in the machining sequence.
9. This system cannot be used for each and every new part.

21.5.2 Generative Process-planning System

1. The component under production is described in terms of:
   (i) Geometry elements, e.g., cylindrical surfaces, plane surfaces, internal and external tapers, holes, splines, keyways, threads, etc.
   (ii) Dimensions
   (iii) Accuracy
   (iv) Surface finish
2. The computer is provided with technological data file is its back-up storage.
3. The tecnological data file contains list of operations and corresponding machine tools, jigs and fixtures and tooling, etc. for machining each geometrical element.
4. The alternative sequences of operations are displayed on the terminal. The optimum sequence is also displayed.
5. Other data such as production time and cost are displayed or printed out.

21.6 PROCESSING OF NC PROGRAMMES

Numerical control machines may be programmed by:

1. Manual methods using basic instruction set of the machine contoller
2. Computer-aided NC processing system.
3. Integrated CAD/CAM system

All methods require a good deal of skill and experience on the part of programmer. The geometry of the part and tool motion is described by the programmer in the language of the system.

This specialized knowledge is covered in chapter 22.

21.7 COMPUTERISED INFORMATION SYSTEM

The computerised information system or information storage and retrieval system for a manufacturing unit should contain the following.

1. Master data files
2. Package of software programmes
3. Data acquisition and updating system
4. Information display system
5. On-line control system

21.7.1 Master Data Files

The following master data files must be created and maintained.

1. **Engineering data master file:** Design data of components and assemblies: dimensions, tolerances, surface finish, material specifications, heat and surface treatments, classification into families.
2. **Process data master file:** Different sequences of operation, dimensional accuracy, surface finish, machine tools, tools, jigs and fixtures, machining conditions for different families or group of components.
3. **Inventory master file:** Data on new materials, bought-out items, tools, machines, consumables in the stock, on order, stocks under process, stocks allocated to job, list of suppliers, price, purchase conditions.
4. **Sales master file:** Order in hand, quantities delivery dates, forecast and history of sales.
5. **Schedule master file:** Detailed schedule for machines, work centres, shops: quantities, completion dates for machines, shops, component, assemblies, etc., schedules and quantities produced.
6. **Current status master file:** Updated data on employees, production facilities, machines under repair, etc.

21.7.2 Software Programmes

1. **Design office:** Part analysis, product and component design.
2. **Process planning:** Part programmers for NC machines.
3. **Sales deptt:** Planning, forecasting and market research.
4. **Production planning and control:** Linear programming, simulation and OR programmers.
5. **Accounts deptt:** Data processing programmes for recording transactions and preparation of statement of accounts.

21.7.3 Data Acquisition and Updating System

All data files should be speedily updated. Examples are:

1. New product design and processes. Data should be entered into respective files by remote terminals.
2. Stores transactions, issues, allocations, receipts, available stocks should be updated.
3. New order and deliveries, pending orders should be entered in sales files.
4. Worker attendance and machine status should be entered in current status file.

21.7.4 Information Display System

The manager should be able to obtain the required information on his personal computer from control computer to make decisions urgently. There may be changes made by other managers or departments in response to unforeseen situations. These changes must be known to the manager taking decisions. There should be a proper networking between central computer, peripheral computers and personal computers of the managers.

21.7.5 On-line Control System

The central computer should transmit commands to appropriate peripheral computers regarding production schedules and current plant status from master files. The peripheral computer has a library of part programmes as back-up storage. The peripheral computers can obey the central computer and transmit commands to production machines, robots, material handling system, assembly, inspection and testing machines.

The peripheral computers receives feedback from production machines through appropriate sensors. The feedback consists of parts completed, quality and machine status. It analyses the information and modifies future manufacturing instructions to the production machines. There is automatic on-line control system.

QUESTION BANK NO. 21

1. Explain product cycle and discuss the scope of computerisation of product cycle.
2. Write a note on CIM.
3. What is process planning? What is the scope of computerization?
4. What are different CAPP systems? List the merits and limitations of each system.
5. Explain the computerized information system for manufacturing unit.

22

NC/CNC PART PROGRAMMING

Introduction; Manual part programming; Computer-aided NC processing; Numerical control processing languages; NC programming using APT-based languages; Programming example in APT, Graphic-based NC processing system, Question Bank.

22.1 INTRODUCTION

Numerical control of machine tools is a method of automation by controlling the various functions of machine tools by a programme fed into it. The programme consists of precise instructions on manufacturing process and movements of machine tool. The machines can be used for any part and are controlled electronically, hydraulically or pnematically. The following functions are controlled:

1. Starting and stopping of spindle of machine tool.
2. Control of spindle speed.
3. Positioning the tool tip at desired locations and guiding along desired path by automatic control of motion of the slides.
4. Controlling feed rate by movement of rate of tool tip.
5. Change of tools in the spindle.

Part programmes are carried out manually for simple components and computer-aided programming is used for complex and repetitive operations.

The following steps are involved in the development of NC/CNC part programmers:

1. Process planning.
2. Selection of axes.
3. Selection tool.
4. Planning of cutting process parameters.
5. Planning of jobs and tool set-up.
6. Planning of machining paths.
7. Writing of part programme.
8. Testing of part programme.

The following documentation is essential for part programming:

1. Part drawing.
2. *Process planning* sheet containing sequence of operations, the name of machine tool, tools with speeds, feeds, etc.
3. *Tool cards* with details of tools, collects, chucks, etc.
4. *Setting card* showing positions of tools on machines and compensation values.
5. Programming sheet.
6. Punched paper tape where required.

22.2 MANUAL PART PROGRAMMING

Numerical control machines respond to a series of coded instructions stored in the controller. Earlier coded instructions were fed by means of punched paper tape. In modern applications, NC programmes are input to the machine controller electronically. The format of basic instructions is standardized in *word address format*.

Each set of instruction codes is called NC block and is terminated by an end of block character (eob). This means the carriage returns. The NC machine acts on each block of coded instructions in turn.

Example: A typical block of instructions is given below:

*n*001 *g*80 *x*75.0 *y*650 *z*10.0 *f*60.0 *s*2000 *t*001 *m*03 *eob*.

Each instruction consists of a letter code which indicates the address followed by data. Therefore, this is called word address format. The explanation for each word and code are given below:

1. **Sequence number (*n*-codes):** This is an identification number for each block of instructions and increases sequentially through the programme.
2. **Preparatory function (*g*-codes):** These are the main motion commands of the machine and are identified by the prefix *g*. These instructions tell the controller what type of motion or action is to be carried out. These codes are standardized and given in Table 22.1.

   g 02 indicates that next motion will be circular interpolation in a clockwise direction.
   g 80 indicates fixed cycle cancel.
3. **Coordinate data (*x*, *y*, *z*-codes, etc.):** These indicate the coordinates for the tool to move to (the end of the motion) is the manner indicated by the preparatory function (*g*-code) currently in force. The *x*, *y*, *z*-codes indicate the conventional cartesian coordinates. In four and five-axis machines other codes (*a*, *b*, *c*) may be used to indicate angular positions. For circular interpolation *i*, *j*, *k*-codes may be used to indicate the centre of the circular arc to be followed.

Table 22.1 Preparatory Functions (EIA Stndard RS-273).

Code	*Function*
*g*00	Point to point positioning
*g*01	Linear interpolation (straight line motion)
*g*02	Circular interpolation arc CW
*g*03	Circular interpolation arc CCW
*g*04	Dwell
*g*05	Hold
*g*06 & *g*07	Unassigned
*g*08	Acceleration
*g*09	Deceleration
*g*10	Linear interpolation (long dimensions)
*g*11	Linear interpolation (short dimensions)
*g*12	Unassigned
*g*13-*g*16	Axis selection
*g*17	xy plane selection
*g*18	zx plane selection
*g*19	yz plane selection
*g*20	Circular interpolation arc CW (long dimensions)
*g*21	Circular interpolation arc CW (short dimensions)
*g*22-*g*24	Unassigned
*g*25-*g*29	Permanently unassigned
*g*30	Circular interpolation arc CCW (long dimensions)
*g*31	Circular interpolation arc CCW (short dimensions)
*g*32	Unassigned
*g*33	Thead cutting constant lead
*g*34	Thead cutting, increasing lead
*g*35	Thead cutting, decreasing lead
*g*36-*g*39	Reserved for control use only
*g*40	Cutter compensation cancel
*g*41	Cutter compensation-left
*g*42	Cutter compensation-right
*g*43-*g*49	Cutter compensation if used; otherwise unassigned
*g*50-*g*59	Unassigned
*g*60-*g*79	Reserved for positioning only
*g*80	Fixed cycle cancel
*g*81	Fixed cycle 1
*g*82	Fixed cycle 2
*g*83	Fixed cycle 3
*g*84	Fixed cycle 4
*g*85	Fixed cycle 5
*g*86	Fixed cycle 6
*g*87	Fixed cycle 7
*g*88	Fixed cycle 8
*g*89	Fixed cycle 9
*g*90-*g*99	Unassigned

4. **Feed rate (*f*-code):** This specifies the feed rate for the operation in millimetres per minute.
5. **Spindle speed (*s*-code):** This specifies the spindle speed to be used for the operator in rpm.
6. **Tool number (*t*-code):** This indicates which tool is being used for the operation.
7. **Miscellaneous functions (*m*-code):** These codes programme various auxiliary functions on the machine tools. These are standardized and given in Table 22.2.

Table 22.2 Miscellaneous Functions (EIA Standard RS-273).

Code	*Function*
*m*00	Program stop
*m*01	Optional (planned) stop
*m*02	End of program
*m*03	Spindle on CW
*m*04	Spindle on CCW
*m*05	Spindle off
*m*06	Tool change
*m*07	Coolant No.2 on
*m*08	Coolant No.1 on
*m*09	Coolant off
*m*10	Clamp
*m*11	Unclamp
*m*12	Unassigned
*m*13	Spindle on CW & coolant on
*m*14	Spindle on CCW & coolant on
*m*15	Motion +
*m*16	Motion –
*m*17-*m*24	Unassigned
*m*25-*m*29	Permanently unassigned
*m*30	End of tape (tile)
*m*31	Interlock bypass
*m*32-*m*35	Constant cutting speed
*m*36-*m*39	Unassigned
*m*40-*m*45	Gear changes if used; otherwise unassigned
*m*46-*m*49	Reserved for control use only
*m*50-*m*99	Unassigned

*m*03: Start the spindle with clockwise rotation

Note: (1) With word address format, information need not be repeated in successive blocks if it is to remain the sames for subsequent block. For example,

*n*001 g01 *x*75.0 *y*65.5 *z*10.0 *f*60.0 *s*2000 *m*05 *eob*

*n*002 *x*87.5 *y*75.0 *eob*. The g–, *f*–, *s*– and *m*– code remain the same for the second block also

(2) For safety reasons, some preparaotry functions are cancelled after the block is acted upon and must be repeated in a subsequent block, if required. For example, for circular interpolations (*g*02, *g*03) and codes for thread cutting.

22.2.1 Fixed or Canned Cycles

Preparatory codes *g*80 and above are used to fixed or canned cycle programmes. The examples are: drilling cycle, tapping cycle, $2^1/_2$ axis milling machines. Users own fixed cycles for common operations.

(i) **Drilling cycle (*g*81):** The use of preparatory function *g*81 will cause a drill to move at rapid transverse above the work surface to a specified *x*, *y* position, then to drill down to specified *z* depth at the specified feed rate. This will be followed by rapid retraction to just above the work surface.

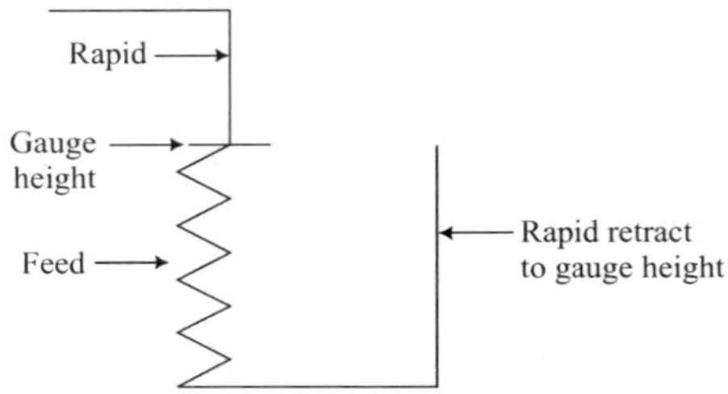

Fig. 22.1 Drilling cycle (*g*81).

(ii) **Tapping cycle (*g*84):** This cycle is similar to drilling cycle. The spindle is reversed and the tool feeds out at the appropriate tapping feed.

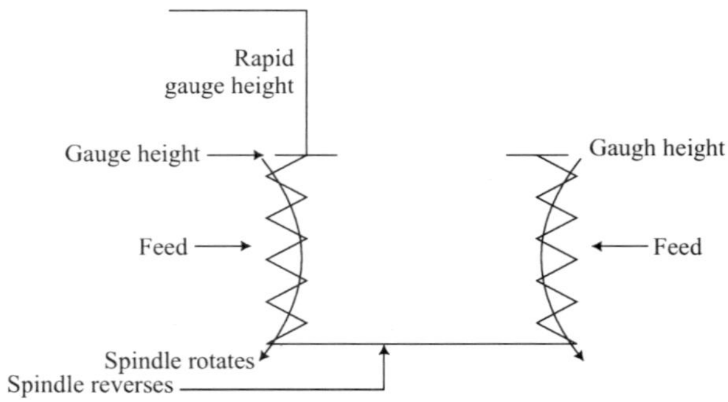

Fig. 22.2 Tapping cycle (*g*84).

22.2.2 Tool Path Coordinates

Determination of tool path coordinates is an important part of NC programming. It involves many repetitive calculations, some of which may be highly complex.

The tool paths are offset from the surface of workpiece. The programme point of the tool is driven along the tool path to machine the workpiece. Each path is made up of linear or circular arc (circular interpolation) line segments. The coordinates of the change points between each of line segments must be determined.

Example: Tool paths segment change points for a milling operation in which an end mill machines the outside profile of the workpiece are shown in Fig. 22.3. The change point coordinates to be determined are shown as the centres of the dotted circles which represent the tool positions. Additionally, the coordinates of the centres of the circular arcs in the profile have to be deterrmined.

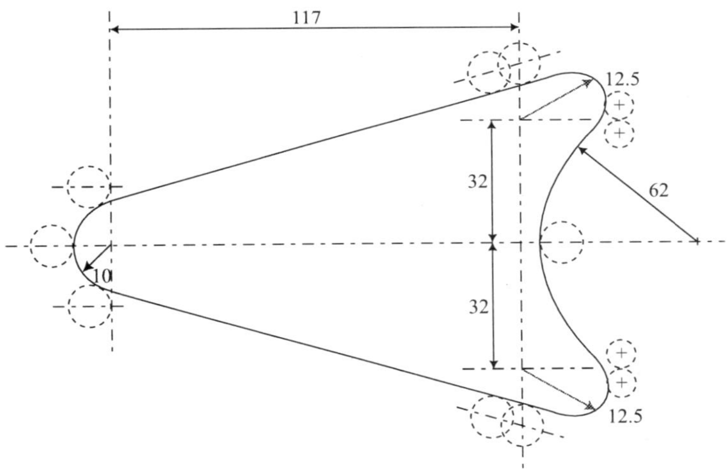

Fig. 22.3 Milling of outside profile.

This is a simple example which requires quite complicated calculations to determine the tool path coordinates. Determination of tool paths for complex three-dimensional surfaces cannot be done manually.

22.3 COMPUTER-AIDED NC PROCESSING

Computer-aided NC processing system have been developed to carry out pogramming of NC machines using stored data on tools, materials, machines and geometric algorithms, from an input of the geometric description of the components to be machined. These are economical for a company having many NC machine tools. Also complex workpieces with three-dimensional surfaces require NC processing systems.

22.3.1 Processing Stages

Computer-aided NC processing is carried out in two or more stages as shown in Fig. 22.4.

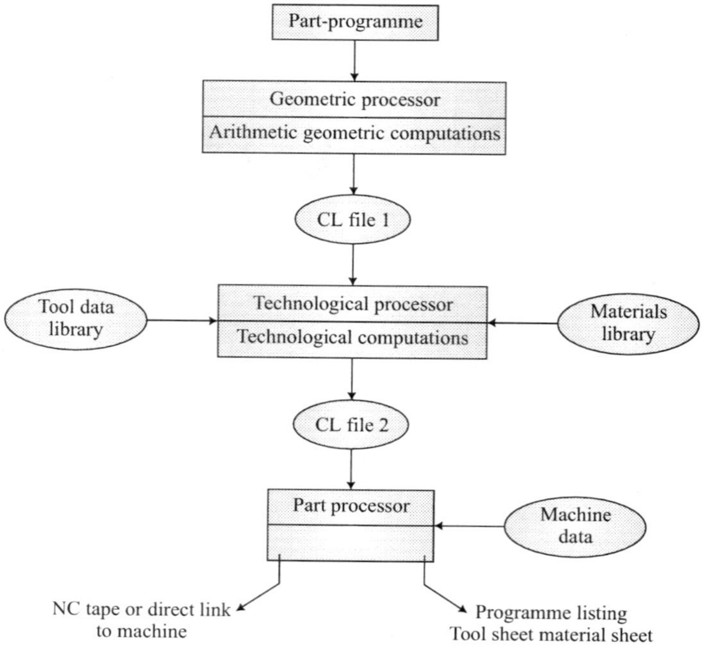

Fig. 22.4 Processing stages for computer-aided programme preparation.

1. NC language is used to describe the geometry of the part.
2. A processor carries out the main geometric calculations.
3. The NC processor has two stages.
   (i) *Geometric processing stage:* To determine the offset tool path positions to machine the workpiece.
   (ii) *Technological processing stage:* To select the tools and determine proper feeds and speeds. Files of standard tool data and machined data are acessed.
4. The output from the processor is Cutter Location (CL) file which contains the coordinates of tool positions in a standard format.
5. A post-processor programme is used to convert general CL data into specific input instructions for the NC machine to be used.
6. A tape image file is the output of post processor.

22.3.2 Functions of Portprocessor Programme

The main functions of portprocessor programe are:

1. Reading of CL data.
2. Conversion of above CL data into machine tool coordinate systems.
3. Conversion of data into absolute coordinates or increment coordinates as required.
4. Checking of machine tool limitations, i.e., allowable feeds and speed, slide travel, interference, etc.
5. Development or selection of feeds and speeds for the given machine.
6. Development of motion commands for the given machine and controller.
7. Allowance for the type of interpolator.
8. The output of machine tool programme should be in proper block format and code set, i.e., *g*-godes.
9. Printing out of porgramme for checking.
10. Priting out of error codes and diagnosis.

22.4 NUMERICAL CONTROL PROCESSING LANGUAGES

A large number of NC processing languages have been developed. These can be classified based on input information level and similarites of language structure.

22.4.1 Input Information Level

The languages are classified into three groups

1. **Group I:** The finished part geometry is used as basic input for the determination of cutter locations data. All motions are specified and feeds, speeds, tools, etc. are selected. ADAPT and APT processor belong to this group.
2. **Group II:** The initial workpiece geometry along with finished part geometry are used as basic input: The programme contains routines for automatic processing of tool path for area clearance, pocket milling, machining patterns of hole. Feeds, speeds and tools are selected by the programmer. API, NELAPT, IFAPT and COMPACT II languages belong to this goup.
3. **Group III:** Based on files of tool and material data, tool, feed, speed selection and other technological decisions are made by the programmer. The finished part geometry, initial workpiece geometry and workpiece material are used as basic input. Sequences of tools and operations are automaticaly determined by the processor. EXAPT and MITURN languages fall wihin this group.

22.4.2 Similarity of Language Structure

NC processing languages can also be classified as per similarity of language structure into:

1. Free-Format languages
2. Fixed-Format languages

1. **Free-format languages**: These are similar in concept to high-level computer programming languages such as BASIC or FORTRAN. These languages are flexible to use, easy to add more facilities and used for different types of machine tools. A vocabulary of words are used to describe part geometry and tool motions.
2. **Fixed-format languages:** A series of numbers and codes are entered in a fixed sequence and used for particular types of machine tools. These are easy-to-use languages and can be programmed without much training. However, these languages are not flexible and difficult to expand their facilities.

22.4.3 Division of Languages

NC programming languages can also be divided into APT-based and Non-APT languages.

1. ***APT*-based languages:** These are free-format languages based on APT language structure. There are very commonly used languages and have become part of modern *CAM* systems. These languages are very powerful for generation of NC programmes for very complex parts using multi-axis machine tools. This group of languages include APT, ADAPT, NELAPT, IFAPT, EXAPT, etc.
2. **Non-APT languages:** This group includes COMPACT–H, AUTO SPOT, SPLIT, etc.

22.5 NC PROGRAMMING USING APT-BASED LANGUAGES

Automatically Programmed Tools (APT) processor can deal with programming of machine tools with five or more axes of control

1. **Geometric statements:** APT language has more than 300 English-like words to describe component geometry and tool motion. The geometry of component and workpiece is described by points, lines, circles, surfaces, etc. Each geometric stratment is of the form:

   SYMBOL (NAME) = WORD/MODIFIER, MODIFIER

Example 1 A point is defined as:

$$P = \text{POINT}/b, 20, 40$$

Example 2 A circle is defined as:

$$CIRC1 = CIRCLE/\text{CENTER}, P, \text{RADIUS}, 57.75$$

2. **Motion statements:** These are commands which describe the method of machining of previously defined geometry. Each statement has the following forms

   Motion command/descriptive data (modifier)

   (i) *Point-to-point motion statement:* A simple point-to-point motion statement is:

   $GOTO/P1.$

   The tool will move from current position to point $P1$ for which coordinates have already been defined.

   (ii) *Incremental motion:* These are specified by the statement *GO DLTA*.

   Example: GODLTA/ 2.0, 3.0, 4.0 moves the tool by 2.0, 3.0 and 4.0 increments in *X*, *Y*, *Z* dimetions, respectively.

   (iii) *Contouring operation:* The general method for specifying motion is in the following three intersecting surfaces:

   1. Drive surface: The tool moves parallel to drive surface during cutting and side of the cutter remains in contact with this surface.
   2. Part surface: The end of the tool remains in contact with part surface during cutting.
   3. Check surface: The tool is brought upto check surface to end the cutting.

      The drive, part and check surfaces can be orthogonal or three-dimensional curved surfaces. The tool follows drive surface leaving the part surface until it reaches the check surface.

   (iv) APT *Motion Commands:* The following motion commands are specified as illustrated in Fig. 22.5.

   GOLEFT, GOFWD, GOUP, GORCHT, GOBACK, GODOWN

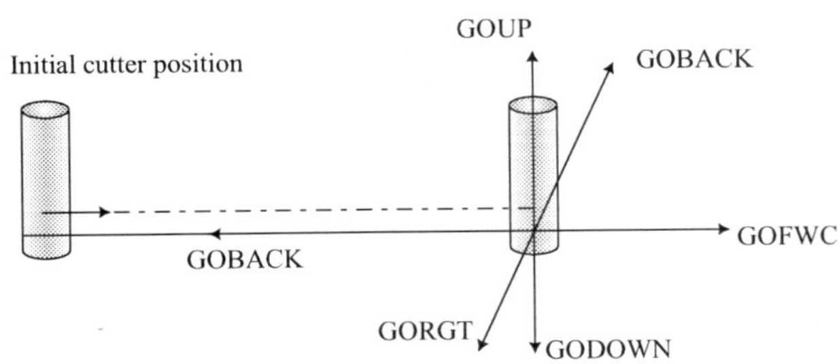

Fig 22.5 APT Motion commands.

   (v) *Motion command modifiers:* Motion command modifiers are used for finishing position at the check surface. The main modifiers are:

   TO, ON, PAST, TANTO. Their effects are shown in Fig. 22.6

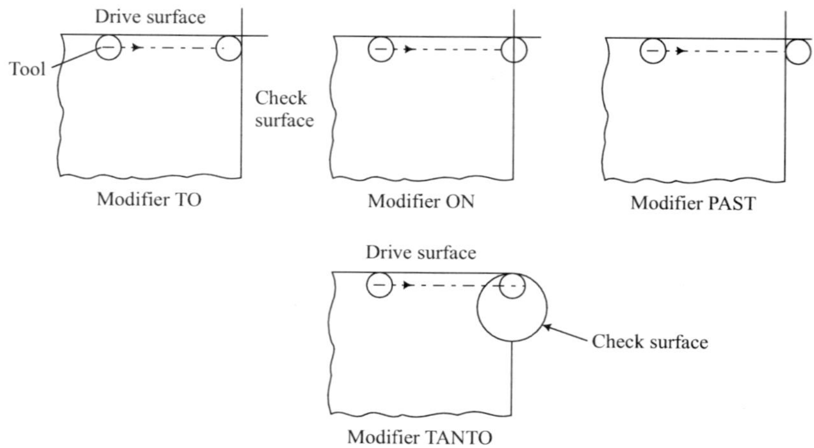

Fig. 22.6 Motion command modifiers for the APT systems.

(vi) **Auxiliary Statements:** The auxiliary statements used are:

PARTO, INTOL/, OUTTOL/, CUTTER/, etc.

(i) *Cutter specification:* Tool geometry must be known for determination of offset tool paths by APT processor. The cutter definition for milling and drilling operations are shown in Fig. 22.7. Example cutter *D*, *r*, *E*, *F*, α, β, *h*.

CUTTER/12.5 defines a 12.5 mm-diamenter end-milling cutter as other dimensions are zero by default.

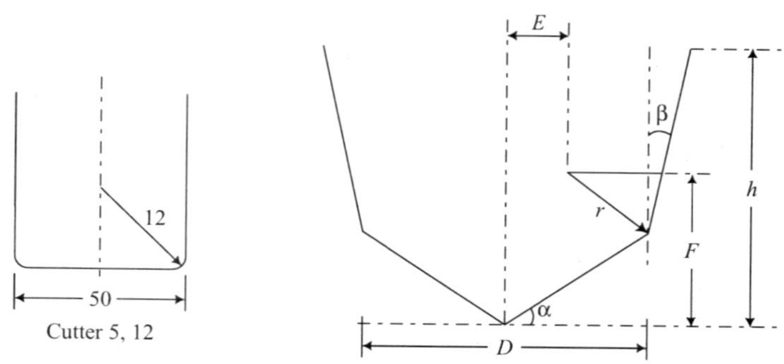

Fig. 22.7 Cutter definition in APT system.

(ii) *Statements directly processed to port processor:* These statements in APT language include:

COOLNT/, RAPID, SPINDL/, FEDRAT/, TURRET/, MACHIN;

(iii) *Other statements:*

FEDRAT : Feed rate

SPINDLE : Spindle speed.

22.6 PROGRAMMING EXAMPLE IN APT

The component shown in Fig. 22.3 is to be machined with a 25-mm-diameter end-milling cutter.

The APT programme to generate NC programme to machine the component is given in Table 22.3 where statements 1 to 9 are header and postprocessor statements. The tool, feed and speed are selected by the programmer and specified by these commands. Statements 10 to 21 are geometry statements and programme has decided to use to define the geometry of the component.

The geometry specified by geometry statements is shown in Fig. 22.8. Statements 22 to 31 of table 22.3 are the motion commands. The effect of statement 27 is shown in Fig. 22.9.

12 – 19 Remark geometric statements follow

*P*1 = POINT/0, 0, 6

*P*2 = POINT/ 117, 32, 6

*P*3 = POINT/117, –32, 6

*C*1 = CIRCLE/CENTER, *P*1, RADIUS, 10

*C*2 = CIRCLE/CENTER, *P*2, RADIUS, 12.5

*C*3 = CIRCLE/CENTER, *P*3, RADIUS, 12.5

*L*1 = LINE/RIGHT, TANTO, *C*1, RIGHT, TANTO, *C*3

*L*2 = LINE/LEFT, TANTO *C*1, LEFT, TANTO, *C*2

*C*4 = CIRCLE, *X* LARGE, OUT, *C*2, OUT, *C*3, RADIUS, 62

SP = POINT/ –25, –25, 25

*PL*1 = PLANE/*P*1, *P*2, *P*3

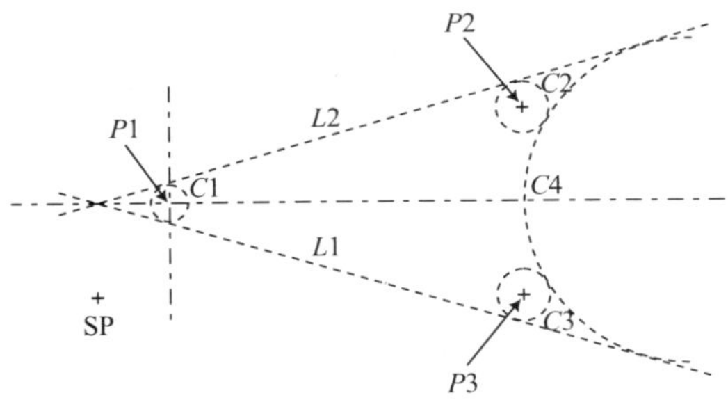

Fig. 22.8 Geometric statements in APT for the milling operation in Figure 12.12.

Table 22.3 APT Program for part shown in Figure 22.3.

SEQ No.	*INPUT STATEMENT*	
1	PART No. STOP PLATE (DRG. No. AJ001X7)	Header
2	REMARK HEADING STATEMENTS FOLLOW	
3	NOPOST	
4	FULIST/ON	
5	CLPRNT	
6	CUTTER/25	
7	SPINDL/350	
8	TOLER/0.03	
9	FEDRAT/15	
10	REMARK GEOMETRIC STATEMENTS FOLLOW	Geometry statements
11	P1 = POINT/0, 0, 6	
12	P2 = POINT/117, 32, 6	
13	P3 = POINT/117, -32, 6	
14	C1 = CIRCLE/CENTER, P1, RADIUS, 10	
15	C2 = CIRCLE/CENTER, P2, RADIUS, 12.5	
16	C3 = CIRCLE/CENTER, P3 RADIUS, 12.5	
17	L1 = LINE/RIGHT, TANTO, C1, RIGHT, TANTO, C3,	
18	L2 = LINE/LEFT, TANTO, C1, LEFT, TANTO, C2	
19	C4 = CIRCLE/XLARGE, OUT, C2 OUT, C3, RADIUS, 62	
20	SP = POINT/-25, -25, 25	
21	PL1 = PLANE/P1, P2, P3	
22	REMERK MOTION STATEMENTS FOLLOW	Motion statement
23	FROM/SP	
24	GO/TO, L1, TO, PL2	
25	GORGT/L1, TANTO, C3	
26	GOFWD/C3, TANTO, C4	
27	GOFWD/C4, TANTO, C2	
28	GOFWD/C2, TANTO, L2	
29	GOFWD/L2, TANTO, C1	
30	GOFWD/C1, TANTO, L1	
31	GOTO/SP	
32	STOP	
33	PRINT/3, ALL	
34	FINI	

```
FROM/SP
GO/TO L1, TO P1
GO RGT/L1, TANTO, C3
GO
GO FWD/C3, TANTO, C4
GO FWD/C4, TANTO, C2
```

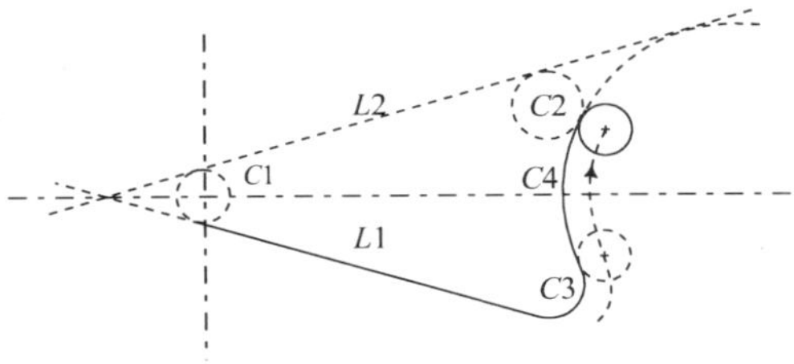

Fig 22.9 Motion statements convert the unbounded geometry into the required part profile.

22.7 GRAPHICS-BASED NC PROCESSING SYSTEMS

Graphics-based NC processing systems are replacing language based systems. These systems can be

(i) Integral part of comprehensive *CAD* system and use solid modeling of part geometry. The tool paths generated are associated with *CAD* geometric model.

(ii) Standalone system which processes NC programme for output of *CAD* systems. A number of commercial systems are available that run on graphics workstation or *PCs*.

1. Tool and materials database maintained by user can be incorporated.
2. The part geometry can be entred into the system with the help of a special purpose graphics-based user interface.
3. Geometry processes can be read into the system from an available *CAD* system directly.
4. The routines for tool part generation used in APT language processes are incorporated into the software.
5. Simple editing routines are included for quality postprocessors for new machines and controllers.
6. A library of postprocessors of popular machine tools are available in the market.

QUESTION BANK NO. 22

1. What steps are involved in the development of NC/CNC part programmes?
2. Explain step by step manual part programming.
3. Explain briefly computer-aided NC processing.
4. Write a short note on graphics-based NC processing systems.

Part V

Annexures

Annexure I

MACHINING DATA

Table I.1 Cutting speeds and feeds for turning operation.

Work Material	*Tool material*	*Cutting Speeds*			
		Depth of cut in mm			
		5 - 10	*2.5 - 5*	*0.5 - 2.5*	*0.1 - 0.5*
		Feeds in mm/rev.			
		0.5 - 0.6	*0.3 - 0.5*	*0.25 - 0.35*	*0.05 - 0.25*
Free Machining Steel	HSS Carbide	20 - 50 100 - 150	50 - 70 125 - 175	40 - 50 150 - 250	50 - 120 200 - 500
Mild Steel	HSS Carbide	25 - 35 60 - 120	35 - 50 85 - 150	40 - 60 120 - 200	45 - 80 150 - 450
Medium Carbon Steel	HSS Carbide	15 - 25 50 - 110	25 - 45 60 - 120	25 - 50 100 - 150	35 - 70 120 - 300
Alloy Steel	HSS Carbide	15 - 20 30 - 65	15 - 25 40 - 80	15 - 36 65 - 100	120 - 300 80 - 180
Tool Steel	HSS Carbide	15 - 20 50 - 110	20 - 25 60 - 120	25 - 30 100 - 150	30 - 60 120 - 300
Stainless Steel	HSS Carbide	15 - 20 35 - 60	20 - 25 40 - 70	20 - 30 50 - 80	25 - 50 50 - 90
Cast Irons Grey, ductile and malleable	HSS Carbide	20 - 25 60 - 90	25 - 30 70 -100	30 - 45 85 - 110	40 - 60 80 - 120
Alluminium Alloys	HSS Carbide	40 - 75 60 - 150	75 -100 90 - 180	90 - 120 90 - 450	100 - 120 150 - 600
Copper Alloys	HSS Carbide	40 - 60 50 - 110	60 - 100 60 - 150	90 - 120 100 - 180	120 - 200 120 - 310
Magnesium Alloys	HSS Carbide	40 - 75 60 - 150	70 - 100 90 - 180	90 - 120 90 - 450	100 - 200 150 - 600
Titanium Alloys	HSS Carbide	10 - 15 15 - 30	15 - 30 30 - 90	30 - 60 60 - 90	60 - 90 60 -120

Table I.2 Speeds and feeds used for milling operations.

Work Material	*Tool Material*	*Cutting Speed m/min*	*Feed mm per tooth*				
			Face Milling	*Slab Milling*	*Slotting, side and face milling*	*End milling*	*Form milling*
Free Machining Steel	HSS Carbide	35 - 40 90 - 120	0.3	0.25	0.175	0.15	0.1
Mild Steel	HSS Carbide	20 - 40 85 - 130	0.25	0.2	0.15	0.125	0.1
Medium Carbon Steel	HSS Carbide	15 - 30 60 - 100	0.2	0.15	0.125	0.1	0.075
Alloy Steel	HSS Carbide	10 - 25 40 - 60	0.15	0.1	0.075	0.06	0.05
Tool Steel	HSS Carbide	10 - 25 60 - 85	0.2	0.15	0.1	0.08	0.05
Stainless Steel	HSS Carbide	10 - 20 25 - 60	0.15	0.1	0.1	0.0.8	0.05
Cast Irons*	HSS Carbide	30 - 35 65 - 100	0.35	0.3	0.2	0.15	0.1
Aluminium Alloys	HSS Carbide	55 - 100 65 - 180	0.5	0.4	0.3	0.2	0.18
Copper Alloys	HSS Carbide	40 - 80 60 - 100	0.3	0.25	0.2	0.2	0.15
Magnesium Alloys	HSS Carbide	60 - 100 60 - 180	0.5	0.4	0.25	0.2	0.18
Titanium Alloys	HSS Carbide	10 - 30 25 - 50	0.15	0.125	0.1	0.1	0.05

*Grey, ductile and malleable

Table I.3 Speeds for drilling, reaming and tapping operations.

Work Material	*Cutting Speed-m/min*		
	Drilling	*Reaming*	*Tapping*
Free Machining Steels	20 - 30	11 - 20	8 - 12
Mild Steel	20 - 25	11 -15	10 - 12
Medium Carbon Steel	15 - 20	10 - 15	8 - 11
Alloy Steel	15 - 22	10 - 15	8 - 12
Tool Steel	5 - 8	3 - 5	3 - 5
Stainless Steel	10 - 15	9 - 25	8 - 10
Cast Irons*	20 - 25	10 - 17	9 - 12
Aluminium Alloys	30 - 55	25 - 30	15 - 18
Copper Alloys	25 - 45	20 - 40	10 - 12
Magnesium Alloys	60 - 111	30 - 45	15 - 25
Titanium Alloys	10 - 15	10 - 25	8 - 9

Table I.4 Feeds for drilling and reaming operations.

Hole Diameter mm	*Feed in mm/rev*		*Hole diameter mm*	*Feed in mm/ rev*	
	Drilling	*Reaming*		*Drilling*	*Reaming*
1 - 2 2.5 - 4	0..4 - 0.06 0.05 0.1	0.08 - 0.12 0.1 - 0.2	15 - 18 18 - 21	0.18 - 0.25 0.2 - 0.3	0.5 -0.6 0.5 - 0.7
4.5 - 6	0.05 - 0.15	0.15 - 0.3	21 - 24	0.22 - 0.3	0.56 - 0.75
6.5 - 8.5	0.1 - 0.2	0.2 -0.4	25 - 3	0 0.25 - 0.35	0.61 - 0.80
9 - 12	0.145 - 0.2	0.3 - 0.5	32 - 38	0.28 - 0.4	0.71 -0.9
12 - 14	0.15 -0.25	0.4 - 0.6	Over 40	0.3 - 0.4	0.8 - 1.0

Table I.5 Rise per tooth and cutting speeds for broaching operation.

Work Material	*Rise per tooth in mm*	*Cutting Speed in m/min.*
Free Machining Steel	0.1	10
Mild Steel	0.1	6 - 10
Medium Carbon Steel	0.08	4 - 8
Alloy Steel	0.08	5 - 10
Tool Steel	0.05	3 - 7
Stainless Steel	0.075	4 - 7
Cast Irons*	0.1	7 - 10
Aluminium Alloys	0.15	10 -15
Copper Alloys	0.125	8 - 10
Magnesium Alloys	0.15	8 -15
Titanium Alloys	0.025	1.5 - 4

Table I.6 Wheel speeds and feeds for cylindrical grinding.

Work Material	*Wheel speed in m/sec.*	*Work speed in m/min.*	*Infeed mm/pass on diameter*	*Transverse feed rate/rev*	*Wheel speci-fication*
Free Machining Steel & Mild Steel	25 - 35	18 - 30	0.05-RC 0.0125 FC	0.5 × w RC 0.16 × w FC	A60KV
Medium Carbon Steel	25 - 32	18 - 30	0.05 RC 0.0125 FC max	0.25 × w RC 0.125 × w FC	A60KV
Alloy Steel	25 - 32	18 - 30	do	0.5 × w RC 0.16 × w FC	A60KV
Stainless Steel	25 - 32	18 - 30	do	0.5 × w RC 0.16 × w FC	A60KV
Cast Irons, grey, ductile & Smalleable	25 -32	20 30	do	0.5 × w RC 0.16 × w FC	A60LV
Aluminium Alloys	25 - 32	15 - 45	do	0.5 × w RC 0.16 × w FC	A46HV
Copper Alloys	25 - 32	20 - 30	do	0.3 × w RC 0.16 × w FC	A60NV A60LV
Magnesium Alloys	25 - 32	20 - 40	do	0.3 × wRC 0.16 × w FC	C46JV
Titanium Alloys	10 - 30	15 - 30	0.025 RC 0.0125 FC	0.2 × w RC 0.1 × w FC	C60JV

*RC = Rough Cut; FC = Finishing Cut: w = Width of wheel

Table I.7 Wheel speeds, feeds and wheel specification for internal cylindrical grinding.

Work Material	*Wheel speed in m/sec.*	*Work speed in m/min.*	*Infeed in mm/pass on diameter*	*Transverse feed rate/ in multiple of width*	*Wheel specification*
Free Machining Steel & Mild Steel	25 - 35	20 - 60	0.0125 RC 0.005 FC max	0.3 w 0.16 w	A60MV A60KV
Medium Carbon Steel	do	do	do	do	A60KV A60MV
Alloy Steel	do	do	do	do	A60KV
Tool Steel	do	do	do	do	A80KV
Stainless Steel	do	do	do	do	A60JV
Cast Iron	do	do	do	do	A60JV
Copper Alloys	do	do	do	do	C60KV
Aluminium Alloys	do	do	0.075 RC 0.005 FC	do	C60KV
Magnesium Alloys	do	20 - 60	do	do	A60KV
Titanium Alloys	10 - 30	15 - 45	0.0125 RC 0.005 FC max	do	A60JV

w = width of wheel; RC - Rough Cutting; FC - Finish Cutting

Table I.8 Speeds, feeds and wheel specification for surface grinding on machines with horizontal spindle and reciprocating tables.

Work Material	*Wheel Speed in m/sec.*	*Table Speed in m/min.*	*Down feed/ pass mm*	*Transverse feed rate/*	*Wheel speci-fication*
Free Machining Steels and Mild Steel	25 - 30	15 - 30	Rough 0.075 Finish 0.025	w/4	A4JV
Medium Carbon Steel	25 - 30	15 - 30	Rough 0.05 Finish 0.0125	w/6 to w/4	A46JV
Alloy Steel	25 - 30	15 - 30	Rough 0.075 Finish 0.0125	w/4	A46JV
Stainless Steel	25 - 30	15 - 30	Rough 0.05 Finish 0.0125	w/4	A46JV
Cast Irons	25 - 30	0	do	w/5	A46JV
Aluminium Alloys	25 - 30	15 - 30	Rough 0.05 Finish 0.025	w/3	C46KV
Copper Alloys	25 - 30	15 - 30	Rough 0.075 Finish 0.0125	w/3	C46JV A46KV
Magnesium Alloys	25 - 30	15- 30	Rough 0.05 Finish 0.025	w/3	C46KV
Titanium Alloys	25 - 30	10	Rough 0.025 Finish 0.0125	W/10	C46KV

Table I.9 Speeds, feeds and wheel specifications for surface grinding on machines with vertical spindle and rotary table

Work Material	*Wheel Speed m/sec*	*Table Speed m/min*	*Down feed mm/rev of table*	*Wheel specifications*
Free Machining Steels and Mild Steel	15 - 25	25 - 60	0.025 - 0.125	A80I or A80H
Medium Carbon Steel	15 - 25	25 - 60	0.025 - 0.075	A80I or A8oH
Alloy Steel	15 - 25	25 - 60	0.025 - 0.075	A80I or A80H
Stainless Steel	15 - 25	30 - 100	0.025 - 0.075	A100H or A100G
Cast Irons	15 - 25	30 - 100	0.075 - 0.2	A80I or C80I
Aluminium Alloys	15 - 25	15 - 75	0.025 - 0.1	A80I or C80I

Annexure II

MAIN CHARACTERISTICS OF MACHINE TOOLS

Table II.1 Characteristics of grinding machines.

Machine	*Primary*				*Motion Feed Motion*			*Power P_e*	*Overall Efficiency*	*Maximum Size of Workpiece*	*Machining Accuracy*
	Type	*Axis*	*Applied to*	*Range of Speed*	*Type*	*Applied to*	*Range of Feed (f)*				
Centre lathe	Rotary	Horizontal	Workpiece	0.3 -27 s^{-1} (20 - 1600 rpm)	Continuous	Tool	0.05 - 2.5 mm	4 kW	70	l_w < 760 mm d_w < 360 mm	0.013 mm
Vertical Boring Machine	Rotary	Vertical	Workpiece	0.07 - 3.3 s^{-1} (4 - 200 rpm)	Continuous	Tool	0.05 - 2.5 mm	15 kW	70	d_w < 1.2 m	0.025 mm
Horizontal Boring Machine	Rotary	Horizontal	Tool	0.3 - 27 s^{-1} (20 - 1600 rpm)	Continuous	Tool	0.05 - 0.75 mm	7.5 kW	70	l_w < 1.2 m	0.025 mm
Shaper	Linear	Horizontal	Tool	To 0.75 m/s (150 ft/min)	Intermittent	Workpiece	0.05 - 0.75 mm	4 kW	70	l_w < 410 mm b_w < 410 mm	0.075 mm
Planer	Linear	Horizontal	Workpiece	To 1.5 m/s (300 ft/min)	Intermittent	Tool	0.05 - 2.5 mm	30 kW	80	l_w < 7.6 m b_w <1.2 m	0.064 mm

Key: bw = width of workpiece: dw = diameter of workpiece: lw = length of workpiece

Table II.2 Charateristics of Machines using multipoint tools.

Machine	Primary				Motion Feed Motion				Overall Efficiency n_m	Maximum Size of Workpiece	Maximum size of tool	Machining Accuracy
	Type	Axis	Applied to	Range of Speed	Type	Applied to	Range of Feed v_f or f	Power P_e				
Upright drill press	Rotary	Vertical	Tool	1 - 10 s^{-1} (60 - 600 rpm)	Continuous	Tool	0.15 - .064 mm	3.7 kW	70	600 × 600 mm	25 mm dia. drill	0.13 mm
Radial-arm drill	Rotary	Vertical	Tool	1 - 10 s^{-1} (60 - 600 rpm)	Continuous	Tool	0.15 - 0.64 mm	3.7 kW	70	1.2 × 1.2 m	25 mm dia. drill	0.13 mm
Bench drill press	Rotary	Vertical	Tool	12 - 83 s^{-1} (700 - 5000 rpm)	Hand operated	Tool	–	250 W	85	250 × 250 mm	8 mm dia drill	0.05 mm
Horizontal miller	Rotary	Horizontal	Tool	0.3 - 8.3 s^{-1} (20 - 500 rpm)	Continuous	Workpiece	0.2 - 8.5 mm/s	3.7 kW	50	l_w < 915 mm	200 mm dia	10.076 mm
Vertical knee-type miller	Rotary	Vertical	Tool	0.3 - 8.3 s^{-1} (20 - 500 rpm)	Continuous	Workpiece	0.2 - 8.5 mm/s	3.7 kW	50	l_w < 915 mm b_w < 250 mm	150 mm dia	0.076 mm
Vertical bed-type miller	Rotary	Vertical	Tool	0.3 - 8.3 s^{-1} (20 - 500 rpm)	Continuous	Workpiece	0.2 - 8.5 mm/s	3.7 kW	50	760 × 760 mm	250 mm dia	0.076 mm
Broacher	Linear	Horizontal or Vertical	Tool	To 0.5 m/s (100 ft/min)	Provided by cutting edges	Tool	0.013 - 0.13 mm	19 W	80	l_w < 460 mm	1500 mm	0.13 mm

Key: b_ω = width of workpiece; d_ω = diameter of workpiece; l_ω = length of workpiece

Table II.3 Characteristics of grinding machines.

Machine	Rotary primary motion to applied to wheel		Traverse			Rotary motion of workpiece		Intermittent feed		Power P_e	Overall efficiency η_m	Maximum size of workpiece	Machining accuracy
	Axis	Speed	Axis	Applied to	Maximum feed	Axis	Speed	Applied to	Maximum feed f				
Horizontal spindle surface grinder	Horizontal	33 s^{-1} (2000 rpm)	Horizontal	Workpiece	0.75 m/s	–	–	Workpiece	6.4 mm	1 kW	85	l_w < 600 mm b_w < 200 mm	0.0025 mm
Vertical spindle surface grinder	Vertical	33 s^{-1} (2000 rpm)	Horizontal	Workpiece	0.75 m/s	–	–	Wheel	1.5 mm	3.7 kW	85	l_w < 760 mm b_w <250 mm	0.025 mm
Vertical spindle surface grinder with rotary table	Vertical	33 s^{-1} (200 rpm)	–	–	–	Vertical	1.7 s^{-1} (100 rpm)	Wheel	0.25 mm	1 kW	85	400 mm dia. table	0.0025 mm
808 Cylindrical grinder	Horizontal (2000 rpm)	33 s^{-1} (100 rpm)	Horizontal	Workpiece	0.5 m/s	Horizontal	1.7 s^{-1} (100 rpm)	Wheel	0.4 mm	1 kW	85	l_w < 600 mm d_w < 250 mm	0.0025 mm
Internal grinder	Horizontal	1700 s^{-1} (100000 rpm)	Horizontal	Wheel	0.75 m/s	Horizontal	3.3 s^{-1} (200 rpm)	Wheel	0.025 mm	1 kW	85	l_w < 200 mm d_w < 150 mm	0.0025 mm

Key: b_ω = width of workpiece; d_ω = diameter of workpiece; l_ω = length of workpiece

Annexure III

CLASSIFICATION OF MACHINE TOOLS

1. Lathes

(a) Centre lathe
(b) Tool room lathe
(c) Special purpose lathes
(d) Copying lathe
(e) Gap bed lathe
(f) Capstan and turret lathes
(g) Vertical turning lathe
(h) Automatic lathes
(i) N.C. and C.N.C. lathes

2. Milling machines

(a) Horizontal milling machine
(b) Vertical milling machine
(c) Universal milling machine
(d) Rotary-type milling machine
(e) Planer-type milling machine

3. Broaching machines

(a) Horizontal broaching machine
(b) Vertical broaching machine
(c) Continuous broaching machine

4. Grinding machines

(a) Horizontal spindle surface grinder
(b) Vertical spindle surface grinder
(c) Cylindrical grinding machine
(d) Centreless grinding machine

5. Drilling machines

(a) Upright Drill press
(b) Hand-feed drilling machine
(c) Radial drilling machine
(d) Multiple spindle drilling machines
(e) Gang drilling machine
(f) Turret-type drilling machine

6. Reciprocating machines
   (a) Shaper
   (b) Planer
   (c) Slotter

7. Hole making machines
   (a) Drilling machines
   (b) Reamer
   (c) Borer
   (d) Lapping machine
   (e) Honing machine

8. Other machines
   (a) Sawing machines
   (b) Lapping machine
   (c) Gear cutting machines

INDEX